Carbohydrate Metabolism in Health and Disease

Carbohydrate Metabolism in Health and Disease

Special Issue Editor

Javier T. Gonzalez

MDPI • Basel • Beijing • Wuhan • Barcelona • Belgrade

MDPI

Special Issue Editor
Javier T. Gonzalez
University of Bath
UK

Editorial Office
MDPI
St. Alban-Anlage 66
Basel, Switzerland

This is a reprint of articles from the Special Issue published online in the open access journal *Nutrients* (ISSN 2072-6643) from 2017 to 2018 (available at: http://www.mdpi.com/journal/nutrients/special_issues/Carbohydrate_Metabolism_Health_Disease)

For citation purposes, cite each article independently as indicated on the article page online and as indicated below:

LastName, A.A.; LastName, B.B.; LastName, C.C. Article Title. *Journal Name* **Year**, *Article Number*, Page Range.

ISBN 978-3-03842-999-9 (Pbk)
ISBN 978-3-03897-000-2 (PDF)

Cover image courtesy of Javier T. Gonzalez

Contents

About the Special Issue Editor

Javier T. Gonzalez, After completing a BSc (Hons) in Sport and Exercise Science and an MRes in Exercise Physiology, Javier Gonzalez completed his PhD in 2013, in the areas of human nutrition, exercise metabolism, and appetite. Following this, he completed a post-doc, studying the effects of nutrition on liver and muscle metabolism. Javier is now a Senior Lecturer (Associate Professor) in the Department for Health at the University of Bath, UK. Javier's research seeks to understand the interactions between nutrition and exercise in the context of human health and disease and he has published over 50 peer-reviewed articles and book chapters in this area. One strand of this work is to explore the role of carbohydrate availability in the regulation of energy balance, metabolic health and sports performance. A second strand aims to uncover new dietary approaches to influence the production of hormones from the gut, and thereby regulate appetite and energy expenditure.

Preface to "Carbohydrate Metabolism in Health and Disease"

Carbohydrates are the most important energy yielding substrate in biology. For most organisms, from bacteria to humans, carbohydrates are a principal cellular fuel source. In addition, carbohydrates also serve as basic constituents of DNA and form a key constituent of biological structures. In humans, carbohydrates are not only important fuel sources, but also act as signalling molecules regulating physiological processes at the transcriptome through to the proteome. In this collection of papers, the role of carbohydrate metabolism in health and disease is tackled from a number of angles. Collectively, these articles provide insights into the effects of altering the type and amount of carbohydrate consumed at rest, and before, during and after exercise, and the implications these may have for health and/or performance. A number of other studies discuss the role of carbohydrates in certain chronic conditions such as Type 1 Diabetes, and other articles cover the role of non-carbohydrate dietary factors that influence carbohydrate metabolism. There have been many important advances in the area of carbohydrate metabolism in recent years, but many questions still remain. It is clear that this is an exciting and rapidly advancing area of research.

Javier T. Gonzalez
Special Issue Editor

nutrients

MDPI

Article

Post-Exercise Carbohydrate-Energy Replacement Attenuates Insulin Sensitivity and Glucose Tolerance the Following Morning in Healthy Adults

Harry L. Taylor [1], Ching-Lin Wu [2], Yung-Chih Chen [1], Pin-Ging Wang [2], Javier T. Gonzalez [1],*and James A. Betts [1],*

[1] Department for Health, University of Bath, Bath BA2 7AY, UK; taylor210@googlemail.com (H.L.T.); y.chen2@bath.ac.uk (Y.-C.C.)

[2] Graduate Institute of Sports and Health Management, National Chung Hsing University, Taichung 402, Taiwan; wuchinglin@icloud.com (C.-L.W.); pspgw@dragon.nchu.edu.tw (P.-G.W.)

* Correspondence: j.t.gonzalez@bath.ac.uk (J.T.G.); j.betts@bath.ac.uk (J.A.B.); Tel.: +44-122-538-5518 (J.T.G.); +44-122-538-3448 (J.A.B.)

Received: 17 December 2017; Accepted: 24 January 2018; Published: 25 January 2018

Abstract: The carbohydrate deficit induced by exercise is thought to play a key role in increased post-exercise insulin action. However, the effects of replacing carbohydrate utilized during exercise on postprandial glycaemia and insulin sensitivity are yet to be determined. This study therefore isolated the extent to which the insulin-sensitizing effects of exercise are dependent on the carbohydrate deficit induced by exercise, relative to other exercise-mediated mechanisms. Fourteen healthy adults performed a 90-min run at 70% $\dot{V}O_2$max starting at 1600–1700 h before ingesting either a non-caloric artificially-sweetened placebo solution (CHO-DEFICIT) or a 15% carbohydrate solution (CHO-REPLACE; 221.4 ± 59.3 g maltodextrin) to precisely replace the measured quantity of carbohydrate oxidized during exercise. The alternate treatment was then applied one week later in a randomized, placebo-controlled, and double-blinded crossover design. A standardized low-carbohydrate evening meal was consumed in both trials before overnight recovery ahead of a two-hour oral glucose tolerance test (OGTT) the following morning to assess glycemic and insulinemic responses to feeding. Compared to the CHO-DEFICIT condition, CHO-REPLACE increased the incremental area under the plasma glucose curve by a mean difference of 68 mmol·L^{-1} (95% CI: 4 to 132 mmol·L^{-1}; $p = 0.040$) and decreased the Matsuda insulin sensitivity index by a mean difference of −2 au (95% CI: −1 to −3 au; $p = 0.001$). This is the first study to demonstrate that post-exercise feeding to replaceme the carbohydrate expended during exercise can attenuate glucose tolerance and insulin sensitivity the following morning. The mechanism through which exercise improves insulin sensitivity is therefore (at least in part) dependent on carbohydrate availability and so the day-to-day metabolic health benefits of exercise might be best attained by maintaining a carbohydrate deficit overnight.

Keywords: insulin sensitivity; exercise; carbohydrate metabolism; oral glucose tolerance test

1. Introduction

Physical activity is a powerful tool to improve insulin sensitivity and glycemic control [1]. Accordingly, increasing physical activity is a crucial counter-measure in reducing T2D prevalence [2,3]. On an acute level, single bouts of exercise consistently enhance insulin sensitivity and muscle glucose uptake in both insulin-resistant [4] and healthy individuals [5], although the acute effects of exercise on glucose *tolerance* are less clear [6,7].

Metabolic flexibility is a key aspect of insulin sensitivity and reflects the ability to switch between substrate sources for oxidation according to availability. Individuals with robust metabolic flexibility display high rates of fat oxidation in the fasted state, switching to high rates of carbohydrate oxidation in the fed (or insulin-stimulated) state. It has been suggested that impaired metabolic flexibility may be an early cause of insulin resistance via lipid accumulation [8,9], although the direction of the relationship between metabolic flexibility and insulin sensitivity remains unclear. Notwithstanding this, substrate selection in the fasted and fed state may be an important mechanism by which exercise alters insulin sensitivity and postprandial glycemia.

The beneficial effects of exercise on glycemic control and insulin sensitivity are thought to be mediated, at least partly, by whole-body carbohydrate status. The increased insulin action after a single bout of exercise, or during the early phases (6 d) of moderate-intensity exercise training (walking), are attenuated when the energy and/or carbohydrate utilized by exercise is replaced by dietary intake [10–12]. These responses appear to be largely driven by carbohydrate status rather than energy status, since an exercise-induced carbohydrate deficit increases insulin sensitivity even in the presence of energy balance, whereas an exercise-induced energy deficit in the presence of high muscle glycogen does not [13]. Furthermore, in rodents, adrenaline-induced muscle glycogen depletion enhances insulin sensitivity of the muscle to glucose transport [14], highlighting the role of muscle carbohydrate status in insulin sensitivity, independent of exercise.

Whilst a number of studies have assessed post-exercise glucose metabolism and/or insulin action under conditions of carbohydrate and/or energy replacement, these studies have all employed intravenous tests of glucose metabolism and insulin sensitivity. Notwithstanding the strengths of tightly-controlled intravenous tests, there is a need to understand postprandial responses with the oral ingestion of glucose. This is particularly important since the acute exercise-induced increases in glucose disposal can be offset by increases in glucose appearance rates, thereby altering glucose tolerance [6,15]. Furthermore, it has been suggested that carbohydrate status around the exercise period may be a further mediator of glucose tolerance post-exercise [16]. Therefore, to translate these mechanistic findings for application, an understanding of metabolic responses to food ingestion is needed.

The purpose of this study was to investigate the role of replacing post-exercise carbohydrate on glucose tolerance and insulin sensitivity using an oral glucose tolerance test (OGTT). It was hypothesized that post-exercise replacement of the carbohydrate utilized during 90 min of treadmill running at ~70% $\dot{V}O_2$max exercise would impair glucose tolerance and insulin sensitivity the following morning during an OGTT, compared to when the exercise-induced carbohydrate deficit is maintained overnight.

2. Materials and Methods

This study was approved by the University of Bath Research Ethics Approval Committee for Health (REACH—EP 15/16 182). Fourteen healthy participants (11 men and 3 women) were recruited for the study (Table 1). Written informed consent was obtained from participants after confirming their understanding of the study design and possible risks. Nine participants (6 men and 3 women) were tested in the United Kingdom and were native to the UK, and five participants (all men) were tested in Taiwan and were native to Taiwan. None of the participants self-reported as smokers. The insulin sensitivity responses to the intervention did not differ between locations (data not shown).

Table 1. Summary of participant characteristics, $n = 14$.

	Mean $\pm$ SD
Age (years)	24 $\pm$ 5
Height (m)	1.76 $\pm$ 0.06
Mass (kg)	71.1 $\pm$ 9.0
BMI (kg/m^2)	23.6 $\pm$ 4.5
$\dot{V}O_2$max (mL·kg^{-1}·min^{-1})	56 $\pm$ 10

2.1. Experimental Design

This study was a dual-center (University of Bath and National Chung Hsing University), randomized, double-blind crossover design with two treatment arms. All participants underwent two 2-day experimental trials with a 7-day washout period between trials. On day 1, between 1600–1700 h, participants were asked to run on a treadmill at 70% $\dot{V}O_2$max for 90 min. After exercise, subjects were immediately given either a carbohydrate replacement drink (CHO-REPLACE) or placebo drink (CHO-DEFICIT). A low carbohydrate pack-dinner was provided in both trials. The following morning after an overnight fast, participants were asked to perform an oral glucose tolerance test (OGTT; Figure 1). In order to standardize metabolic parameters prior to trials, participants were asked to record their diet 3 days before the first trial and repeated the same diet before the next trial. In addition, they were asked to refrain from smoking and ingesting alcohol- and caffeine-containing beverages for 24 h before the OGTT. All participants reported successful replication of lifestyle prior to trials.

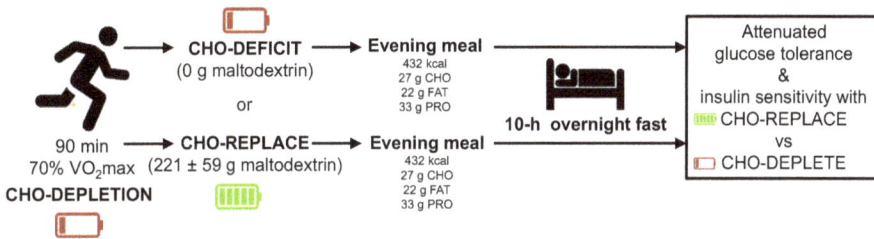

Figure 1. Schematic of the study design. CHO, carbohydrate; FAT, fat; PRO, protein; OGTT, oral glucose tolerance test.

2.2. Preliminary Tests

Subjects completed two preliminary tests: a maximal oxygen uptake test ($\dot{V}O_2$max) and a familiarization test, at least 1 week before the main trial. The $\dot{V}O_2$max test protocol has been described previously [17]. At least 3 days prior to the first main trial, a 60-min familiarization run was performed to accustom participants to an extended period of treadmill-running. The session was also used to re-affirm the appropriate treadmill speed required to achieve an exercise intensity of 70% $\dot{V}O_2$max. Accordingly, heart-rate, RPE, and expired air samples were collected and analyzed at 15 min intervals.

2.3. Main Trials

Day 1—90-min run: Participants were asked to arrive at the laboratory between 15:30–15:45, and body-mass and stature were recorded. Participants were then fitted with a heart-rate monitor and the 90-min run at 70% $\dot{V}O_2$max initiated at ~16:00. After the run, a test drink containing either carbohydrate or placebo was ingested immediately post-exercise and participants were asked to ingest the drink within one hour. Finally, a standardized dinner (432 kcal; 27 g carbohydrate (23 g of which sugars), 22 g fat, 33 g protein) was also provided, to be consumed between 19:30 and 20:00. Participants were then asked to abstain from consuming any further food or drink other than water.

Day 2—Oral Glucose Tolerance Test (OGTT): Participants attended the laboratory between 07:30–07:45, to perform a two-hour OGTT having completed an overnight fast (>10 h) and refrained from further exercise.

Upon arrival, participants rested for 15-min in a semi-recumbent position with their hand placed in a hot-box set to 55 °C [17], with a subsequent five-minute (baseline) expired air sample collected from a subgroup ($n = 7$). Following this, a cannula was inserted into an antecubital vein of the participant's forearm and a 5-mL (baseline) blood sample drawn. A 75 g glucose load was then administered orally (113 mL Polycal: Nutricia, UK (mixed with 87 mL water)) and blood samples were collected at 30 min intervals over 2-h. 5-min expired air samples were also collected at 25 to 30, 55 to 60, 85 to 90, and 115 to 120-min from a subgroup (n = 7). Since there were some slight differences in protocols between University of Bath and the National Chung Hsing University (namely, Bath protocols included a hot-box for blood sampling and expired breath analysis), the data from both institutions were initially analyzed separately to check that responses were similar. Since both protocols produced similar overall responses for glycaemia, insulinemia, and insulin sensitivity indices, the data were combined for the present manuscript (n = 14).

2.4. Test Drink

During the CHO-REPLACE trial, a 15% Maltodextrin solution (MyProtein, Cheshire, UK; Batch No.: L626929168) was ingested to precisely replace carbohydrate oxidized during the preceding run. The total amount of carbohydrate utilized during exercise were determined via indirect calorimetry from expired gases collected every 15-min during exercise. The amount of carbohydrate replacement for the CHO-REPLACE trial was 221.4 ± 59.3 g. Conversely, a (0 g carbohydrate) 1.5% artificially-sweetened placebo solution (Truvia, Silver Spoon, Peterborough, UK) was ingested during the CHO-DEFICIT trial.

2.5. Blood and Expired Air Samples Collection and Analysis

2.5.1. Blood Sample Collection and Analysis

The arterialized blood samples were obtained via a cannula inserted into antecubital vein of each subject's forearm [17]. A non-heparinized tube was used to collect 2 mL of blood sample, and it was allowed to stand for 1 h to wait for the blood to coagulate. Another tube containing ethylenediaminetetraacetic acid (EDTA; BD, Oxford, UK) was used to collect 3 mL of blood sample. The collected sample was then centrifuged (Eppendorf 5810, Hamburg, Germany) in 4 °C at 2500 g for 10 min. The extracted serum and plasma samples were stored at −80 °C before later analysis. Serum insulin concentrations were analyzed using enzyme-linked immunosorbent assays (ELISA, Mercodia AB, Uppsala, Sweden), following the manufacturer's instructions. Minimal detectable concentrations for serum insulin were set at 18 pmol·L^{-1} and intra-plate coefficients of variation were <4.4%. Plasma glucose concentrations were analyzed using a spectrophotometric analyzer (Randox Daytona, Randox Laboratories Ltd., Crumlin, UK). Due to a cannula blockage on one trial for one participant, data for blood-based variables are (n = 13).

2.5.2. Expired Gas Samples Collection and Analysis

The Douglas bag method was used to assess substrate metabolism at rest and during exercise. For all samples, participants were provided the mouthpiece before gas collections for a stabilization period. At rest, the stabilization and gas collection periods were each 5 min, whereas during exercise the stabilization and gas collection periods were each 1 min. Samples were collected in 200 L Douglas bags (Hans Rudolph, Kansas City, MO, USA) through falconia tubing (Baxter, Woodhouse and Taylor Ltd., Macclesfield, UK). Expired O_2 and CO_2 concentrations were measured in a known volume of each sample, using paramagnetic and infrared transducers, respectively (Mini HF 5200, Servomex Group Ltd., Crowborough, East Sussex, UK). The sensor was calibrated using known concentrations of low (99.998% Nitrogen, 0% O_2 and CO_2) and high (balance nitrogen mix, 16.04% O_2, 5.06% CO_2) calibration gases (both BOC Industrial Gases, Linde AG, Munich, Germany). Substrate utilization was determined during exercise using the equations of Jeukendrup and Wallis (2005) [18], whilst Frayn's (1983) [19] equations were used for samples collected at rest as follows (where $\dot{V}O_2$ and $\dot{V}CO_2$ are expressed in L/min):

$$\text{Fat utilisation at rest and during exercise (g/min)} = \left(1.695 \times \dot{V}O_2\right) - \left(1.701 \times \dot{V}CO_2\right) \quad (1)$$

$$\text{Carbohydrate utilisation at rest (g/min)} = \left(4.585 \times \dot{V}CO_2\right) - \left(3.226 \times \dot{V}O_2\right) \qquad (2)$$

$$\text{Carbohydrate utilisation during exercise (g/min)} = \left(4.210 \times \dot{V}CO_2\right) - \left(2.962 \times \dot{V}O_2\right) \qquad (3)$$

2.6. Sample Size Estimation

The sample size estimation was performed using data on insulin concentrations during steady-state intravenous glucose infusion following exercise training with, or without, carbohydrate and energy replacement. In the absence of carbohydrate replacement, insulin concentrations were ~225 $\pm$ 141 pmol$\cdot$L^{-1}, compared to ~345 $\pm$ 85 pmol$\cdot$L^{-1} when the carbohydrate and energy utilized during exercise was replaced. Based on this effect size (d = 1.03), 12 participants should provide more than a 90% chance of detecting such an effect with an alpha level of 0.05.

2.7. Statistical Analysis

Incremental area under the curve (iAUC; divided by 120 min to provide time average values) and Matsuda Insulin sensitivity index (Matsuda index; [18]) were calculated from plasma glucose and serum insulin data using Microsoft Excel (Version 15.26, Microsoft, Redmond, WA, USA). The updated homeostasis assessment model of insulin resistance (HOMA2-IR; [19]) was calculated using freely available online software (https://www.dtu.ox.ac.uk/homacalculator/). Statistical analyses were performed using GraphPad Prism v7 (GraphPad Software, San Diego, CA, USA). Differences between trials in time-dependent variables (glucose and insulin concentrations, and carbohydrate and fat oxidation rates) were analyzed by a two-way ANOVA with repeated measures. For non-time dependent variables, paired t-tests were applied. A p-value of $\leq$0.05 was considered statistically significant. Data are presented in the body of the text as mean $\pm$ SD, whereas error bars on figures are confidence intervals normalized to remove between subject variance, consistent with this within-subject design [20]. With this approach, any error bars that do not overlap the mean of their respective comparison can be considered to have a significance level of <0.05.

3. Results

3.1. Indirect Calorimetry during the Treadmill Run

The mean rates of oxygen consumption and carbon dioxide production were 2.65 $\pm$ 0.43 L$\cdot$min^{-1} and 2.65 $\pm$ 0.48 L$\cdot$min^{-1}, respectively. This resulted in a respiratory exchange ratio of 1.00 $\pm$ 0.12 ($\dot{V}O_2$:$\dot{V}CO_2$). Based on data collected at Bath, the total amount of carbohydrate oxidized during the run in CHO-DEPLETE was 205 $\pm$ 58 g vs. 220 $\pm$ 52 g in the CHO-REPLACE trial.

3.2. Glycemia, Insulinemia and Insulin Sensitivity

Pre-OGTT glucose concentrations were 4.56 $\pm$ 0.55 mmol$\cdot$L^{-1} during the CHO-REPLACE trial and 4.36 $\pm$ 0.51 mmol$\cdot$L^{-1} during the CHO-DEFICIT trial (p = 0.149). Following ingestion of the OGTT, plasma glucose concentrations rose to a greater extent in CHO-REPLACE versus CHO-DEFICIT (Figure 2A,B; p = 0.040), whereby peak glucose concentrations were 9.74 $\pm$ 1.22 mmol$\cdot$L^{-1} on the CHO-REPLACE and 8.33 $\pm$ 1.76 mmol$\cdot$L^{-1} on the CHO-DEFICIT trial. Repeated measures ANOVA revealed main effects of time (p < 0.001) and treatment (p = 0.007), but no time–treatment interaction effect (p = 0.170).

Pre-OGTT insulin concentrations were higher during the CHO-REPLACE trial compared to the CHO-DEFICIT trial (36 $\pm$ 31 pmol$\cdot$L^{-1} compared to 30 $\pm$ 20 pmol$\cdot$L^{-1}, respectively p = 0.023). Following ingestion of the OGTT, the increase in serum insulin concentrations was greater with CHO-REPLACE versus CHO-DEFICIT (Figure 2C,D), whereby peak insulin concentrations were 337 $\pm$ 107 pmol$\cdot$L^{-1} during CHO-REPLACE, compared to 260 $\pm$ 101 pmol$\cdot$L^{-1} during CHO-DEFICIT

($p = 0.012$). Repeated measures ANOVA revealed main effects of time ($p < 0.001$) and treatment ($p = 0.028$), but no time—treatment interaction effect ($p = 0.158$).

The HOMA2-IR was ~16% higher with CHO-REPLACE versus CHO-DEFICIT (Figure 3A; $p = 0.015$), whereas the Matsuda insulin sensitivity index was ~25% lower with CHO-REPLACE vs. CHO-DEFICIT (Figure 3B; $p = 0.001$).

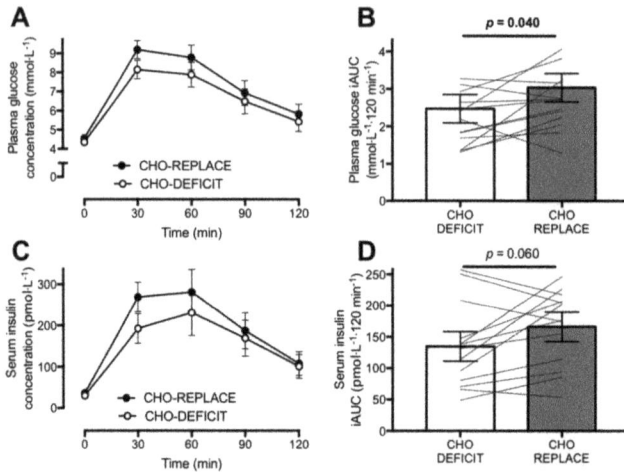

Figure 2. Postprandial glycaemia (**A,B**) and insulinemia (**C,D**) expressed as absolute concentrations (**A,C**) or as the incremental time-averaged area under the curve (iAUC; **B,D**) during the oral glucose tolerance test conducted ~16 h after exercise with either carbohydrate replacement (CHO-REPLACE) or a maintenance of the exercise-induced carbohydrate deficit (CHO-DEFICIT). $n = 13$. Data are means ± normalized 95% CI.

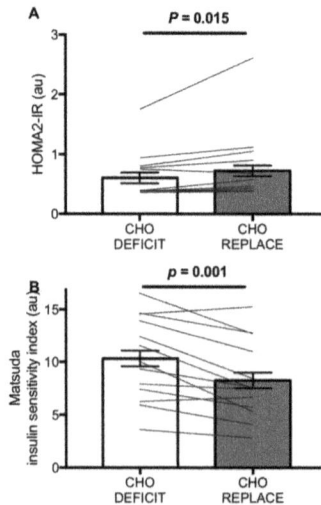

Figure 3. Homeostasis model of insulin resistance (HOMA2-IR) (**A**) and the Matsuda insulin sensitivity index (**B**) during the oral glucose tolerance test conducted ~16 h after exercise with either carbohydrate replacement (CHO-REPLACE) or a maintenance of the exercise-induced carbohydrate deficit (CHO-DEFICIT). $n = 13$. Data are means ± normalized 95% CI.

3.3. Whole-Body Substrate Utilisation

Pre-OGTT, whole-body carbohydrate utilization was 0.08 ± 0.05 g·min^{-1} during the CHO-REPLACE trial and 0.06 ± 0.05 g·min^{-1} during the CHO-DEFICIT trial ($p = 0.639$). Following ingestion of the OGTT, carbohydrate utilization increased ~2-fold in both trials (Figure 4A; time effect: $p < 0.001$), with no differences between trials (treatment effect: $p = 0.378$; time–treatment interaction effect: $p = 0.099$).

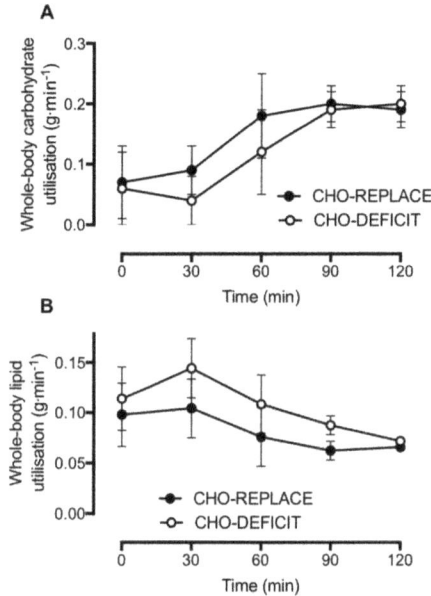

Figure 4. Whole-body carbohydrate (**A**) and lipid utilization (**B**) during an oral glucose tolerance test conducted ~16 h after exercise with either carbohydrate replacement (CHO-REPLACE) or a maintenance of the exercise-induced carbohydrate deficit (CHO-DEFICIT). $n = 7$. Data are means ± normalized 95% CI.

Pre-OGTT, whole-body lipid utilization was 0.10 ± 0.03 g·min^{-1} during the CHO-REPLACE trial and 0.11 ± 0.03 g·min^{-1} during the CHO-DEFICIT trial ($p = 0.350$). Following ingestion of the OGTT, lipid utilization was suppressed in both trials (Figure 4B; time effect: $p < 0.001$), but to a greater extent in CHO-REPLACE vs. CHO-DEFICIT (Figure 4B; treatment effect: $p = 0.033$; time–treatment interaction effect: $p = 0.048$).

4. Discussion

The present study demonstrates that replacement of the carbohydrate utilized during a single bout of exercise impairs both insulin sensitivity and glucose tolerance by ~20–25% the following morning, relative to when the exercise-induced carbohydrate deficit is maintained. Importantly, these changes were most clearly apparent in the postprandial state. Furthermore, postprandial fat oxidation was suppressed by post-exercise replacement of carbohydrate use.

Previous work has demonstrated that, whilst exercise is a potent method of stimulating muscle glucose uptake and insulin sensitivity, the carbohydrate deficit induced by exercise is key factor that mediates these responses. However, previous work has primarily used intravenous methods of assessing insulin sensitivity and/or action, which do not necessarily translate into the tolerance of ingested nutrients.

It has been suggested that the degree of whole-body carbohydrate depletion is a key mediator of exercise induced-increases in insulin action. Indeed, a positive relationship has been reported between post-exercise carbohydrate depletion and the change in insulin action assessed during intravenous glucose infusion, whereby a carbohydrate deficit of greater than 90 g was associated with an increase in insulin action [21]. In the present study, the mean carbohydrate deficit was 221 ± 59 g; all participants had a carbohydrate deficit of at least 99 g and there was a clear ~25% increase in insulin sensitivity as assessed by the Matsuda index, which was apparent in 12 of 14 individuals. Furthermore, whilst we did not have an energy-matched, low-carbohydrate condition to isolate the effect of carbohydrate versus energy-replacement, it has previously been demonstrated that re-feeding fat post-exercise does not influence glucose tolerance or insulin sensitivity the following morning [22]. Therefore, our data extend those findings demonstrating that post-exercise carbohydrate re-feeding does influence glucose tolerance and insulin sensitivity the following morning. Taken together, the evidence suggests that whole-body carbohydrate depletion induced by exercise is a key mediator of the enhanced insulin sensitivity and glucose control induced by exercise.

We report that both HOMA2-IR and the Matsuda insulin sensitivity index indicated an impairment in insulin sensitivity with post-exercise carbohydrate replacement, by ~16% and ~25%, respectively. Fasting concentrations of insulin and glucose primarily reflect changes in hepatic insulin sensitivity, whereas at postprandial concentrations, hepatic glucose production is negligible [18,23]. On this basis, it has been suggested that HOMA2-IR is primarily reflective of hepatic insulin sensitivity, whereas the Matsuda insulin sensitivity index is more heavily influenced by peripheral insulin sensitivity [24]. The finding that postprandial metabolic responses are more clearly altered than fasted measures by post-exercise carbohydrate replacement suggests that peripheral insulin sensitivity was more heavily influenced by replacement of carbohydrate compared to hepatic insulin sensitivity. It should be acknowledged that the present data does not allow for interpretation of insulin secretion to be assessed. Therefore, the reduction in glucose tolerance with carbohydrate replacement compared to the maintenance of the carbohydrate deficit could be due, in part, to an inability of the pancreas to secrete sufficient insulin to compensate for the change in insulin sensitivity. Furthermore, it has previously been shown that the timing of carbohydrate re-feeding post-exercise can alter insulin action the following day [11]. Delaying the re-feeding of carbohydrate by 3 h results in lower insulin action compared to immediate post-exercise carbohydrate refeeding. Therefore, the immediate re-feeding in the present study may result in a lower-bound estimate of the impairment in insulin sensitivity with carbohydrate replacement in the hours following exercise.

During a prolonged bout (90 min) of moderate-to-high intensity exercise (70% $\dot{V}O_2$peak), both muscle and liver glycogen concentrations can be expected to be depleted by ~60% [25–27]. In the absence of meaningful quantities of carbohydrate or glycogenic amino acid ingestion, only negligible net quantities of muscle and liver glycogen will be synthesized. Therefore, prior to the OGTT we can be confident that both muscle and liver glycogen stores would have been depleted with the carbohydrate restriction trial, compared to the carbohydrate replacement trial. Accordingly, the observed suppression of insulin sensitivity with replacement of carbohydrate is likely to represent depletion of all major glycogen stores. This is important, since hepatic and muscle insulin sensitivity appear to respond differentially to carbohydrate status during acute (3-d) overfeeding [28].

Increased insulin sensitivity after exercise does not always translate into changes in glucose control after the ingestion of nutrients. Post-exercise increases in glucose disposal can be offset by changes in the rate of appearance of glucose from both endogenous and exogenous sources [6], leading to either no change or even a worsening of glucose tolerance after a single bout of exercise [7]. This highlights the importance of complementing mechanistic studies of insulin sensitivity that involve intravenous infusion methods, with oral ingestion of nutrients. In the present study, we employed an oral glucose tolerance test and demonstrated that restoring carbohydrate balance via post-exercise feeding increases postprandial glycaemia compared to a maintenance of the exercise-induced carbohydrate deficit.

We also observed a difference in postprandial substrate metabolism, whereby whole-body lipid utilization was suppressed when the carbohydrate deficit of prior exercise had been replaced. This suppression of postprandial lipid utilization is consistent with findings from others in which the energy utilized by exercise was replaced [12] and further highlights the role of carbohydrate balance in the regulation of whole-body lipid utilization. Furthermore, the greater suppression of postprandial lipid utilization with carbohydrate replacement is consistent with the high fasting and postprandial insulinemia that we observed, since insulin is a potent inhibitor of adipose tissue lipolysis in vivo [29].

The present study design does not allow for inferences to be drawn about the effects of exercise on insulin sensitivity and glycemic control, since there was no non-exercise trial. Therefore, the carbohydrate replacement in this study could be either: (1) partly attenuating the effects of exercise; (2) completely reversing the effects of exercise; or (3) superseding the effects of exercise. However, the effects of exercise on glycemia and insulin sensitivity are well-characterized, and the aim of the present investigation was specifically to establish the degree to which the carbohydrate deficit of exercise alters glycemia and insulin sensitivity. By comparing the carbohydrate replacement trial with the maintenance of the exercise-induced carbohydrate deficit, we are able to establish the extent to which the whole-body carbohydrate deficit alters postprandial glycaemia and insulin sensitivity. Furthermore, the findings of this study will need further work to provide greater generalizability and to further characterize the underlying mechanisms. In order to be able to generalize the findings to people at risk of metabolic disease, this work should be repeated in overweight/obese people, and at lower exercise intensities. Some disease states and lower exercise intensities would reduce the reliance on glycogen use during exercise, thereby altering the nutrition interaction with exercise. Furthermore, to firmly establish the underlying mechanisms, isotopic tracers and euglycemic hyperinsulinemic clamps could establish rates of appearance and disappearance of glucose, and peripheral insulin sensitivity, respectively. Nonetheless, the present study provides the first evidence, as proof of principle, that replacing the carbohydrate deficit induced by exercise has the capacity to reduce postprandial glycemic control and insulin sensitivity the following morning.

5. Conclusions

This study is the first to show that feeding carbohydrate to replace that utilized during exercise can reduce insulin sensitivity and glucose tolerance the next morning in healthy adults, when compared to a preservation of the exercise-induced carbohydrate deficit. Furthermore, carbohydrate replacement suppresses subsequent postprandial fat utilization. The mechanism through which exercise improves insulin sensitivity and glucose control is therefore (at least partly) dependent on carbohydrate availability, and so the day-to-day metabolic health benefits of exercise might be best attained by maintaining a carbohydrate deficit overnight.

Acknowledgments: The study was partially funded by Ministry of Science and Technology in Taiwan (MOST 105-2918-I-005-004). We appreciate the technical support provided by the Sport Science Research Center of National Taiwan University of Sport.

Author Contributions: C.-L.W., J.A.B. and J.T.G. conceived and designed the experiments and wrote the manuscript; H.L.T. and P.-G.W. performed the experiments; Y.-C.C. ran assays and analyzed the data; C.-L.W., H.L.T., J.A.B. and J.T.G analyzed the data. All authors contributed to editing the manuscript.

Conflicts of Interest: The authors declare no conflict of interest.

References

1. Edinburgh, R.M.; Betts, J.A.; Burns, S.F.; Gonzalez, J.T. Concordant and divergent strategies to improve postprandial glucose and lipid metabolism. *Nutr. Bull.* **2017**, *42*, 113–122. [CrossRef]
2. Maarbjerg, S.J.; Sylow, L.; Richter, E.A. Current understanding of increased insulin sensitivity after exercise—Emerging candidates. *Acta Physiol.* **2011**, *202*, 323–335. [CrossRef] [PubMed]

3. Pan, X.R.; Li, G.W.; Hu, Y.H.; Wang, J.X.; Yang, W.Y.; An, Z.X.; Hu, Z.X.; Lin, J.; Xiao, J.Z.; Cao, H.B.; et al. Effects of diet and exercise in preventing NIDDM in people with impaired glucose tolerance. The Da Qing IGT and diabetes study. *Diabetes Care* **1997**, *20*, 537–544. [CrossRef] [PubMed]

4. Rabol, R.; Petersen, K.F.; Dufour, S.; Flannery, C.; Shulman, G.I. Reversal of muscle insulin resistance with exercise reduces postprandial hepatic de novo lipogenesis in insulin resistant individuals. *Proc. Natl. Acad. Sci. USA* **2011**, *108*, 13705–13709. [CrossRef] [PubMed]

5. Mikines, K.J.; Sonne, B.; Farrell, P.A.; Tronier, B.; Galbo, H. Effect of physical exercise on sensitivity and responsiveness to insulin in humans. *Am. J. Physiol.* **1988**, *254*, E248–E259. [CrossRef] [PubMed]

6. Rose, A.J.; Howlett, K.; King, D.S.; Hargreaves, M. Effect of prior exercise on glucose metabolism in trained men. *Am. J. Physiol. Endocrinol. Metab.* **2001**, *281*, E766–E771. [CrossRef] [PubMed]

7. Gonzalez, J.T.; Veasey, R.C.; Rumbold, P.L.; Stevenson, E.J. Breakfast and exercise contingently affect postprandial metabolism and energy balance in physically active males. *Br. J. Nutr.* **2013**, *110*, 721–732. [CrossRef] [PubMed]

8. Kelley, D.E.; Mandarino, L.J. Fuel selection in human skeletal muscle in insulin resistance: A reexamination. *Diabetes* **2000**, *49*, 677–683. [CrossRef] [PubMed]

9. Galgani, J.E.; Moro, C.; Ravussin, E. Metabolic flexibility and insulin resistance. *Am. J. Physiol. Endocrinol. Metab.* **2008**, *295*, E1009–E1017. [CrossRef] [PubMed]

10. Black, S.E.; Mitchell, E.; Freedson, P.S.; Chipkin, S.R.; Braun, B. Improved insulin action following short-term exercise training: Role of energy and carbohydrate balance. *J. Appl. Physiol.* **2005**, *99*, 2285–2293. [CrossRef] [PubMed]

11. Stephens, B.R.; Sautter, J.M.; Holtz, K.A.; Sharoff, C.G.; Chipkin, S.R.; Braun, B. Effect of timing of energy and carbohydrate replacement on post-exercise insulin action. *Appl. Physiol. Nutr. Metab.* **2007**, *32*, 1139–1147. [CrossRef] [PubMed]

12. Burton, F.L.; Malkova, D.; Caslake, M.J.; Gill, J.M. Energy replacement attenuates the effects of prior moderate exercise on postprandial metabolism in overweight/obese men. *Int. J. Obes.* **2008**, *32*, 481–489. [CrossRef] [PubMed]

13. Newsom, S.A.; Schenk, S.; Thomas, K.M.; Harber, M.P.; Knuth, N.D.; Goldenberg, N.; Horowitz, J.F. Energy deficit after exercise augments lipid mobilization but does not contribute to the exercise-induced increase in insulin sensitivity. *J. Appl. Physiol.* **2010**, *108*, 554–560. [CrossRef] [PubMed]

14. Nolte, L.A.; Gulve, E.A.; Holloszy, J.O. Epinephrine-induced in vivo muscle glycogen depletion enhances insulin sensitivity of glucose transport. *J. Appl. Physiol.* **1994**, *76*, 2054–2058. [CrossRef] [PubMed]

15. Gonzalez, J.T. Paradoxical second-meal phenomenon in the acute post-exercise period. *Nutrition* **2014**, *30*, 961–967. [CrossRef] [PubMed]

16. Gonzalez, J.T.; Stevenson, E.J. Assessment of the post-exercise glycemic response to food: Considering prior nutritional status. *Nutrition* **2014**, *30*, 122–123. [CrossRef] [PubMed]

17. Edinburgh, R.M.; Hengist, A.; Smith, H.A.; Betts, J.A.; Thompson, D.; Walhin, J.P.; Gonzalez, J.T. Prior exercise alters the difference between arterialised and venous glycaemia: Implications for blood sampling procedures. *Br. J. Nutr.* **2017**, *117*, 1414–1421. [CrossRef] [PubMed]

18. Matsuda, M.; DeFronzo, R.A. Insulin sensitivity indices obtained from oral glucose tolerance testing: Comparison with the euglycemic insulin clamp. *Diabetes Care* **1999**, *22*, 1462–1470. [CrossRef] [PubMed]

19. Levy, J.C.; Matthews, D.R.; Hermans, M.P. Correct homeostasis model assessment (HOMA) evaluation uses the computer program. *Diabetes Care* **1998**, *21*, 2191–2192. [CrossRef] [PubMed]

20. Loftus, G.R.; Masson, M.E. Using confidence intervals in within-subject designs. *Psychon. Bull. Rev.* **1994**, *1*, 476–490. [CrossRef] [PubMed]

21. Holtz, K.A.; Stephens, B.R.; Sharoff, C.G.; Chipkin, S.R.; Braun, B. The effect of carbohydrate availability following exercise on whole-body insulin action. *Appl. Physiol. Nutr. Metab.* **2008**, *33*, 946–956. [CrossRef] [PubMed]

22. Fox, A.K.; Kaufman, A.E.; Horowitz, J.F. Adding fat calories to meals after exercise does not alter glucose tolerance. *J. Appl. Physiol.* **2004**, *97*, 11–16. [CrossRef] [PubMed]

23. Radziuk, J. Homeostatic model assessment and insulin sensitivity/resistance. *Diabetes* **2014**, *63*, 1850–1854. [CrossRef] [PubMed]

24. Gonzalez, J.T.; Barwood, M.J.; Goodall, S.; Thomas, K.; Howatson, G. Alterations in whole-body insulin sensitivity resulting from repeated eccentric exercise of a single muscle group: A pilot investigation. *Int. J. Sport Nutr. Exerc. Metab.* **2015**, *25*, 405–410. [CrossRef] [PubMed]

25. Gonzalez, J.T.; Fuchs, C.J.; Betts, J.A.; van Loon, L.J. Liver glycogen metabolism during and after prolonged endurance-type exercise. *Am. J. Physiol. Endocrinol. Metab.* **2016**, *311*, E543–E553. [CrossRef] [PubMed]

26. Gonzalez, J.T.; Fuchs, C.J.; Smith, F.E.; Thelwall, P.E.; Taylor, R.; Stevenson, E.J.; Trenell, M.I.; Cermak, N.M.; van Loon, L.J. Ingestion of glucose or sucrose prevents liver but not muscle glycogen depletion during prolonged endurance-type exercise in trained cyclists. *Am. J. Physiol. Endocrinol. Metab.* **2015**, *309*, E1032–E1039. [CrossRef] [PubMed]

27. Alghannam, A.F.; Jedrzejewski, D.; Tweddle, M.G.; Gribble, H.; Bilzon, J.; Thompson, D.; Tsintzas, K.; Betts, J.A. Impact of muscle glycogen availability on the capacity for repeated exercise in man. *Med. Sci. Sports Exerc.* **2016**, *48*, 123–131. [CrossRef] [PubMed]

28. Lundsgaard, A.M.; Sjoberg, K.A.; Hoeg, L.D.; Jeppesen, J.; Jordy, A.B.; Serup, A.K.; Fritzen, A.M.; Pilegaard, H.; Myrmel, L.S.; Madsen, L.; et al. Opposite regulation of insulin sensitivity by dietary lipid versus carbohydrate excess. *Diabetes* **2017**, *66*, 2583–2595. [CrossRef] [PubMed]

29. Jensen, M.D.; Caruso, M.; Heiling, V.; Miles, J.M. Insulin regulation of lipolysis in nondiabetic and IDDM subjects. *Diabetes* **1989**, *38*, 1595–1601. [CrossRef] [PubMed]

nutrients

MDPI

Review

Regulation of Muscle Glycogen Metabolism during Exercise: Implications for Endurance Performance and Training Adaptations

Mark A. Hearris, Kelly M. Hammond, J. Marc Fell and James P. Morton *

Research Institute for Sport & Exercise Sciences, Liverpool John Moores University, Liverpool L3 3AF, UK;
M.Hearris@2014.ljmu.ac.uk (M.A.H.); K.M.Hammond@2009.ljmu.ac.uk (K.M.H.);
J.M.Fell@2015.ljmu.ac.uk (J.M.F.)
* Correspondence: J.P.Morton@ljmu.ac.uk; Tel.: +44-151-904-6233

Received: 9 January 2018; Accepted: 27 February 2018; Published: 2 March 2018

Abstract: Since the introduction of the muscle biopsy technique in the late 1960s, our understanding of the regulation of muscle glycogen storage and metabolism has advanced considerably. Muscle glycogenolysis and rates of carbohydrate (CHO) oxidation are affected by factors such as exercise intensity, duration, training status and substrate availability. Such changes to the global exercise stimulus exert regulatory effects on key enzymes and transport proteins via both hormonal control and local allosteric regulation. Given the well-documented effects of high CHO availability on promoting exercise performance, elite endurance athletes are typically advised to ensure high CHO availability before, during and after high-intensity training sessions or competition. Nonetheless, in recognition that the glycogen granule is more than a simple fuel store, it is now also accepted that glycogen is a potent regulator of the molecular cell signaling pathways that regulate the oxidative phenotype. Accordingly, the concept of deliberately training with low CHO availability has now gained increased popularity amongst athletic circles. In this review, we present an overview of the regulatory control of CHO metabolism during exercise (with a specific emphasis on muscle glycogen utilization) in order to discuss the effects of both high and low CHO availability on modulating exercise performance and training adaptations, respectively.

Keywords: muscle glycogen; glucose; athletes; train-low

---

1. Introduction

The study of carbohydrate (CHO) metabolism in relation to the field of sport and exercise is an area of investigation that is now over 100 years old. Almost a century ago, Krogh and Lindhard [1] reported the efficiency of CHO as a fuel source during exercise and demonstrated that fatigue occurs earlier when subjects consume a low CHO diet (as compared with a high CHO diet) in the days preceding an exercise bout undertaken at a fixed workload. Levine et al. [2] also observed that runners who completed the 1923 Boston marathon exhibited hypoglycemia (<4 mmol·L^{-1}) immediately post-exercise, thus suggesting that low CHO availability may be linked to fatigue. These early studies provided the initial evidence that CHO was an important fuel source for sustaining exercise performance.

Nonetheless, much of the foundation of our understanding of CHO metabolism was developed by Scandinavian researchers in the late 1960s with the introduction of the muscle biopsy technique [3–6]. These researchers provided the platform for modern day sports nutrition practice in a series of studies that collectively demonstrated that (1) muscle glycogen is depleted during exercise in an intensity dependent manner; (2) high CHO diets increase muscle glycogen storage and subsequently improve exercise capacity and (3) muscle glycogen storage is acutely enhanced following prior glycogen depletion (i.e., the super-compensation effect), the magnitude of which is dependent on high CHO availability [4]. This body of work remains some of the most highly cited papers in the field and is referenced accordingly in the most recent sport nutrition guidelines [7].

The field continued to develop throughout the 1980s and 1990s with the consistent finding that CHO feeding during exercise also improved exercise performance and capacity [8–12]. Such studies relied on the use of stable isotope methodology (to quantify exogenous CHO oxidation) as well as magnetic resonance imaging to quantify liver glycogen depletion during exercise [13]. As such, it is now generally accepted that liver glycogen depletion is also a major contributing cause of fatigue during endurance exercise. It is noteworthy, however, that CHO feeding can also improve physical performance via non-metabolic effects through modulating regions of the brain associated with reward and motor control [14,15].

In addition to a simple "fuel store", our understanding of CHO metabolism has advanced considerably in recent years with the use of more sophisticated molecular biology techniques. In this regard, it is now accepted that glycogen is more than a store [16], acting as a regulator of many key cell signaling pathways related to promoting the oxidative phenotype, insulin sensitivity, contractile processes, protein degradation and autophagic processes [16,17]. When considered this way, it is remarkable that whole body storage of only 500 g of substrate can exert such profound effects on multiple tissues, organs and systems, the results of which have considerable effects on human health and performance.

The aim of this review is to, therefore, present a contemporary discussion of our understanding of CHO metabolism (focusing on muscle glycogen metabolism) with specific reference to sport and exercise. We begin by presenting an overview of CHO storage followed by outlining regulatory steps in the control of both muscle glycogen metabolism and muscle glucose uptake. We then proceed to discuss how manipulating substrate availability (i.e., CHO availability itself) and alterations to specifics of the exercise protocol (e.g., intensity, duration) and training status of the athlete can all affect the magnitude of CHO utilized during exercise. The previous section, therefore, provides the platform to discuss the well-documented effects of both endogenous (i.e., liver and muscle glycogen) and exogenous (i.e., CHO feeding during exercise) CHO availability on exercise performance. Finally, we close by discussing the role of CHO availability on modulating aspects of training adaptation, a field of research that has grown rapidly in the last decade.

2. Overview of CHO Storage

Carbohydrate is predominantly stored as glycogen in both the liver (approximately 100 g) and muscle (approximately 400 g) with 5 g also circulating in the blood stream as glucose. In skeletal muscle, glycogen is typically expressed as $mmol \cdot kg^{-1}$ of dry muscle (d.w.) where concentrations in whole muscle homogenate can vary from 50 to 800 $mmol \cdot kg^{-1}$ d.w., depending on training status, fatigue status and dietary CHO intake (see Figure 1). Muscle glycogen can also be expressed as $mmol \cdot kg^{-1}$ of wet muscle (ww), as commonly reported in the early classical literature, where values range between 10 to 180 $mmol \cdot kg^{-1}$ ww [3–6].

The glycogen granule itself is essentially a tiered assembly of glucose units (i.e., polymers) that is formed in a branch-like structure via 1:4 and 1:6 α-glycosidic bonds. Glycogen granules are formed on the protein, glycogenin, and can be as large as 42 nm in diameter as well as potentially having 12 tiers. At its maximal size, the granule can consist of as much as 55,000 glucosyl units [18]. Nonetheless, the majority of glycogen granules in human skeletal muscle are reported to be 25 nm in diameter with approximately eight tiers [19]. Although muscle glycogen has traditionally been quantified through acid hydrolysis in whole muscle homogenate, it is, of course, apparent that glycogen is expressed and utilized in fibre type specific patterns as well as being located in specific intracellular locations within muscle cells themselves. Using histochemical techniques, it has typically been reported that resting glycogen content is not apparently different between type I and type II fibres [20–22]. Nonetheless, using biochemical quantification (a more quantitative measure), it has been reported that type II fibres may contain 50–100 $mmol \cdot kg^{-1}$ d.w. more glycogen than type I fibres [9,23]. Regardless of method of quantification, glycogen depletion during exercise is dependent on fibre type recruitment patterns, according to the specifics of the exercise protocol. For example, during prolonged steady state type protocols, type I fibres show preferential depletion, whereas during near maximal or supra-maximal type activity, type II fibres become recruited and show considerable glycogen depletion [24]. In activities involving high-intensity intermittent exercise (e.g., a soccer match), considerable glycogen depletion is observed in both muscle fibre types (54% and 46% of type I and II fibres classed as completely or almost empty, respectively) thus reflecting recruitment patterns to support both moderate and high-intensity running speeds [25].

The use of transmission electron microscopy (TEM) has also revealed that glycogen is stored in three distinct sub-cellular pools contained in the myofibrils (intra-myofibrillar glycogen, 5–15% of total glycogen pool), between myofibrils (inter-myofibrallar glycogen, 75% of total glycogen pool) and beneath the sarcolemmal region (sub-sarcolemmal glycogen, 5–15% of total glycogen pool). In endurance trained athletes, it appears that both intra-myofibrillar and sub-sarcolemmal glycogen stores are greater in type I fibres compared with type II fibres, whereas inter-myofibrillar glycogen storage is greater in type II fibres [26]. In relation to acute exercise itself, it is also apparent that intra-myofibrillar glycogen stores show preferential depletion [27], and the failure to restore this specific pool in the immediate hours after exercise is associated with impaired Ca^{2+} release from the sarcoplasmic reticulum [28,29]. Clearly, our understanding of muscle glycogen storage has advanced considerably in recent years, and there remains a definitive need to further quantify intracellular glycogen utilization in a variety of exercise settings, according to training status, age and gender.

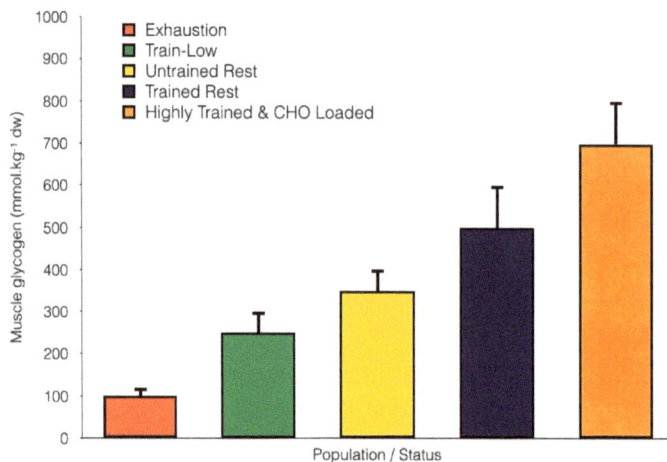

Figure 1. Variations in muscle glycogen storage according to fatigue status, training status and dietary carbohydrate (CHO) intake (data are compiled from males only and from several studies including Taylor et al. [30]; Bartlett et al. [31]; Arkinstall et al. [32]; Gollnick et al. [24]; Coyle et al. [8]).

3. Regulation of CHO Metabolism

There are a number of potential sites of control that can regulate the interaction of CHO and lipid metabolism during endurance exercise (see Figure 2). These include the availability of intra-muscular and extra-muscular substrate (controlled by diet and the action of key hormones such as the catecholamines and insulin), the abundance of transport proteins involved in transporting substrates across both the plasma and mitochondrial membranes and, of course, the activity of the key regulatory enzymes involved in the metabolic pathways. The activity of regulatory enzymes can be modified acutely through covalent modification (i.e., phosphorylation and dephosphorylation, largely under hormonal control) and/or allosteric regulation via important signalling molecules that are produced in the muscle as a result of contraction (e.g., adenosine diphosphate (ADP), adenosine monophosphate (AMP), inosine monophosphate (IMP), inorganic phosphate (Pi), calcium (Ca^{2+}), hydrogen ion (H^+). Enzyme activity can also be modified through substrate activation or product inhibition such that increasing the substrate concentration increases catalysis, whereas increased product concentration may inhibit the reaction. Finally, enzyme activity can be regulated chronically through increasing the muscle cell's content of the actual enzyme protein (i.e., more of the enzyme is actually present) as would occur with endurance training. Clearly, muscle cells possess a highly coordinated and regulatory network of signalling and feedback pathways which function to ensure ATP demand is matched by ATP synthesis. From a physiological perspective, key factors, such as exercise intensity, duration, nutritional status, training status etc., can all regulate substrate utilization during exercise, largely through influencing the potential regulatory control points discussed above. This section will outline the regulation of CHO utilization during endurance exercise, where we pay particular attention to what are currently considered the predominant sites of regulation that are relevant to the specific situation.

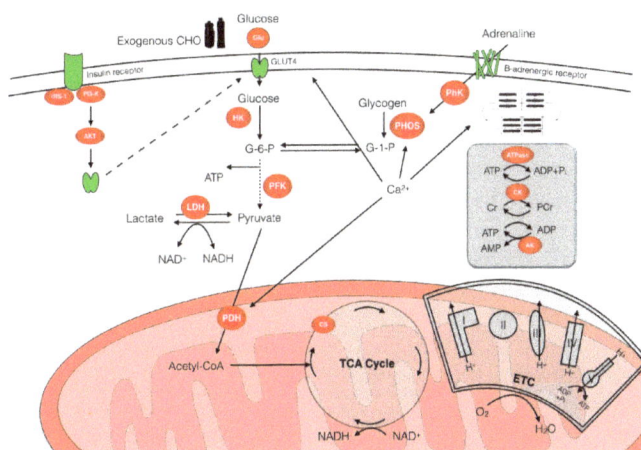

Figure 2. Overview of CHO metabolism and main control points. Key regulatory enzymes are well recognized as phosphorylase (PHOS), hexokinase (HK), phosphofructokinase (PFK), lactate dehydrogenase (LDH) and pyruvate dehydrogenase (PDH). Additionally, the rate of muscle glucose uptake can also determine the flux through glycolysis. Abbreviations: ADP, Adenosine diphosphate; AK, Adenylate kinase; AKT, Protein kinase B; AMP, Adenosine monophosphate; ATP, Adenosine triphosphate; Ca^{2+}, Calcium; CHO, Carbohydrate; CK, Creatine kinase; Cr, Creatine; CS, Citrate synthase; ETC, Electron transport chain; G-1-P, Glucose-1-phosphate; G-6-P, Glucose-6-phosphate; Glu, Gluose; GLUT4, Glucose transporter 4; H^+, Hydrogen ion; H_2O, water; IRS-1, Insulin receptor substrate 1; HK, Hexokinase; LDH, Lactate dehydrogenase; O_2, Oxygen; NAD, Nicotinamide adenine dinucleotide; TCA cycle, Tricarboxylic acid cycle; P_i, phosphate; PCr, Phosphocreatine; PFK, Phosphofructokinase; PhK, Phosphorylase kinase; PHOS, Glycogen phosphorylase; PI3-K, Phosphoinositide 3-kinase.

3.1. Effects of Exercise Intensity and Duration

As exercise intensity progresses from moderate (i.e., 65% VO_{2max}) to high-intensity (85% VO_{2max}), muscle glycogenolysis, liver glycogenolysis and glucose uptake increase such that CHO metabolism predominates [33,34]. In contrast, there is a reduction in whole body lipid oxidation due to a reduction in both plasma free fatty acids (FFA) and intramuscular triglyceride oxidation. Maximal rates of lipid oxidation are considered to occur around 65% VO_{2max}, though this is dependent on a number of other factors, such as training status, gender and diet [35].

The breakdown of muscle glycogen to glucose 1 phosphate is under the control of glycogen phosphorylase, and this reaction requires both glycogen and Pi as substrates. Phosphorylase, in turn, exists as a more active *a* form (which is under the control of phosphorylation by phosphorylase kinase) and also as a more inactive *b* form (which exists in a dephosphorylated form due to the action of protein phosphatase 1). Given that phosphorylase can be transformed via covalent modification (i.e., phosphorylation by phosphorylase kinase) mediated through epinephrine, it would be reasonable to expect that greater phosphorylase transformation from *b* to *a* may be one mechanism to explain the increased glycogenolysis evident with increasing exercise intensity. This would also be logical given that sarcoplasmic Ca^{2+} levels would be increased with high-intensity exercise (given the need for more rapid cross-bridge cycling) and that Ca^{2+} is a potent positive allosteric regulator of phosphorylase kinase through binding to the calmodulin subunit [36]. However, the percentage of phosphorylase in the more active *a* form does not appear to be increased with exercise intensity and, in actual fact, is decreased after only 10 min of high intensity exercise, which may be related to the reduced pH associated with intense exercise [37]. Whereas this mechanism of transformation (mediated by Ca^{2+} signalling) may be in operation within seconds of the onset of contraction [38], it appears that post-transformational mechanisms are in operation during more prolonged periods of high-intensity exercise given that glycogenolysis still occurs despite reduced transformation. In this regard, vital signals related to the energy status of the cell play more prominent roles. Indeed, as exercise intensity progresses from moderate to high-intensity exercise, the rate of ATP hydrolysis increases so much that there is a greater accumulation of ADP, AMP and Pi. In this way, the increased accumulation of Pi as a result of increased ATP hydrolysis can increase glycogenolysis, as it provides the increased substrate required for the reaction. Furthermore, greater accumulations of free ADP and AMP can also subsequently fine-tune the activity of phosphorylase *a* through allosteric regulation [37].

In addition to muscle glycogen, the contribution of plasma glucose to ATP production also increases with exercise intensity. The most likely explanation for this is due to increased muscle blood flow (and hence substrate delivery) in addition to increased muscle fibre recruitment [39]. The delivery of glucose to the contracting muscle is, of course, also a reflection of increased rates of liver glycogenolysis in accordance with increases in exercise intensity [40]. The regulation of liver metabolism during exercise is beyond the scope of the present review, and the reader is directed to the comprehensive review by Gonzalez et al. [34]. Although glucose uptake is also regulated by GLUT4 content, GLUT4 is unlikely to play a role in this situation given that GLUT4 translocation to the plasma membrane is not increased with exercise intensity [41]. Once glucose is transported into the cytosol, it is phosphorylated to glucose-6-phosphate under the control of hexokinase. Evidence suggests that hexokinase activity is also not limiting given that patients with type 2 diabetes (who have reduced maximal hexokinase activity) display normal patterns of exercise-induced glucose uptake likely due to normal perfusion and GLUT4 translocation [42]. In contrast, during intense exercise at near maximal or supra-maximal intensity, glucose phosphorylation may be rate limiting to glucose utilization given that high rates of glucose-6-phosphate, secondary to muscle glycogen breakdown, can directly inhibit hexokinase activity [43]. Once glucose enters the glycolytic pathway, the rate limiting enzyme to glycolysis is considered to be phosphofructokinase (PFK). PFK is allosterically activated by ADP, AMP and Pi, and this mechanism is likely to explain high rates of glycolysis during intense exercise even in the face of metabolic acidosis when PFK could be inhibited.

In contrast to high intensity exercise, prolonged steady state exercise lasting several hours is characterized by a shift towards increased lipid oxidation and reduced CHO oxidation rates [44]. This shift in oxidation rates is accompanied by an increased contribution of plasma FFA towards energy expenditure and a decreased reliance on both muscle glycogen and intramuscular triglycerides (IMTGs) [44]. Studies examining the regulatory mechanisms underpinning this shift in substrate utilization have suggested that a reduction in muscle glycogen availability (due to progressive glycogen depletion and hence a reduced glycolytic flux) down-regulate pyruvate dehydrogenase (PDH) activity, thereby leading to reduced CHO oxidation. In addition, progressive increases in plasma FFA availability (due to continual lipolysis in adipose tissue) stimulate lipid oxidation. The down-regulation of PDH activity as exercise duration progresses may be due to reduced pyruvate flux, therefore, reducing the substrate production required for the PDH reaction [44]. In addition, more recent data demonstrate an up-regulation of PDH kinase activity during exercise which would, therefore, directly inhibit PDH activity [45]. Taken together, these data are consistent with the many observations that increasing or decreasing substrate availability is one of the most potent regulators of fuel utilization patterns during exercise, and this concept is discussed in the next section.

3.2. Effects of Substrate Availability

Modifying substrate availability through dietary manipulation (such as loading regimens, pre-exercise meals or providing enhanced substrate availability during exercise) has been consistently shown to alter metabolic regulation during endurance exercise through various control points. Increasing muscle glycogen concentration enhances glycogenolysis during exercise [46] by enhancing phosphorylase activity given that glycogen is a substrate for phosphorylase. The enhanced glycogenolysis with elevated glycogen stores does not appear to affect muscle glucose uptake [46,47]. In addition to glycogenolysis, muscle glycogen also appears to be a potent regulator of PDH activity (and thus CHO oxidation) during exercise. Indeed, commencing exercise with reduced muscle glycogen attenuates the exercise-induced increase in PDH activity and vice versa [48], likely due to reduced glycolytic flux as well as an increased resting content of pyruvate dehydrogenase kinase 4 (PDK4, the kinase responsible for deactivating PDH) when glycogen concentrations are low. PDH regulation appears particularly sensitive to nutritional status, even at rest. In fact, just 3 days of a low CHO (but increased fat) diet up-regulates PDH kinase activity and down-regulates PDH activity [49].

Although the effects of exercise intensity on substrate utilization were discussed previously, it appears that muscle glycogen availability can influence fuel metabolism over and above that of exercise intensity. Indeed, Arkinstall et al. [47] observed that glycogen utilization was enhanced during exercise at 45% VO_{2max} that was commenced with high glycogen (591 mmol·kg^{-1} d.w.) as opposed to exercise at 70% VO_{2max} commenced with low glycogen concentration (223 mmol·kg^{-1} d.w.), despite the higher intensity. In contrast to glycogen utilization and CHO oxidation rates, lipid oxidation was highest when exercise was commenced with reduced glycogen stores. The shift towards fat oxidation when pre-exercise muscle glycogen is low is likely mediated by a number of contributing factors. Firstly, a reduced glycogen availability is associated with an increased epinephrine concentration and plasma FFA availability, thus favoring conditions for augmented lipolysis and lipid oxidation, respectively, compared with conditions of high glycogen concentration [47]. However, when a pre-exercise meal is ingested, and glucose is infused during glycogen depleted exercise such that minimal differences exist between plasma FFA and epinephrine, lipid oxidation is still augmented [50]. In such circumstances, available evidence points to regulation within the muscle cell itself and more specifically, a carnitine mediated increase in lipid oxidation. Carnitine is important given its role as a substrate for carnitine palmitoyltransferase 1 (CPT-1) activity, the rate limiting enzyme that facilitates long chain fatty acid entry to the mitochondria. Indeed, these researchers observed lower PDH activity and acetyl CoA and acetyl carnitine content, and an increased free carnitine concentration during glycogen depleted exercise when compared with glycogen loaded conditions. Interestingly, acetyl CoA carboxylase (ACC) phosphorylation increased and malonyl CoA decreased similarly in both conditions, despite higher AMP activated protein kinase (AMPK) activity when glycogen was reduced. Such data provide further support for a critical role of carnitine in regulating the interaction between CHO and lipid utilization [51] and suggest that malonyl CoA (as an allosteric inhibitor of CPT-1 activity) does not play a major role in fine-tuning fat oxidation in human skeletal muscle. Nonetheless, it is acknowledged that the apparent disconnect between malonyl CoA, CPT-1 activity and in vivo lipid oxidation may be dependent on palmitoyl-CoA concentrations [52].

In addition to muscle glycogen availability, consuming CHO in a pre-exercise meal and/or during exercise also induces potent effects on the regulation of CHO and lipid metabolism during exercise. Indeed, CHO ingestion during exercise appears to suppress or even abolish hepatic glucose output during exercise, thus attenuating the decline in liver glycogen content [53]. In relation to lipid metabolism, CHO feeding attenuates lipolysis in adipose tissue (as mediated via anti-lipolytic effects of insulin) such that plasma FFA availability is reduced during exercise undertaken in CHO fed conditions [54]. Horowitz et al. [54] studied male participants during 60 min of exercise at 45% VO_{2max} in fasted conditions, or 1 h after consuming 0.8 $g \cdot kg^{-1}$ of glucose (to induce a high insulin response) or 0.8 $g \cdot kg^{-1}$ fructose (to induce a low insulin response), or in an additional glucose trial during which intralipid and heparin were infused so as to maintain plasma FFA availability in the face of high insulin. In accordance with the insulin response in CHO fed conditions, lipolysis (as indicated by rate of appearance of glycerol), FFA availability and lipid oxidation were reduced. However, when intralipid and heparin were infused during the additional glucose trial, lipid oxidation rates were enhanced by 30% (4.0 $\mu mol \cdot kg^{-1} \cdot min^{-1}$) compared with the glucose only trial (3.1 $\mu mol \cdot kg^{-1} \cdot min^{-1}$) but were still not restored to levels occurring during fasted exercise (6.1 $\mu mol \cdot kg^{-1} \cdot min^{-1}$). Taken together, whilst these data suggest that only small elevations in insulin can attenuate lipolysis (i.e., 10–30 $\mu U/mL$), they also demonstrate a limitation within the muscle cell itself during CHO fed conditions.

In an effort to ascertain the source of limitation to lipid oxidation within the muscle following CHO feeding, Coyle et al. [55] infused octanoate (a medium chain fatty acid, MCFA) or palmitate (a long chain fatty acid, LCFA) during 40 min of exercise at 50% VO_{2max} after an overnight fast or 60 min after ingesting 1.4 $g \cdot kg^{-1}$ of glucose. As expected (based on the previously discussed study), plasma FFA and lipid oxidation were higher in the fasted trials whilst CHO oxidation was lower compared with the glucose trials. However, the major finding of this study was that the percentage of palmitate oxidized during the glucose trial was reduced compared with fasting (70% vs. 86%, respectively) whereas octanoate was unaffected (99% vs. 98%, respectively). These data, therefore, suggest that LCFA uptake into the mitochondria is reduced with CHO feeding. When taken in the context of previous sections, it becomes apparent that any condition which accelerates glycolytic flux (e.g., increased intensity, muscle glycogen, glucose feeding) can down-regulate intramuscular lipid metabolism. Furthermore, the increased insulin and decreased epinephrine levels which accompany glucose ingestion during exercise appear to result in the attenuation of intra-muscular hormone sensitive lipase (HSL) activity [56], thus highlighting an additional point of control.

3.3. Effects of Training Status

Endurance training results in a number of profound physiological and metabolic adaptations which function to reduce the degree of perturbations to homeostasis for a given exercise intensity and ultimately, delay the onset of fatigue. Adaptations to endurance training are most recognized functionally by an increase in maximal oxygen uptake as well as a rightward shift in the lactate threshold. From a metabolic perspective, the most prominent adaptation is an increase in the size and number of mitochondria (i.e., mitochondrial biogenesis), which essentially permits a closer matching between ATP requirements and production via oxidative metabolism. The adaptive response of muscle mitochondria is also accompanied by increases in capillary density, substrate transport proteins and increased activity of the enzymes involved in the main metabolic pathways. In addition, endurance training increases the capacity for skeletal muscle, but not liver [34], to store glycogen and triglycerides, thereby increasing substrate availability. In relation to substrate utilization during exercise following endurance training, the most notable response is a reduction in CHO utilization with a concomitant increase in lipid oxidation [57].

For a given exercise intensity, glycogen utilization is reduced with exercise training [58], an effect that is confined locally to the actual muscles that were trained [59]. The reduced glycogenolysis observed after training is not due to any change in phosphorylation transformation but rather, allosteric mechanisms [60,61]. Indeed, exercise in the trained state is associated with reduced contents of ADP, AMP and Pi, thereby providing a mechanism leading to reduced phosphorylase activity. Le Blanc et al. [61] also observed reduced pyruvate and lactate production during exercise undertaken in the trained state as well as reduced PDH activity. As a result of the reduced CHO flux, it is, therefore, likely that the attenuated pyruvate production (in addition to reduced ADP accumulation) may have attenuated PDH activity. In addition to training-induced reductions in muscle glycogenolysis, several investigators have observed that training reduces exercise-induced liver glycogenolysis [34]. In this way, trained individuals will have more liver glycogen available late in exercise to thereby maintain the plasma glucose concentration and hence, support maintenance of the desired exercise intensity. There is some evidence that endurance training also increases gluconeogenesis following training [62]. In accordance with reduced rates of glucose production, muscle glucose uptake is reduced when exercise is undertaken at the same absolute workload following a period of endurance training [63].

Despite the fact that training increases total muscle GLUT4, the reduction in exercise-induced muscle glucose uptake is likely caused by a reduced translocation of GLUT4 to the sarcolemma following training, thereby reducing the capacity to transport glucose [64]. One particular study utilized a knee extensor training and exercise model where only one limb was trained but yet both limbs performed the exercise protocol before and after training. In this way, training-induced alterations in hormonal and cardiovascular status were minimized and the reduced glucose uptake and GLUT4 translocation were likely mediated by local contractile factors. In summarizing the link between liver glucose production and muscle glucose uptake, it is generally accepted that training-induced changes in hormone concentrations such as epinephrine, insulin and glucagon are unable to explain all of the effects [65]. Rather, it is possible that the actual rate of muscle glucose uptake acts as a feedback signal to regulate glucose output from the liver [65].

4. CHO and Exercise Performance

Given the effects of exercise intensity, duration and training status on muscle glycogen utilization, it follows that glycogen depletion (in both muscle and liver) is a major cause of fatigue in both endurance and high-intensity (intermittent) type activities. As such, traditional nutritional advice for these types of activities (whether it is competitive situations or training sessions) is to ensure high daily CHO intake before, during and after the activity so as to promote both performance and recovery.

4.1. Muscle Glycogen and CHO Loading

The basic principles of CHO loading were developed in the late 1960s where it was identified that a period of exhaustive exercise followed by several days of high dietary CHO intake induces a super-compensation effect so that glycogen storage is augmented [3,4]. A less extreme form of CHO loading was developed in the 1980s where Sherman et al. [66] observed that a simple exercise taper in conjunction with several days of increased dietary CHO intake was also sufficient to increase glycogen storage. It is now generally accepted that trained athletes can increase glycogen storage in both type I and II fibres within 24–48 h of increased CHO intake [67]. In relation to practical application, it is also suggested that high glycemic foods are superior to low glycemic foods [68] in augmenting glycogen storage and that dietary intakes of 8–12 $g \cdot kg^{-1}$ per day are likely required to "maximize" glycogen storage [7]. The general consensus from the wealth of studies undertaken in the last 40 years is that CHO loading can improve performance and capacity when the exercise is greater than 90 min in duration [69]. The enhanced performance effect is likely initially mediated by a delay in the time-point at which energy availability becomes limiting to the maintenance of the desired workload, which, in the case of race pace, is dependent on sustained and high rates of CHO oxidation [70,71]. Indeed, in reviewing the literature, Hawley et al. [69] cited that CHO loading can improve exercise capacity by approximately 20%, and time trial performance can increase by 2–3%. In addition to providing substrate availability for ATP production, it is now recognized that glycogen availability (especially the intramyofibrillar storage pool) can directly modulate contractile function. Indeed, a series of studies from Ørtenblad and colleagues [28,29,72] have collectively shown preferential utilization of this storage pool during exercise in a manner that also correlates with impaired Ca^{2+} release from the sarcoplasmic reticulum. Such impaired excitation–contraction coupling is likely to be of particular importance during situations where higher power outputs and sprint finishes are required in the very late and finishing stages of races.

4.2. Pre-Exercise CHO Availability

Whereas the 1960s and 1970s focused on CHO loading studies, research in the next two decades examined the effects of pre-exercise feeding as well as consuming additional CHO during exercise. Pre-exercise feeding (i.e., 3–4 h before competition) is not only advantageous as it can lead to further elevations in muscle glycogen content [73] but can also restore liver glycogen content, which is usually depleted after an overnight fast. The latter is particularly important given that liver glycogen content is also related to exercise capacity [13]. Sherman et al. [74] observed that time trial performance after 90 min of steady state exercise at 70% VO2max was greater when 150 g of CHO was consumed before exercise compared with 75 g of CHO, both of which were greater than no meal. The enhanced performance effect was associated with the maintenance of blood glucose concentration late into exercise, which is important because liver glucose production and muscle glucose uptake and oxidation become more important when muscle glycogen concentrations begin to decline. In a further study, the same authors also observed that performance can be further increased when CHO is ingested during exercise in addition to a pre-exercise meal [75]. As such, current CHO guidelines for pre-exercise feeding advise an intake of 1–4 $g \cdot kg^{-1}$ body mass, 3–4 h prior to exercise [7].

4.3. CHO Feeding during Exercise

In addition to high endogenous pre-exercise muscle and liver glycogen stores, it is widely accepted that exogenous CHO feeding during exercise also improves physical elements of performance [76]. Whereas it was generally accepted that exogenous CHO oxidation rates were limited at approximately 1 $g \cdot min^{-1}$ due to saturation of intestinal glucose transporters, it is now known that exogenous CHO oxidation rates can increase to 1.8 $g \cdot min^{-1}$ with the addition of sucrose or fructose to the CHO blend [77]. When taken together, it is currently thought that CHO feeding during exercise may, therefore, augment exercise performance via multiple mechanisms, consisting of muscle glycogen sparing [22], liver glycogen sparing [53] and maintenance of plasma glucose and CHO oxidation rates [8]. The role of CHO feeding in reducing liver glycogen breakdown as a performance enhancing mechanism is gaining increasing recognition [78]. In simple terms, a liver glycogen sparing effect ensures that more liver glycogen is, therefore, available late in exercise, thereby maintaining plasma glucose availability and delivery to the muscle to meet the CHO oxidation rates necessary to sustain the required workload.

It is noteworthy that exogenous CHO feeding during exercise also improves performance [11] when exercise duration is <60 min (i.e., where muscle and liver glycogen availability is not likely limiting), an effect that is not apparent when glucose is directly infused to the bloodstream during exercise [79]. Such data suggest that CHO feeding may also improve exercise performance via non-metabolic effects but through direct effects on the central nervous system [14]. To this end, the last decade of research has resulted in a growing body of literature demonstrating that simply "rinsing" CHO in the oral cavity (for 10-s periods every 5–10 min during exercise) is also ergogenic to performance [80], an effect that is independent of sweetness [15] and that is especially apparent in the absence of a pre-exercise CHO meal [81] and low pre-exercise muscle glycogen [82], although this effect is not always evident [83].

The conventional approach to CHO fueling during exercise is to consume 6–8% CHO beverages, although relying solely on this approach does not allow for flexibility in terms of individual variations in body mass or actual fluid requirements given variations in ambient conditions [84]. As such, many athletes rely on a CHO fueling approach that is based on a combination of solids (e.g., bars), semi-solids (e.g., gels) and fluids (e.g., sports drinks) so as to collectively meet their personalized exogenous CHO targets, typically in the region of 30–90 $g \cdot h^{-1}$, depending on exercise duration.

Nevertheless, although there is little difference in exogenous CHO oxidation rates (albeit in fluid matched conditions) between the aforementioned sources [85,86], it is noteworthy that many athletes experience gastrointestinal discomfort when attempting to hit these targets, possibly related to extreme differences in osmolality between commercially available CHO gels [87] as well as the presence of fibre, fat and protein in energy bars [88]. As such, it is now advised that athletes should clearly practice their approach to in-competition fueling during training sessions of a similar intensity and duration as competition. As a general rule of thumb, it is suggested that 30–60 g·h^{-1} of CHO (glucose polymers e.g., maltodextrin) is consumed during events lasting < 60–150 min [78], whereas in events > 2.5–3 h, 60–90 g·h^{-1} (glucose/fructose blends) is the recommended rate [7]. Whilst beyond the scope of the present review, it is noteworthy that CHO ingestion (in either drink or gel format) during team sport type activity (i.e., <90 min duration) can also improve performance of technical skills (see reference [89] for an extensive review on this topic), thus providing further evidence for the ergogenic properties of CHO feeding during exercise.

5. CHO and Training Adaptations

5.1. Overview of Molecular Regulation of Training Adaptations

Skeletal muscle is a highly malleable tissue that has the ability to undergo major adaptations and alter its phenotype in response to exercise stimuli. In relation to endurance training, prominent adaptations include increased mitochondrial biogenesis, lipid oxidation and angiogenesis, recognized functionally by a rightward shift of the lactate threshold curve [90]. Upon the onset of muscle contraction, the accumulation of multiple metabolic signals generated during exercise (i.e., increased AMP/ATP ratio, Ca^{2+} flux, lactate, hypoxia and energy availability) initiates a cascade of events that activate or suppress specific signalling pathways that regulate gene expression and protein translation [91–94]. The dynamic fluctuation in content and subcellular location of metabolites activates regulatory cell signalling kinases that converge on nuclear and mitochondrial transcription factors, and co-activators to induce a co-ordinated up-regulation of both nuclear and mitochondrial genomes [95]. It is the combination of transient increases in gene expression and protein content that ultimately form the molecular basis of training adaptations. Many exercise regulated signalling pathways are also sensitive to nutrient availability [17], and a schematic overview of the potential exercise and nutrient interactions that modify the early signalling events regulating mitochondrial biogenesis is displayed in Figure 3.

As previously discussed, the principle of promoting high CHO availability before, during and after exercise is the foundation on which traditional sports nutrition guidelines are based. Although this is essential for promoting competition performance and ensuring adequate recovery, accumulating data now suggest that restricting CHO before, during, and in recovery from endurance-based exercise augments the cell signaling and gene expression responses associated with oxidative adaptations in human skeletal muscle. Indeed, both acute and training based studies have collectively observed that the reduction of both endogenous and/or exogenous CHO promotes an increase in mitochondrial enzyme activity and protein content, increases both whole body [96] and intramuscular [97] lipid metabolism and can improve both exercise capacity [98] and performance [99]. This approach to CHO periodization has been termed "train-low, compete-high", a model which promotes carefully scheduled periods of CHO restricted training for augmenting adaptation, but ensures high CHO availability prior to and during competition in order to promote maximal performance.

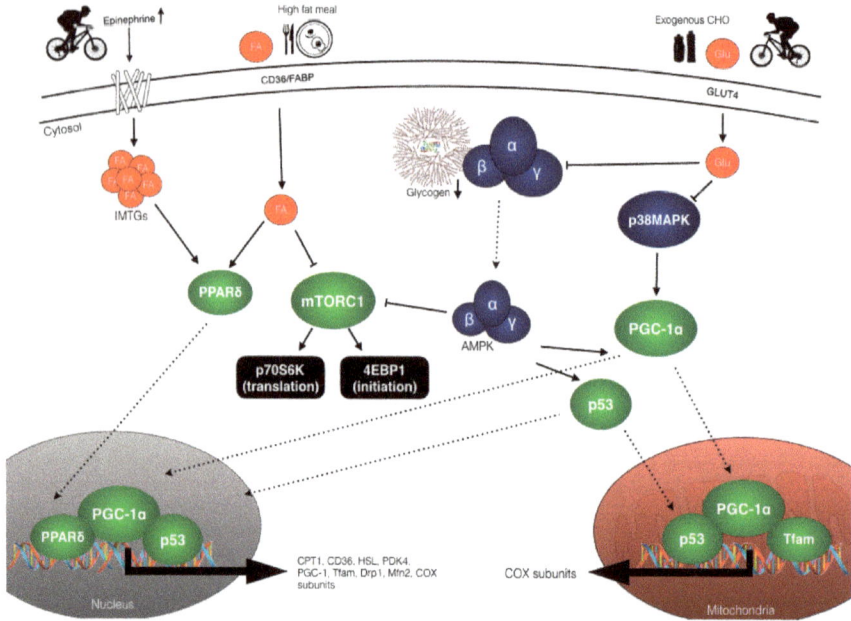

Figure 3. Schematic overview of the potential exercise-nutrient sensitive cell signalling pathways regulating the enhanced mitochondrial adaptations associated with training with low CHO availability. Reduced muscle glycogen and exogenous glucose availability enhance both AMPK and p38MAPK phosphorylation which results in activation and translocation of PGC-1α and p53 to the mitochondria and nucleus. Upon entry into the nucleus, PGC-1α co-activates additional transcription factors, (i.e., NRF1/2) to increase the expression of COX subunits and Tfam as well as auto-regulating its own expression. In the mitochondria, PGC-1α co-activates Tfam to coordinate the regulation of mtDNA and induces expression of key mitochondrial proteins of the electron transport chain, e.g., COX subunits. Similar to PGC-1α, p53 also translocates to the mitochondria to modulate Tfam activity and mtDNA expression and to the nucleus where it functions to increase the expression of proteins involved in mitochondrial fission and fusion (DRP-1 and MFN-2) and electron transport chain protein proteins. Exercising in conditions of reduced CHO availability increases adipose tissue and intramuscular lipolysis via increased circulating epinephrine concentrations. The resulting elevation in FFA activates the nuclear transcription factor, PPARδ, to increase expression of proteins involved in lipid metabolism, such as CPT-1, PDK4, CD36 and HSL. However, consuming pre-exercise meals rich in CHO and/or CHO during exercise can down-regulate lipolysis (thereby negating FFA mediated signalling) as well as reducing both AMPK and p38MAPK activity, thus having negative implications for downstream regulators. High fat feeding can also modulate PPARδ signalling and up-regulate genes with regulatory roles in lipid metabolism (and down-regulate CHO metabolism), though high fat diets may also reduce muscle protein synthesis via impaired mTOR-p70S6K signalling, despite feeding leucine rich protein. Abbreviations: 4EBP1; eukaryotic translation initiation factor 4E-binding protein 1, AMPK; AMP-activated protein kinase, CHO; carbohydrate, COX; cytochrome c oxidase, CPT-1; carnitine palmitoyltransferase 1, Drp1; dynamin-related protein 1, FA; fatty acid, FABP; fatty acid binding protein, GLU; glucose, HSL; hormone sensitive lipase, IMTG; intramuscular triglycerides, Mfn2; mitofusion-2, mTORC1; mammalian target of rapamycin complex 1, p38 mitogen-activated protein kinase, p70S6K; ribosomal protein S6 kinase, PDK4; pyruvate dehydrogenase kinase 4, PGC-1α; peroxisome proliferator-activated receptor gamma coactivator 1-alpha, PPARδ; peroxisome proliferator-activated receptor, Tfam; mitochondrial transcription factor A.

5.2. Low Muscle Glycogen Availability and Twice per Day Training Models

The altered metabolic milieu created through exercising with low glycogen availability has a direct impact on molecular signalling events controlling muscular adaptations. The notion that CHO restriction augments markers of training adaptation emerged from an initial investigation from Pilegaard et al. [100] who observed an enhanced expression of genes involved in mitochondrial biogenesis and substrate utilization when exercise was undertaken with reduced muscle glycogen. Indeed, exercise induced PDK4 and UCP3 gene expression were both augmented with low (240 ± 38 mmol·kg^{-1} d.w.) pre-exercise muscle glycogen when compared to normal (398 ± 52 mmol·kg^{-1} d.w.) glycogen levels. On the basis of the molecular evidence derived from acute studies, initial training studies adopted a "training twice every second day versus once daily" model. Using this model, Hansen et al. [98] subjected seven untrained males to 10 weeks of knee extensor exercise training under conditions of either high or low muscle glycogen. Subjects trained both legs according to two different schedules, whereby one leg was trained every day (HIGH) whilst the contralateral leg was trained twice a day, every other day (LOW). Exercise during the twice per day sessions was interspersed with 2 h of recovery, during which time CHO intake was restricted. As such, one leg (LOW) commenced 50% of training sessions under conditions of low glycogen, while the other leg performed each session under conditions of high muscle glycogen (HIGH), whilst allowing the matching of total work done between legs. Following 10 weeks of training, the leg that commenced 50% of training sessions with low muscle glycogen demonstrated a superior increase in the maximal activity of both citrate synthase (CS) and β-hydroxyacyl-CoA dehydrogenase (β-HAD) when compared with the contralateral leg. Furthermore, the LOW leg also demonstrated an almost two-fold increase in exercise capacity when compared with the HIGH leg, thus demonstrating a potent effect of altering substrate availability on exercise-induced adaptation and subsequent performance.

Yeo and colleagues [96] subsequently adopted a "real world" design more applicable to elite athletes. Using well trained male cyclists completing a 3-week training block, cyclists trained six times per week, either once every day with high muscle glycogen availability or twice every other day, so the second session was undertaken with reduced levels of muscle glycogen. In the "high" group, cyclists alternated between steady-state and high-intensity training (HIT) each day, whereas in the "low" group, steady-state exercise was performed in the morning and HIT exercise was performed after a 1–2 h recovery period during which time CHO was restricted. Before and after this training block, muscle biopsies were obtained to assess markers of adaptation, and a time trial was completed to examine performance improvements in each group. Despite significant increases in CS and β-HAD activity, cytochrome c oxidase subunit IV (COXIV) protein content and rates of fat oxidation in the "low" group following training, training-induced improvements in time trial performance were comparable between groups. Interestingly, the enhanced adaptive responses occurred in the "low" group despite cyclists having to reduce exercise intensity during the HIT training session. These findings suggest that even when overall training intensity is reduced, reduced CHO availability is associated with an adaptive response. In a similar study design, Hulston and colleagues [97] reported greater increases in intra-muscular lipid oxidation and the expression of CD36 and β-HAD activity following "training low" compared with "training high".

5.3. Fasted Training

Performing endurance training in the fasted state represents a simpler model of "training low" where exercise is performed prior to breakfast. Although pre-exercise muscle glycogen is not altered as a result of the overnight fast, liver glycogen remains lower whilst FFA availability is increased [54] compared with when breakfast is fed. Exercising in the fasted state increases post-exercise AMPK activity [101] and mRNA of genes controlling substrate utilization (PDK4, GLUT4, CD36, CPT-1) and mitochondrial function (UCP3) [102,103] compared with when CHO is fed before and during exercise. Accordingly, chronic periods of fasted training elicit similar adaptations to those observed when training with low muscle glycogen. Nybo et al. [104] demonstrated that 8 weeks of endurance training (50–90 min of high-intensity intervals at 70–85% VO_{2max}) in the fasted state enhanced training induced increases in β-HAD activity and basal muscle glycogen content compared with when CHO was fed before and during training. Similarly, Van Proeyen et al. [105] observed augmented CS and β-HAD activity when regular steady state (1–1.5 h cycling at 70% VO_{2max}) cycling was performed in the fasted state compared to when breakfast was fed. Nonetheless, the augmented biochemical adaptations did not translate to improved exercise performance.

5.4. Post-Exercise CHO Restriction

In addition to restricting CHO prior to endurance exercise training, data also demonstrate beneficial adaptive responses when restricting CHO during the post-exercise recovery period. Indeed, Pilegaard et al. [106] explored this idea with participants completing 75 min of cycling at 75% VO_{2max} followed by the consumption of a diet either high or low in CHO for the next 24 h. These authors observed that although the mRNA expression of pyruvate dehydrogenase kinase 4 (PDK4), lipoprotein lipase (LPL), uncoupling protein 3 (UCP3), and carnitine palmitoyltransferase 1 (CPT1) increased in response to exercise, expression levels were sustained at 24 h post-exercise in the low CHO group only. In a twice per day, 6-week training study, it was also observed that when glucose is consumed during recovery from the first session, the enhanced oxidative adaptations are blunted compared to when CHO is restricted, despite reduced levels of muscle glycogen [107]. When taken together, data from these studies suggest that reducing CHO availability in the recovery period also modulates the muscle adaptive process.

5.5. Sleep-Low, Train-Low Models

More recent train-low investigations have adopted a "sleep-low, train-low" model, whereby participants perform an evening training session, restrict CHO overnight and complete a training session the subsequent morning under levels of low muscle glycogen availability. Acute studies using whole body exercise [31] demonstrate that commencing HIT running with low muscle glycogen (as a result of glycogen depleting exercise the prior evening) leads to significant phosphorylation of both ACC and p53 alongside enhanced gene expression of PGC-1α, COXIV, Tfam and PDK4. In contrast, when exercise was commenced with high muscle glycogen and exogenous CHO was provided before, during and after exercise, the aforementioned activation of signalling kinases and gene expression were completely abolished. In a subsequent study, Lane et al. [108] manipulated the timing of daily CHO ingestion to elicit sleeping with reduced muscle glycogen. In this way, subjects consumed either 8 g·kg^{-1} CHO prior to a bout of evening HIT (and subsequently restricted CHO intake overnight) or consumed 4 g·kg^{-1} CHO prior to the evening HIT and 4 g·kg^{-1} CHO post-exercise in order to replenish muscle glycogen. Although fat oxidation during morning exercise and post-exercise PDK4 mRNA expression were significantly greater in the sleep-low group, the genes involved in the regulation of mitochondrial biogenesis displayed similar exercise induced increases in both groups [108]. Using a similar sleep-low model as Lane and colleagues [108], Marquet et al. [99] observed that 3 weeks of sleep-low training in elite triathletes and cyclists improved cycling efficiency (3.1%), 20 km cycling time trial performance (3.2%) and 10 km running performance (2.9%) compared with traditional high CHO approaches.

While the mechanisms underpinning the aforementioned adaptive responses to both acute and chronic exercise are still not fully understood, they are likely mediated by upstream signalling kinases. Indeed, AMPK has the capacity to be modulated by the glycogen status of the muscle through a glycogen-binding domain on the β-subunit [109]. Wojtaszewski et al. [92] demonstrated that when pre-exercise muscle glycogen levels are reduced, AMPKα2 activity and ACC^{Ser221} phosphorylation are significantly elevated following steady state cycling compared to when muscle glycogen is high. In a subsequent study, Chan et al. [110] also observed a significantly greater nuclear abundance of p38MAPK, both pre and post exercise, when muscle glycogen levels were low. In a twice per day, train-low model, Cochran et al. [93] also reported significantly greater elevations in p38MAPK phosphorylation following the second exercise session when participants consumed no CHO during recovery. These data are highly suggestive of both AMPK and p38MAPK being nutrient sensitive, and thus likely regulating the downstream events (e.g., p53 and PGC-1α activation) that co-ordinate mitochondrial biogenesis.

5.6. Critical Limitations of Train-Low Models

Despite the theoretical rationale for train-low protocols, it is noteworthy that the augmented cell signalling responses, enzymatic changes and improved performance outcomes are not always apparent or consistent between studies (see Impey et al. [111] for an extensive review of this topic). Indeed, recent work from Gejl et al. [112] observed no additional benefit when elite triathletes performed selected training sessions with low CHO availability on both CS and β-HAD maximal activity when compared with traditional high CHO availability approaches. Similarly, Burke et al. [113] observed that periodising CHO availability (to incorporate train low sessions) offered no additional benefit to 10 km race walking performance in Olympic standard race walkers when compared with high CHO availability. However, given that the enhanced training response associated with train-low is potentially mediated by muscle glycogen availability, close examination of the available glycogen data between studies may explain the lack of molecular, biochemical and performance changes reported within some train-low studies. In this regard, we recently suggested the concept of a glycogen threshold, whereby the beneficial adaptations associated with train-low paradigms are especially apparent when the exercise session is commenced with muscle glycogen concentrations < 300 mmol·kg^{-1} d.w. [111]. To this end, there is, therefore, a definitive need to better understand the exercise and nutrient conditions that truly constitute train-low conditions.

5.7. Practical Applications

There are also a number of potential limitations to this type of training that can make it difficult for exercise physiologists and nutritionists to optimally periodise this type of training into an elite athlete's training schedule. For example, reduced CHO availability impairs acute training intensity [96,97] and hence, if performed long-term, may actually lead to a de-training effect. Additionally, given the role of CHO in preventing immunosuppression, it has been suggested that repeated high intensity training under conditions of low CHO increases susceptibility to illness and infection [114]. Nonetheless, proxy markers of immunity and the incidence of upper respiratory tract infections (URTI) appear to be unaltered by 3 weeks of sleep-low training [115]. Restriction of CHO availability has also been shown to increase muscle protein breakdown [116], an effect that, if performed chronically, may lead to muscle mass loss, especially in conditions of both calorie and CHO restriction. Finally, data also demonstrate a reduced ability to oxidize exogenous CHO following regular training with low CHO, which could lead to a negative effect on competition performance [117]. Taking the above limitations into account, it is important to recognize that training with low CHO availability should be carefully periodised in an athlete's training programme. In practice, this approach could represent an amalgamation of train-low paradigms and is perhaps best communicated by the principle of "fuel for the work required" [111]. In this way, athletes could strategically reduce CHO availability prior to completing pre-determined training workloads that can be readily performed with reduced CHO availability, thereby inducing a "work-efficient" approach to training [118]. Alternatively, when the goals of the training session are to complete the highest workload possible over more prolonged durations, then adequate CHO should be provided in the 24 h period prior to and during the specific training session. Careful day-to-day periodization in a meal-by-meal manner (as opposed to chronic periods of CHO restriction) is likely to maintain metabolic flexibility and still allow for the completion of high-intensity and prolonged duration workloads on heavy training days, e.g., interval type workouts undertaken above lactate threshold. Intuitively, train-low sessions may be best left to training sessions that are not CHO dependent and in which the intensity and duration of the session are not likely to be compromised by reduced CHO availability, e.g., steady-state type training sessions performed at intensities below the lactate threshold. Clearly, more studies are required to investigate the optimal practical approach for which to integrate periods of train-low into an elite athlete's training programme.

6. Summary and Future Directions

Despite over 100 years of research, carbohydrate metabolism continues to intrigue muscle biologists and exercise scientists. From its early recognition as a simple fuel store, it is now apparent that the glycogen granule regulates many cell-signaling processes related to both health and human performance. Nonetheless, it is clear that many of the original questions posed in our field are still relevant today, though the array of biochemical tools now at our disposal ensure we are better equipped to answer those questions with greater precision. For example, the storage of the glycogen granule in specific intracellular pools remains a highly active research area. As a related point, the magnitude of exercise-induced utilization of specific storage pools remains to be documented using "real-world" exercise protocols that are relevant to both training and competition scenarios. Whilst the specific regulatory control points of CHO metabolism are now well documented, the precise molecular mechanisms underpinning the regulation of CHO transport, storage and utilization are not yet fully known. Finally, the identification of the glycogen granule as a regulator of training adaptation has opened a new field of study that is likely to dominate the applied nature of sport nutrition research in the coming decade. From the early studies from the pioneers in the field (e.g., Krogh, Lindhard, Bergstrom, Saltin, etc.), it is clear that our field remains as exciting as ever.

Acknowledgments: The authors received no direct funding for preparation of this manuscript. Experimental data on glycogen metabolism from the authors' laboratory have been funded by GlaxoSmithKline, Lucozade Ribena Suntory and Science in Sport.

Author Contributions: All authors contributed equally to the drafting and writing of the paper and all approved the final manuscript.

Conflicts of Interest: The authors declare no conflict of interest.

References

1. Krogh, A.; Lindhard, J. The relative value of fat and carbohydrate as sources of muscular energy: With appendices on the correlation between standard metabolism and the respiratory quotient during rest and work. *Biochem. J.* **1920**, *14*, 290–363. [CrossRef] [PubMed]
2. Levine, S.A.; Gordon, B.; Derick, C.L. Some changes in the chemical constituents of the blood following a marathon race. *J. Am. Med. Assoc.* **1924**, *82*, 1778–1779. [CrossRef]
3. Bergstrom, J.; Hermansen, L.; Hultman, E.; Saltin, B. Diet, muscle glycogen and physical performance. *Acta Physiol. Scand.* **1967**, *71*, 140–150. [CrossRef] [PubMed]
4. Bergstrom, J.; Hultman, E. Muscle glycogen synthesis after exercise: An enhancing factor localized to the muscle cells in man. *Nature* **1966**, *210*, 309–310. [CrossRef] [PubMed]
5. Bergstrom, J.; Hultman, E. The effect of exercise on muscle glycogen and electrolytes in normals. *Scand. J. Clin. Lab. Investig.* **1966**, *18*, 16–20. [CrossRef]
6. Hermansen, L.; Hultman, E.; Saltin, B. Muscle glycogen during prolonged severe exercise. *Acta Physiol. Scand.* **1967**, *71*, 129–139. [CrossRef] [PubMed]
7. Thomas, D.T.; Erdman, K.A.; Burke, L.M. Position of the academy of nutrition and dietetics, dietitians of Canada, and the American College of sports medicine: Nutrition and athletic performance. *J. Acad. Nutr. Diet.* **2016**, *116*, 501–528. [CrossRef] [PubMed]
8. Coyle, E.F.; Coggan, A.R.; Hemmert, M.K.; Ivy, J.L. Muscle glycogen utilization during prolonged strenuous exercise when fed carbohydrate. *J. Appl. Physiol.* **1986**, *61*, 165–172. [CrossRef] [PubMed]
9. Tsinztas, K.; Williams, C.; Boobis, L.; Greenhaff, P. Carbohydrate ingestion and glycogen utilization in different muscle fibre types in man. *J. Physiol.* **1995**, *489*, 243–250.
10. Bosch, A.N.; Dennis, S.C.; Noakes, T.D. Influence of carbohydrate ingestion on fuel substrate turnover and oxidation during prolonged exercise. *J. Appl. Physiol.* **1994**, *76*, 2364–2372. [CrossRef] [PubMed]

11. Jeukendrup, A.; Brouns, F.; Wagenmakers, A.J.; Saris, W.H. Carbohydrate-electrolyte feedings improve 1 h time trial cycling performance. *Int. J. Sports Med.* **1997**, *18*, 125–129. [CrossRef] [PubMed]
12. Jeukendrup, A.E.; Jentjens, R. Oxidation of carbohydrate feedings during prolonged exercise: Current thoughts, guidelines and directions of future research. *Sports Med.* **2000**, *29*, 407–424. [CrossRef] [PubMed]
13. Casey, A.; Mann, R.; Banister, K.; Fox, J.; Morris, P.G.; Macdonald, I.A.; Greenhaff, P.L. Effect of carbohydrate ingestion on glycogen resynthesis in human liver and skeletal muscle, measured by (13) C MRS. *Am. J. Physiol.* **2000**, *278*, E65–E75. [CrossRef]
14. Carter, J.M.; Jeukendrup, A.E.; Jones, D.A. The effect of carbohydrate mouth rinse on 1-h cycle time trial performance. *Med. Sci. Sports Exerc.* **2004**, *36*, 2107–2111. [CrossRef] [PubMed]
15. Chambers, E.S.; Bridge, M.W.; Jones, D.A. Carbohydrate sensing in the human mouth: Effects on exercise performance and brain activity. *J. Physiol.* **2009**, *587*, 1779–1794. [CrossRef] [PubMed]
16. Philp, A.; Hargreaves, M.; Baar, K. More than a store: Regulatory roles for glycogen in skeletal muscle adaptation to exercise. *Am. J. Physiol. Endocrinol. Metab.* **2012**, *302*, E1343–E1351. [CrossRef] [PubMed]
17. Bartlett, J.D.; Hawley, J.A.; Morton, J.P. Carbohydrate availability and exercise training adaptation: Too much of a good thing? *Eur. J. Sport Sci.* **2015**, *15*, 3–12. [CrossRef] [PubMed]
18. Graham, T.E.; Yuan, Z.; Hill, A.K.; Wilson, R.J. The regulation of muscle glycogen: The granule and its proteins. *Acta Physiol.* **2010**, *199*, 489–499. [CrossRef] [PubMed]
19. Marchand, I.; Chorneyko, K.; Tarnopolsky, M.; Hamilton, S.; Shearer, J.; Potvin, J.; Graham, T.E. Quantification of subcellular glycogen in resting human muscle: Granule size, number, and location. *J. Appl. Physiol.* **2002**, *93*, 1598–1607. [CrossRef] [PubMed]
20. Essen, B.; Henriksson, J. Glycogen content of individual muscle fibres in man. *Acta Physiol. Scand.* **1974**, *90*, 645–647. [CrossRef] [PubMed]
21. Essen, B.; Jansson, E.; Henriksson, J.; Taylor, AW.; Saltin, B. Metabolic characteristics of fibre types in human skeletal muscle. *Acta Physiol. Scand.* **1975**, *95*, 153–165. [CrossRef] [PubMed]
22. Stellingwerff, T.; Boon, H.; Gijsen, A.P.; Stegen, J.H.; Kuipers, H.; van Loon, L.J. Carbohydrate supplementation during prolonged cycling exercise spares muscle glycogen but does not affect intramyocellular lipid use. *Pflug. Arch.* **2007**, *454*, 635–647. [CrossRef] [PubMed]
23. Tsintzas, O.K.; Williams, C.; Boobis, L.; Greenhaff, P. Carbohydrate ingestion and single muscle fibre glycogen metabolism during prolonged running in men. *J. Appl. Physiol.* **1996**, *81*, 801–809. [CrossRef] [PubMed]
24. Gollnick, P.D.; Piehl, K.; Saltin, B. Selective glycogen depletion pattern in human muscle fibres after exercise of varying intensity and at varying pedalling rates. *J. Physiol.* **1974**, *24*, 45–57. [CrossRef]
25. Krustrup, P.; Mohr, M.; Steensberg, A.; Bencke, J.; Kjaer, M.; Bangsbo, J. Muscle and blood metabolites during a soccer game: Implications for sprint performance. *Med. Sci. Sports Exerc.* **2006**, *38*, 1165–1174. [CrossRef] [PubMed]
26. Nielsen, J.; Holmberg, H.C.; Schroder, H.D.; Saltin, B.; Ortenblad, N. Human skeletal muscle glycogen utilization in exhaustive exercise: Role of subcellular localization and fiber type. *J. Physiol.* **2011**, *589*, 2871–2885. [CrossRef] [PubMed]
27. Marchand, I.; Tarnopolsky, M.; Adamo, K.B.; Bourgeois, J.M.; Chorneyko, K.; Graham, T.E. Quantitative assessment of human muscle glycogen granules size and number in subcellular locations during recovery from prolonged exercise. *J. Physiol.* **2007**, *580*, 617–628. [CrossRef] [PubMed]
28. Ortenblad, N.; Nielsen, J.; Saltin, B.; Holmberger, H.C. Role of glycogen availability in sarcoplasmic reticulum Ca^{2+} kinetics in human skeletal muscle. *J. Physiol.* **2011**, *589*, 711–725. [CrossRef] [PubMed]
29. Gejl, K.D.; Hvid, L.O.; Frandsen, U.; Jensen, K.; Sahlin, K.; Ortenblad, N. Muscle glycogen content modifies SR Ca^{2+} release rate in elite endurance athletes. *Med. Sci. Sports Exerc.* **2014**, *46*, 496–505. [CrossRef] [PubMed]
30. Taylor, C.; Bartlett, J.D.; van de Graaf, C.S.; Louhelainen, J.; Coyne, V.; Iqbal, Z.; MacLaren, D.P.; Gregson, W.; Close, G.L.; Morton, J.P. Protein ingestion does not impair exercise-induced AMPK signalling when in a glycogen-depleted state: Implications for train-low compete-high. *Eur. J. Appl. Physiol.* **2013**, *113*, 1457–1468. [CrossRef] [PubMed]
31. Bartlett, J.D.; Louhelainen, J.; Iqbal, Z.; Cochran, A.J.; Gibala, M.J.; Gregson, W.; Close, G.L.; Drust, B.; Morton, J.P. Reduced carbohydrate availability enhances exercise-induced p53 signalling in human skeletal muscle: Implications for mitochondrial biogenesis. *Am. J. Physiol. Regul. Integr. Comp. Physiol.* **2013**, *304*, 450–458. [CrossRef] [PubMed]

32. Arkinstall, M.J.; Bruce, C.R.; Nikolopoulos, V.; Garnham, A.P.; Hawley, J.A. Effect of carbohydrate ingestion of metabolism during running and cycling. *J. Appl. Physiol.* **2001**, *91*, 2125–2134. [CrossRef] [PubMed]

33. Van Loon, L.J.; Greenhaff, P.L.; Constantin-Teodosiu, D.; Saris, W.H.; Wagenmakers, A.J. The effects of increasing exercise intensity on muscle fuel utilisation in humans. *J. Physiol.* **2001**, *536*, 295–304. [CrossRef] [PubMed]

34. Gonzales, J.T.; Fuchs, C.J.; Betts, J.A.; van Loon, L.J. Liver glycogen metabolism during and after prolonged endurance-type exercise. *Am. J. Physiol. Endocrinol. Metab.* **2016**, *311*, E543–E553. [CrossRef] [PubMed]

35. Achten, J.; Jeukendrup, A. Optimising fat oxidation through exercise and diet. *Nutrition* **2004**, *20*, 716–727. [CrossRef] [PubMed]

36. Picton, C.; Klee, C.B.; Cohen, P. The regulation of muscle phosphorylase kinase by calcium ions, calmodulin and troponin-C. *Cell Calcium* **1981**, *2*, 281–294. [CrossRef]

37. Howlett, R.A.; Parolin, M.L.; Dyck, D.J.; Jones, N.L.; Heigenhauser, G.J.; Spriet, L.L. Regulation of skeletal muscle glycogen phosphorylase and PDH at varying exercise power outputs. *Am. J. Physiol.* **1998**, *275*, R418–R425. [CrossRef] [PubMed]

38. Parolin, M.L.; Chesley, A.; Matsos, M.P.; Spriet, L.L.; Jones, N.L.; Heigenhauser, G.J.F. Regulation of skeletal muscle glycogen phosphorylase and PDH during maximal intermittent exercise. *Am. J. Physiol.* **1999**, *277*, E890–E900. [CrossRef] [PubMed]

39. Rose, A.J.; Richter, E.A. Skeletal muscle glucose uptake during exercise: How is it regulated. *J. Physiol.* **2005**, *20*, 260–270. [CrossRef] [PubMed]

40. Trimmer, J.K.; Schwarz, J.M.; Casazza, G.A.; Horning, M.A.; Rodriguez, N.; Brooks, G.A. Measurement of gluconeogenesis in exercising men by mass isotopomer distribution analysis. *J. Appl. Physiol.* **2002**, *93*, 233–241. [CrossRef] [PubMed]

41. Kraniou, G.N.; Cameron-Smith, D.; Hargreaves, M. Acute exercise and GLUT4 expression in human skeletal muscle: Influence of exercise intensity. *J. Appl. Physiol.* **2006**, *101*, 934–937. [CrossRef] [PubMed]

42. Martin, I.K.; Katz, A.; Wahren, J. Splanchnic and muscle metabolism during exercise in NIDDM patients. *Am. J. Physiol.* **1995**, *269*, E583–E590. [CrossRef] [PubMed]

43. Katz, A.; Broberg, S.; Sahlin, K.; Wahren, J. Leg glucose uptake during maximal dynamic exercise in humans. *Am. J. Physiol.* **1986**, *251*, E65–E70. [CrossRef] [PubMed]

44. Watt, M.J.; Heigenhauser, G.J.; Dyck, D.J.; Spriet, L.L. Intramuscular triacglycerol, glycogen, and acetyl group metabolism during 4 h of moderate exercise in man. *J. Physiol.* **2002**, *541*, 969–978. [CrossRef] [PubMed]

45. Watt, M.J.; Heigenhauser, G.J.; LeBlanc, P.J.; Inglis, J.G.; Spriet, L.L.; Peters, S.J. Rapid upregulation of pyruvate dehydrogenase kinase activity in human skeletal muscle during prolonged exercise. *J. Appl. Physiol.* **2004**, *97*, 1261–1267. [CrossRef] [PubMed]

46. Hargreaves, M.; McConell, G.; Proletto, J. Influence of muscle glycogen on glycogenolysis and glucose uptake during exercise in humans. *J. Appl. Physiol.* **1995**, *78*, 288–292. [CrossRef] [PubMed]

47. Arkinstall, M.J.; Bruce, C.R.; Clark, S.A.; Rickards, C.A.; Burke, L.M.; Hawley, J.A. Regulation of fuel metabolism by pre-exercise muscle glycogen content and exercise intensity. *J. Appl. Physiol.* **2004**, *97*, 2275–2283. [CrossRef] [PubMed]

48. Kiilerich, K.; Gudmundsson, M.; Birk, J.B.; Lundby, C.; Taudorf, S.; Plomgard, P.; Saltin, B.; Pedersen, P.A.; Wojtaszewski, J.F.P.; Pilegaard, H. Low muscle glycogen and elevated plasma free fatty acid modify but do not prevent exercise-induced PDH activation in human skeletal muscle. *Diabetes* **2010**, *59*, 26–32. [CrossRef] [PubMed]

49. Peters, S.J.; St Amand, T.A.; Howlett, R.A.; Heigenhauser, G.J.F.; Spriet, L.L. Human skeletal muscle pyruvate dehydrogenase kinase activity increases after a low carbohydrate diet. *Am. J. Physiol.* **1998**, *275*, E980–E986. [CrossRef] [PubMed]

50. Roepstorff, C.; Halberg, N.; Hillig, T.; Saha, A.K.; Ruderman, N.B.; Wojtaszewski, J.F.P.; Richter, E.A.; Kiens, B. Malonyl CoA and carnitine in regulation of fat oxidation in human skeletal muscle during exercise. *Am. J. Physiol.* **2005**, *288*, E133–E142. [CrossRef] [PubMed]

51. Wall, B.T.; Stephens, F.B.; Constantin-Teodosiu, D.; Marimuthu, K.; Macdonald, I.A.; Greenhaff, P.L. Chronic oral ingestion of L-carnitine and carbohydrate increases muscle carnitine content and alters muscle fuel metabolism during exercise in humans. *J. Physiol.* **2011**, *589*, 963–973. [CrossRef] [PubMed]

52. Smith, B.K.; Perry, C.G.R.; Koves, T.R.; Wright, D.C.; Smith, J.C.; Neufer, D.P.; Muoio, D.M.; Holloway, G.P. Identification of a novel malonyl-CoA IC$_{50}$ for CPT-1: Implications for predicting in vivo fatty acid oxidation rates. *Biochem. J.* **2012**, *448*, 13–20. [CrossRef] [PubMed]

53. Gonzales, J.T.; Fuchs, C.J.; Smith, F.E.; Thelwall, P.E.; Taylor, R.; Stevenson, E.J.; Trenell, M.I.; Cermak, N.M.; Van Loon, L.J. Ingestion of glucose or sucrose prevents liver but not muscle glycogen depletion during prolonged endurance-type exercise in trained cyclists. *Am. J. Physiol. Endocrinol. Metab.* **2015**, *309*, E1032–E1039. [CrossRef] [PubMed]

54. Horowitz, J.F.; Mora-Rodriguez, R.; Byerley, L.O.; Coyle, E.F. Lipolytic suppression following carbohydrate ingestion limits fat oxidation during exercise. *Am. J. Physiol.* **1997**, *273*, E768–E775. [CrossRef] [PubMed]

55. Coyle, E.F.; Jeukendrup, A.; Wagenmakers, A.J.; Saris, W.H. Fatty acid oxidation is directly regulated by carbohydrate metabolism during exercise. *Am. J. Physiol.* **1997**, *273*, E268–E275. [CrossRef] [PubMed]

56. Watt, M.J.; Steinberg, G.R.; Chan, S.; Garnham, A.; Kemp, B.E.; Febbraio, M.A. Beta-adrenergic stimulation of skeletal muscle HSL can be overridden by AMPK signalling. *FASEB J.* **2004**, *18*, 1445–1446. [CrossRef] [PubMed]

57. Henriksson, J. Training induced adaptation of skeletal muscle and metabolism during submaximal exercise. *J. Physiol.* **1977**, *270*, 661–675. [CrossRef] [PubMed]

58. Karlsson, J.; Nordesjö, L.O.; Saltin, B. Muscle glycogen utilization during exercise after physical training. *Acta Physiol.* **1974**, *90*, 210–217. [CrossRef] [PubMed]

59. Saltin, B.; Nazar, K.; Costill, D.L.; Stein, E.; Jansson, E.; Essen, B.; Gollnick, P.D. The nature of the training response: Peripheral and central adaptations of one-legged exercise. *Acta Physiol. Scand.* **1976**, *96*, 289–305. [CrossRef] [PubMed]

60. Chesley, A.; Heigenhauser, G.J.; Spriet, L.L. Regulation of glycogen phosphorylase activity following short term endurance training. *Am. J. Physiol.* **1996**, *270*, E328–E335. [CrossRef] [PubMed]

61. Leblanc, P.J.; Howarth, K.R.; Gibala, M.J.; Heigenhauser, G.J. Effects of 7 weeks of endurance training on human skeletal muscle metabolism during submaximal exercise. *J. Appl. Physiol.* **2004**, *97*, 2148–2153. [CrossRef] [PubMed]

62. Bergman, B.C.; Horning, M.A.; Casazza, G.A.; Wolfel, E.E.; Butterfiled, G.E.; Brooks, G.A. Endurance training increases gluconeogenesis during rest and exercise in men. *Am. J. Physiol. Endocrinol. Metab.* **2000**, *278*, E244–E251. [CrossRef] [PubMed]

63. Bergman, B.C.; Butterfield, G.E.; Wolfel, E.E.; Lopaschuk, G.D.; Casazza, G.A.; Horning, M.A.; Brooks, G.A. Muscle net glucose uptake and glucose kinetics after endurance training in men. *Am. J. Physiol.* **1999**, *277*, E81–E92. [CrossRef] [PubMed]

64. Richter, E.A.; Jensen, P.; Kiens, B.; Kristiansen, S. Sarcolemmal glucose transport and GLUT4 translocation during exercise are diminished by endurance training. *Am. J. Physiol.* **1998**, *274*, E89–E95. [PubMed]

65. Phillips, S.M.; Green, H.J.; Tarnapolsky, M.A.; Heigenhauser, G.J.; Hill, R.E.; Grant, S.M. Effects of training duration on substrate turnover and oxidation during exercise. *J. Appl. Physiol.* **1996**, *81*, 2182–2191. [CrossRef] [PubMed]

66. Sherman, W.M.; Costill, D.L.; Fink, W.J.; Miller, J.M. Effect of exercise-diet manipulation on muscle glycogen and its subsequent utilisation during performance. *Int. J. Sports Med.* **1981**, *2*, 114–118. [CrossRef] [PubMed]

67. Bussau, V.A.; Fairchild, T.J.; Rao, A.; Steele, P.; Fournier, P.A. Carbohydrate loading in human muscle: An improved 1 day protocol. *Eur. J. Appl. Physiol.* **2002**, *87*, 290–295. [CrossRef] [PubMed]

68. Burke, L.M.; Collier, G.R.; Hargreaves, M. Muscle glycogen storage after prolonged exercise: Effect of the glycemic index of carbohydrate feedings. *J. Appl. Physiol.* **1993**, *75*, 1019–1023. [CrossRef] [PubMed]

69. Hawley, J.A.; Schabort, E.J.; Noakes, T.D.; Dennis, S.C. Carbohydrate loading and exercise performance. *Sports Med.* **1997**, *24*, 73–81. [CrossRef] [PubMed]

70. O'Brien, M.J.; Viguie, C.A.; Mazzeo, R.S.; Brooks, G.A. Carbohydrate dependence during marathon running. *Med. Sci. Sports Exerc.* **1993**, *25*, 1009–1017. [PubMed]

71. Leckey, J.J.; Burke, L.M.; Morton, J.P.; Hawley, J.A. Altering fatty acid availability does not impair prolonged, continuous running to fatigue: Evidence for carbohydrate dependence. *J. Appl. Physiol.* **2016**, *120*, 107–113. [CrossRef] [PubMed]

72. Ortenblad, N.; Westerblad, H.; Nielsen, J. Muscle glycogen stores and fatigue. *J. Physiol.* **2013**, *591*, 4405–4413. [CrossRef] [PubMed]

73. Wee, L.-S.; Williams, C.; Tsintzas, K.; Boobis, L. Ingestion of a high glycemic index meal increases muscle glycogen storage at rest but augments its utilization during subsequent exercise. *J. Appl. Physiol.* **2005**, *99*, 707–714. [CrossRef] [PubMed]

74. Sherman, W.M.; Peden, M.C.; Wright, D.A. Carbohydrate feedings 1 h before exercise improves cycling performance. *Am. J. Clin. Nutr.* **1991**, *54*, 866–870. [CrossRef] [PubMed]

75. Wright, D.A.; Sherman, W.M.; Dernbach, A.R. Carbohydrate feedings before, during or in combination improve cycling endurance performance. *J. Appl. Physiol.* **1991**, *71*, 1082–1088. [CrossRef] [PubMed]

76. Stellingwerff, T.; Cox, G.R. Systematic review: Carbohydrate supplementation on exercise performance or capacity of varying durations. *Appl. Physiol. Nutr. Metab.* **2014**, *39*, 998–1011. [CrossRef] [PubMed]

77. Jeukendrup, A. A step towards personalized sports nutrition: Carbohydrate intake during exercise. *Sports Med.* **2014**, *44*, S25–S33. [CrossRef] [PubMed]

78. Newell, M.L.; Wallis, G.A.; Hunter, A.M.; Tipton, K.D.; Galloway, S.D.R. Metabolic responses to carbohydrate ingestion during exercise: Associations between carbohydrate dose and endurance performance. *Nutrients* **2018**, *10*, 37. [CrossRef] [PubMed]

79. Carter, J.M.; Jeukendrup, A.E.; Mann, C.H.; Jones, D.A. The effect of glucose infusion on glucose kinetics during a 1-h time trial. *Med. Sci. Sports Exerc.* **2004**, *36*, 1543–1550. [CrossRef] [PubMed]

80. Burke, L.M.; Maughan, R.J. The Governor has a sweet tooth-mouth sensing of nutrients to enhance sports performance. *Eur. J. Sport Sci.* **2015**, *15*, 29–40. [CrossRef] [PubMed]

81. Lane, S.C.; Bird, S.R.; Burke, L.M.; Hawley, J.A. Effect of a carbohydrate mouth rinse on simulated cycling time-trial performance commenced in a fed or fasted state. *Appl. Physiol. Nutr. Metab.* **2013**, *38*, 134–139. [CrossRef] [PubMed]

82. Kasper, A.M.; Cocking, S.; Cockayne, M.; Barnard, M.; Tench, J.; Parker, L.; McAndrew, J.; Langan-Evans, C.; Close, G.L.; Morton, J.P. Carbohydrate mouth rinse and caffeine improves high-intensity interval running capacity when carbohydrate restricted. *Eur. J. Sport Sci.* **2015**, *16*, 560–568. [CrossRef] [PubMed]

83. Ali, A.; Yoo, M.J.; Moss, C.; Breier, B.H. Carbohydrate mouth rinsing has no effect on power output during cycling in a glycogen-reduced state. *J. Int. Soc. Sports Nutr.* **2016**, *23*, 13–19. [CrossRef] [PubMed]

84. Lee, M.J.; Hammond, K.M.; Vasdev, A.; Poole, K.L.; Impey, S.G.; Close, G.L.; Morton, J.P. Self-selecting fluid intake while maintaining high carbohydrate availability does not impair half-marathon performance. *Int. J. Sports Med.* **2014**, *35*, 1216–1222. [CrossRef] [PubMed]

85. Pfeiffer, B.; Stellingwerff, T.; Zaltas, E.; Jeukendrup, A.E. CHO oxidation from a CHO gel compared with a drink during exercise. *Med. Sci. Sports Exerc.* **2010**, *42*, 2038–2045. [CrossRef] [PubMed]

86. Pfeiffer, B.; Stellingwerff, T.; Zaltas, E.; Jeukendrup, A.E. Oxidation of solid versus liquid CHO sources during exercise. *Med. Sci. Sports Exerc.* **2010**, *42*, 2030–2037. [CrossRef] [PubMed]

87. Zhang, X.; O'Kennedy, N.; Morton, J.P. Extreme Variation of Nutritional Composition and Osmolality of Commercially Available Carbohydrate Energy Gels. *Int. J. Sport Nutr. Exerc. Metab.* **2015**, *25*, 504–509. [CrossRef] [PubMed]

88. Pfeiffer, B.; Stellingwerff, T.; Hodgson, A.B.; Randell, R.; Pottgen, K.; Res, P.; Jeukendrup, A.E. Nutritional intake and gastrointestinal problems during competitive endurance events. *Med. Sci. Sports Exerc.* **2012**, *44*, 344–355. [CrossRef] [PubMed]

89. Hills, S.P.; Russell, M. Carbohydrates for soccer: A focus on skilled actions and half-time practices. *Nutrients* **2018**, *10*, 22. [CrossRef] [PubMed]

90. Holloszy, J.O.; Coyle, E.F. Adaptations of skeletal muscle to endurance exercise and their metabolic consequences. *J. Appl. Physiol.* **1984**, *56*, 831–838. [CrossRef] [PubMed]

91. Egan, B.; Zierath, J.B. Exercise metabolism and the molecular regulation of skeletal muscle adaptation. *Cell Metab.* **2013**, *17*, 162–184. [CrossRef] [PubMed]

92. Wojtaszewski, J.F.; MacDonald, C.; Nielsen, J.N.; Hellsten, Y.; Hardie, D.G.; Kemp, B.E.; Kiens, B.; Richter, E.A. Regulation of 5'AMP-activated protein kinase activity and substrate utilization in exercising human skeletal muscle. *Am. J. Physiol. Endocrinol. Metab.* **2003**, *284*, 813–822. [CrossRef] [PubMed]

93. Cochran, A.J.; Little, J.P.; Tarnopolsky, M.A.; Gibala, M.J. Carbohydrate feeding during recovery alters the skeletal muscle metabolic response to repeated sessions of high-intensity interval exercise in humans. *J. Appl. Physiol.* **2010**, *108*, 628–636. [CrossRef] [PubMed]

94. Rose, A.J.; Frøsig, C.; Kiens, B.; Wojtaszewski, J.F.P.; Richter, E. Effect of endurance exercise training on Ca^{2+} calmodulin-dependent protein kinase II expression and signalling in skeletal muscle of humans. *J. Physiol.* **2007**, *583*, 785–795. [CrossRef] [PubMed]

95. Perry, C.G.; Lally, J.; Holloway, G.P.; Heigenhauser, G.J.; Bonen, A.; Spriet, L.L. Repeated transient mRNA bursts precede increases in transcriptional and mitochondrial proteins during training in human skeletal muscle. *J. Physiol.* **2010**, *588*, 4795–4810. [CrossRef] [PubMed]

96. Yeo, W.K.; Paton, C.D.; Garnham, A.P.; Burke, L.M.; Carey, A.L.; Hawley, J.A. Skeletal muscle adaptation and performance responses to once versus twice every second day endurance training regimens. *J. Appl. Physiol.* **2008**, *105*, 1462–1470. [CrossRef] [PubMed]

97. Hulston, C.J.; Venables, M.C.; Mann, C.H.; Martin, C.; Philp, A.; Baar, K.; Jeukendrup, A.E. Training with low muscle glycogen enhances fat metabolism in well-trained cyclists. *Med. Sci. Sports Exerc.* **2010**, *42*, 2046–2055. [CrossRef] [PubMed]

98. Hansen, A.K.; Fischer, C.P.; Plomgaard, P.; Andersen, J.L.; Saltin, B.; Pedersen, B.K. Skeletal muscle adaptation: Training twice every second day vs. training once daily. *J. Appl. Physiol.* **2005**, *98*, 93–99. [CrossRef] [PubMed]

99. Marquet, L.A.; Brisswalter, J.; Louis, J.; Tiollier, E.; Burke, L.M.; Hawley, J.A.; Hausswirth, C. Enhanced endurance performance by periodization of carbohydrate intake: "Sleep low" strategy. *Med. Sci. Sports Exerc.* **2016**, *48*, 663–672. [CrossRef] [PubMed]

100. Pilegaard, H.; Keller, C.; Steensberg, A.; Helge, J.W.; Pedersen, B.K.; Saltin, B.; Neufer, P.D. Influence of pre-exercise muscle glycogen content on exercise-induced transcriptional regulation of metabolic genes. *J. Physiol.* **2002**, *541*, 261–271. [CrossRef] [PubMed]

101. Akerstrom, T.C.A.; Birk, J.B.; Klein, D.K.; Erikstrup, C.; Plomgaard, P.; Pedersen, B.K.; Wojtaszewski, J. Oral glucose ingestion attenuates exercise-induced activation of 5′-AMP-activated protein kinase in human skeletal muscle. *Biochem. Biophys. Res. Commun.* **2006**, *342*, 949–955. [CrossRef] [PubMed]

102. Civitarese, A.E.; Hesselink, M.K.; Russell, A.P.; Ravussin, E.; Schrauwen, P. Glucose ingestion during exercise blunts exercise-induced gene expression of skeletal muscle fat oxidative genes. *Am. J. Physiol. Endocrinol. Metab.* **2005**, *289*, 1023–1029. [CrossRef] [PubMed]

103. Cluberton, L.J.; McGee, S.L.; Murphy, R.M.; Hargreaves, M. Effect of carbohydrate ingestion on exercise-induced alterations in metabolic gene expression. *J. Appl. Physiol.* **2005**, *99*, 1359–1363. [CrossRef] [PubMed]

104. Nybo, L.; Pedersen, K.; Christensen, B.; Aagaard, P.; Brandt, N.; Kiens, B. Impact of carbohydrate supplementation during endurance training on glycogen storage and performance. *Acta Physiol.* **2009**, *197*, 117–127. [CrossRef] [PubMed]

105. Van Proeyen, K.; Szlufcik, K.; Nielens, H.; Ramaekers, M.; Hespel, P. Beneficial metabolic adaptations due to endurance exercise training in the fasted state. *J. Appl. Physiol.* **2011**, *110*, 236–245. [CrossRef] [PubMed]

106. Pilegaard, H.; Osada, T.; Andersen, L.T.; Helge, J.W.; Saltin, B.; Neufer, P.D. Substrate availability and transcriptional regulation of metabolic genes in human skeletal muscle during recovery from exercise. *Metabolism* **2005**, *54*, 1048–1055. [CrossRef] [PubMed]

107. Morton, J.P.; Croft, L.; Bartlett, J.D.; Maclaren, D.P.; Reilly, T.; Evans, L.; McArdle, A.; Drust, B. Reduced carbohydrate availability does not modulate training-induced heat shock protein adaptations but does up regulate oxidative enzyme activity in human skeletal muscle. *J. Appl. Physiol.* **2009**, *106*, 1513–1521. [CrossRef] [PubMed]

108. Lane, S.C.; Camera, D.M.; Lassiter, D.G.; Areta, J.L.; Bird, S.R.; Yeo, W.K.; Jeacocke, N.A.; Krook, A.; Zierath, J.R.; Burke, L.M.; et al. Effects of sleeping with reduced carbohydrate availability on acute training responses. *J. Appl. Physiol.* **2015**, *119*, 643–655. [CrossRef] [PubMed]

109. McBride, A.; Ghilagaber, S.; Nikolaev, A.; Hardie, D.G. The glycogen-binding domain on the AMPK beta subunit allows the kinase to act as a glycogen sensor. *Cell Metab.* **2009**, *9*, 23–34. [CrossRef] [PubMed]

110. Chan, S.; McGee, S.L.; Watt, M.J.; Hargreaves, M.; Febbraio, M.A. Altering dietary nutrient intake that reduces glycogen content leads to phosphorylation of nuclear p38 MAP kinase in human skeletal muscle: Association with IL-6 gene transcription during contraction. *FASEB J.* **2004**, *18*, 1785–1787. [CrossRef] [PubMed]

111. Impey, S.G.; Hearris, M.A.; Hammond, K.M.; Bartlett, J.D.; Louis, J.; Close, G.L.; Morton, J.P. Fuel for the work required: A theoretical framework for carbohydrate periodization and the glycogen threshold hypothesis. *Sports Med.* **2018**. [CrossRef] [PubMed]

112. Gejl, K.D.; Tharns, L.B.; Hansen, M.; Rokkedal-Lausch, T.; Plomgaard, P.; Nybo, L.; Larsen, F.J.; Cardinale, D.A.; Jensen, K.; Holmberg, H.C.; et al. No superior adaptations to carbohydrate periodization in elite endurance athletes. *Med. Sci. Sports Exerc.* **2017**, *49*, 2486–2497. [CrossRef] [PubMed]

113. Burke, L.M.; Ross, M.L.; Garvican-Lewis, L.A.; Welvaert, M.; Heikura, I.A.; Forbes, S.G.; Mirtschin, J.G.; Cato, L.E.; Strobel, N.; Sharma, A.P.; et al. Low carbohydrate, high fat diet impairs exercise economy and negates performance benefit from intensified training in elite race walkers. *J. Physiol.* **2017**, *595*, 2785–2807. [CrossRef] [PubMed]

114. Gleeson, M.; Nieman, D.C.; Pedersen, B.K. Exercise, nutrition and immune function. *J. Sports Sci.* **2004**, *22*, 115–125. [CrossRef] [PubMed]

115. Louis, J.; Marquet, L.A.; Tiollier, E.; Bernon, S.; Hausswirth, C.; Brisswalter, J. The impact of sleeping with reduced glycogen stores on immunity and sleep in triathletes. *Eur. J. Appl. Physiol.* **2016**, *116*, 1941–1954. [CrossRef] [PubMed]

116. Howarth, K.R.; Phillips, S.M.; MacDonald, M.J.; Richards, D.; Moreau, N.A.; Gibala, M.J. Effect of glycogen availability on human skeletal muscle protein turnover during exercise and recovery. *J. Appl. Physiol.* **2010**, *109*, 431–438. [CrossRef] [PubMed]

117. Cox, G.R.; Clark, S.A.; Cox, A.J.; Halson, S.L.; Hargreaves, M.; Hawley, J.A.; Burke, L.M. Daily training with high carbohydrate availability increases exogenous carbohydrate oxidation during endurance cycling. *J. Appl. Physiol.* **2010**, *109*, 126–134. [CrossRef] [PubMed]

118. Impey, S.G.; Hammond, K.M.; Shepherd, S.O.; Sharples, A.P.; Stewart, C.; Limb, M.; Smith, K.; Jeromson, S.; Hamilton, D.L.; Close, G.L.; et al. Fuel for the work required: A practical approach to amalgamating train-low paradigms for endurance athletes. *Physiol. Rep.* **2016**, *4*, e12803. [CrossRef] [PubMed]

nutrients

MDPI

Article

Metabolic Responses to Carbohydrate Ingestion during Exercise: Associations between Carbohydrate Dose and Endurance Performance

Michael L. Newell [1],*, Gareth A. Wallis [2], Angus M. Hunter [3], Kevin D. Tipton [3] and Stuart D. R. Galloway [3]

1   Department of Life Sciences, Faculty of Science and Technology University of Westminster, London W1W 6UW, UK
2   School of Sport, Exercise and Rehabilitation Sciences, University of Birmingham, Birmingham B15 2TT, UK; g.a.wallis@bham.ac.uk
3   Physiology, Exercise and Nutrition Research Group, Faculty of Health Sciences and Sport, University of Stirling, Stirling FK9 4LA, UK; a.m.hunter1@stir.ac.uk (A.M.H.); k.d.tipton@stir.ac.uk (K.D.T.); s.d.r.galloway@stir.ac.uk (S.D.R.G.)
*   Correspondence: m.newell@westminster.ac.uk; Tel.: +44-(0)207-911-5000 (ext. 64133)

Received: 30 November 2017; Accepted: 25 December 2017; Published: 3 January 2018

Abstract: Carbohydrate (CHO) ingestion during exercise lasting less than three hours improves endurance exercise performance but there is still debate about the optimal dose. We utilised stable isotopes and blood metabolite profiles to further examine metabolic responses to CHO (glucose only) ingestion in the $20–64$ g·h^{-1} range, and to determine the association with performance outcome. In a double-blind, randomized cross-over design, male cyclists ($n = 20$, mean ± SD, age 34 ± 10 years, mass 75.8 ± 9 kg, peak power output 394 ± 36 W, VO$_{2max}$ 62 ± 9 mL·kg^{-1}·min^{-1}) completed four main experimental trials. Each trial involved a two-hour constant load ride (185 ± 25 W) followed by a time trial, where one of three CHO beverages, or a control (water), were administered every 15 min, providing 0, 20, 39 or 64 g CHO·h^{-1}. Dual glucose tracer techniques, indirect calorimetry and blood analyses were used to determine glucose kinetics, exogenous CHO oxidation (EXO), endogenous CHO and fat oxidation; and metabolite responses. Regression analysis revealed that total exogenous CHO oxidised in the second hour of exercise, and suppression of serum NEFA concentration provided the best prediction model of performance outcome. However, the model could only explain ~19% of the variance in performance outcome. The present data demonstrate that consuming ~40 g·h^{-1} of CHO appears to be the minimum ingestion rate required to induce metabolic effects that are sufficient to impact upon performance outcome. These data highlight a lack of performance benefit and few changes in metabolic outcomes beyond an ingestion rate of 39 g·h^{-1}. Further work is required to explore dose-response effects of CHO feeding and associations between multiple metabolic parameters and subsequent performance outcome.

Keywords: glucose; fat oxidation; exogenous; fatty acids; hepatic glucose output

1. Introduction

During prolonged steady state exercise, endogenous glycogen stores and circulating plasma glucose are key substrates for energy provision. Fatigue is often reported to coincide with the depletion of endogenous carbohydrate (CHO) stores and the dysregulation of circulating plasma glucose concentration [1–3]. Ingesting CHO improves performance and extends exercise duration via a range of proposed mechanisms including: better maintenance of circulating plasma glucose [1], higher rates of exogenous [4] and total CHO oxidation, and endogenous glycogen sparing [5]. These proposed mechanisms do not occur in isolation but occur together facilitating force production and improving performance and exercise capacity.

Early research by Coyle et al. [1] reported that feeding CHO maintained blood glucose concentration and CHO oxidation rates, and in turn, exercise capacity increased 33% (3.02 versus 4.02 h) significantly in comparison to a water control. In a follow-up study [2] participants exercised to exhaustion and were then provided with either no CHO, ingested CHO, or infused CHO. Both CHO provision conditions increased exercise duration on commencement of exercise in comparison to no CHO. However, only the infusion condition maintained blood glucose concentration sufficiently to subsequently extend exercise duration above that of the CHO ingestion trial. The authors concluded that the maintenance of blood glucose concentration was the critical factor for maintaining sufficient CHO oxidation rates to extend exercise capacity.

Further research has indicated that the maintenance of higher CHO oxidation can be primarily explained by an increase in exogenous CHO oxidation rates [4]. An elevation in exogenous CHO oxidation rate and enhanced endurance exercise performance are now believed to be directly associated, despite little systematic evaluation to date. As a result, increasing exogenous CHO oxidation rate is thought to be essential for the enhancement of endurance performance when ingesting CHO throughout a range of 20–100 $g \cdot h^{-1}$. This relationship has led some researchers to hypothesise that maximising exogenous CHO oxidation rate, through use of glucose: fructose combinations at high feeding rates will result in further performance gains [6]. However, most studies examining performance benefits of multiple transportable CHO ingestion have been conducted in comparison to isocaloric single source CHO. The findings using this model are likely to be confounded by gastrointestinal issues when ingesting single source CHO at high feeding rates. At lower feeding rates, Smith et al. [4] demonstrated that the largest improvement in performance occurred when ingesting 60 $g \cdot h^{-1}$ in comparison to 15 or 30 $g \cdot h^{-1}$. The 60 $g \cdot h^{-1}$ ingestion rate also resulted in the highest exogenous CHO oxidation rate. These authors reported a non-significant but 'likely' 2.3% improvement in performance when comparing 60 vs. 30 $g \cdot h^{-1}$ suggesting a dose-response effect of CHO feeding rate. We recently reported that a 2.3% performance gain would not necessarily be 'likely' due to typical variance observed in performance outcome measures [7] when using a more suitably powered design. In addition, there has yet to be an extensive exploration of the association between multiple metabolic variables and subsequent exercise performance outcomes using a dose-response investigation. Thus, more work remains to be done to determine the key factors driving performance improvement in exercise lasting <3 h. In our previous work [7] a lack of any further improvement in performance when feeding 64 $g \cdot h^{-1}$ in comparison to 39 $g \cdot h^{-1}$ suggests that the metabolic alterations with feeding rates as low as 39 $g \cdot h^{-1}$ could be sufficient to maximise performance gains within this feeding rate range. As such, peak exogenous carbohydrate oxidation rate may not be the sole, or key, determining factor for performance enhancement during exercise lasting less than 3 h with the single source CHO doses studied.

Feeding CHO during exercise influences the usage of endogenous glycogen stores. Several studies have assessed endogenous glycogen utilisation using stable isotopes during 1–2 h of moderate intensity exercise. McConnell et al. [8] provided participants with 100 $g \cdot h^{-1}$ of CHO during 2 h of exercise at 69 ± 1% VO_{2peak}. Hepatic glucose output was suppressed in comparison to a control trial and remained close to baseline rates throughout the exercise bout. The authors calculated that a 51% reduction in hepatic glucose production occurred as a result of consuming 100 $g \cdot h^{-1}$ CHO in comparison to the control. Furthermore, Jeukendrup et al. [5] provided 30 and 180 $g \cdot h^{-1}$ of a glucose based CHO beverage during a 2 h moderate intensity exercise bout. They reported reduced fat oxidation rates, increased rate of appearance (Ra) and rate of disappearance (Rd) of glucose, and an increase in the oxidation of exogenous CHO particularly with the higher glucose dose. Endogenous muscle glycogen oxidation rates were not altered with either 30 or 180 $g \cdot h^{-1}$ of CHO in their study. However, liver glycogen breakdown was reduced when consuming 30 $g \cdot h^{-1}$, and completely inhibited when consuming 180 $g \cdot h^{-1}$ of CHO. These observations suggest that only when very high doses of glucose are ingested can hepatic glucose production be completely inhibited. Smith et al. [4] estimated a stepped reduction in the contribution of liver glycogen to total CHO oxidation during the second hour of their submaximal exercise bout whilst consuming 15, 30 and 60 g·h^{-1} of CHO. Interestingly, all three studies indicate that muscle glycogen was not spared with any ingestion rate provided. These data suggest that a focus on hepatic glycogen sparing is required when considering factors likely to influence subsequent performance outcomes.

The amount of CHO to ingest for optimal endurance performance has been widely debated. A consensus has been reached that the maximal exogenous CHO oxidation rate that can be achieved with glucose (single source CHO) ingestion is around ~1 g·min^{-1}. As previously mentioned, Smith et al. [4] suggested the existence of a dose response relationship between CHO ingestion rate and endurance exercise performance enhancement when feeding 0, 15, 30 and 60 g·h^{-1} of glucose. However, their initial study was underpowered. Their study was followed up with a multicentre investigation which presented evidence for a curvilinear dose response relationship with ingestion rates of a multi-source CHO beverage spanning 0 to 120 g·h^{-1} with a statistically optimal ingestion rate reported as 78 g·h^{-1}. However, whether maximal exogenous oxidation rates driven by higher CHO ingestion rates result in optimal performances during endurance tasks requires further metabolic analysis. Until now our previously published work is the most suitably powered and most statistically robust study design to indicate the lack of a clear dose response relationship with ingestion rates between 20 and 64 g·h^{-1} [7]. However, from these data alone we are unable to determine what the underlying physiological explanations were for the plateau in performance. We now present the metabolic data to explore these performance changes more comprehensively.

As such, in the present manuscript we aimed to explore the metabolic responses to submaximal endurance exercise with CHO ingestion rates between 0 and 64 g·h^{-1}. We specifically aimed to: examine glucose kinetics and quantify or estimate the total substrate usage from exogenous and endogenous glycogen stores by utilising stable isotopic tracers; measure key circulating metabolites; quantify the percentage contribution of key substrates throughout the exercise bout. We hypothesised that during exercise lasting <3 h there would be a minimum effective dose of CHO required to result in optimal metabolic responses, and endogenous CHO sparing, linked to improved performance outcome. We also hypothesized that both exogenous CHO oxidation rate and reduction in hepatic glucose production would be the key parameters most closely associated with performance outcomes.

2. Materials and Methods

2.1. Participants

Twenty trained male cyclists were recruited from regional cycling and triathlon clubs. The mean ($\pm$SD) characteristics of the participants were: age 34.0 ($\pm$10.2) years, body mass 74.6 ($\pm$7.9) kg, stature 178.3 ($\pm$8.0) cm, peak power output (PPO) 393 ($\pm$36) W, power output (PO) at lactate threshold 206 ($\pm$30) W and VO$_{2max}$ 62 ($\pm$9) mL·kg^{-1}·min^{-1}. Participants were required to have been training for >6 h/week for >3 years. Each individual had the procedures and associated risks explained prior to providing written informed consent to participate in the study. The study was approved by the University of Stirling, Research Ethics Committee (SSREC number 604) in accordance with the Declaration of Helsinki. In some circumstances, not all participants were included in all datasets. Unfortunately, 2 participants had to be removed from all stable isotope and substrate use data due to measurement errors. Hence, the characteristics of participants included in the stable isotope analyses were: body mass 76.9 ($\pm$8.4) kg, stature 178.7 ($\pm$8.1) cm, PPO 392 ($\pm$34) W, VO$_{2max}$ 61.2 ($\pm$8.2) mL·kg^{-1}·min^{-1} and PO at lactate threshold 206 ($\pm$30) W.

2.2. Pretesting

Following pre-screening, on week one of six, after a ten hour overnight fast, participants performed a two-part incremental cycle test (Lode Excalibur Sport, The Netherlands) to determine lactate threshold (LT), maximal oxygen uptake (VO_{2max}), and peak power output as described previously [7]. The mean $\pm$ SD lactate concentration at LT was 2.1 ± 0.4 mmol·L^{-1} corresponding to an intensity of $52 \pm 6\%$ of PPO for LT which is typical of other studies utilising a similar protocol [9]. The test end time and power output of the final stage was used to calculate peak power output (PPO) using the following Equation (1) [10]:

$$\text{PPO} = W_{final} + ([t/60] \times \text{PI}) \tag{1}$$

where, W_{final} = the power output of the final completed stage in (watts), t = the time spent in the final uncompleted stage (seconds), 60 = the duration of each stage (seconds) and PI = the increase in power output between each stage (W). Maximal oxygen uptake (VO_{2max}) was assessed via an automated online gas analyser (Oxycon Pro, Jaeger, Wuerzerberg, Germany). VO_{2max} was determined as the highest average VO_2 captured over a 30 s period.

2.3. Design

In a double blind, placebo controlled, randomised cross-over study design participants visited the laboratory for 5 experimental trials (1 preliminary and 4 intervention) over a five-week period. They completed one visit per week commencing each trial on the same day of the week and at the same time of day. On the first of these trial visits participants completed a full familiarisation. The familiarisation trial and the four subsequent intervention trials consisted of a 120 min steady state submaximal cycle ergometer ride at 95% lactate threshold (185 ± 25 W). Participants were asked to record their habitual dietary intake for 48 h prior to visit one and replicate this dietary intake for the two days prior to each subsequent visit. Additionally, participants were asked to arrive at the laboratory following a ~10 h overnight fast. Water was ingested before and during the familiarisation trial and was consumed at a rate of 1 L·h^{-1}. Thereafter, participants consumed in a counterbalanced randomized cross-over design either: 0%, 2%, 3.9% or 6.4% CHO solutions before and during exercise at a fluid ingestion rate of 1 L·h^{-1}. The 0% trial was a water control trial. Blood samples, expired gas collection and subjective measures were obtained every 15 min throughout the submaximal ride.

2.4. Experimental Trials

On arrival at the laboratory, participants emptied their bladder and bowel prior to nude body mass measurement. Participants then changed into cycling clothing. Teflon catheters were placed into an antecubital vein in each arm. One catheter was attached to a three way stop cock to enable stable isotopic tracer infusion. The second was attached to a 10-cm extension line for multiple venous blood sampling. The sampling line was kept patent with a sterile saline solution flush (2 mL) following each sample collection. A baseline blood sample was drawn (10 mL) prior to commencing the primed (18.54 μmol·kg^{-1}) continuous (0.32 μmol·kg^{-1}·min^{-1}) infusion of $6,6,^2H_2$ glucose via a calibrated syringe pump (Asena GS Syringe Pump; Alaris Medical Systems, Basingstoke, UK) over 60 min at rest. Further blood samples were drawn at 30 min prior to and at the start of exercise for later determination of isotopic enrichments. The concentration of isotopic tracer in the infusate and the pre and post syringe weights were both determined to confirm the actual infusion rate achieved.

2.5. Immediately Pre Exercise

Five minutes prior to the start of exercise, a resting breath sample was collected into an expired gas-sampling bag (Quintron QT00892 GaSampler, Milwaukee, WI 53215, USA). 10 mL gas samples were immediately drawn into a 10 mL syringe from the bag and secured with a three way stop cock. Samples were then extracted with a 21G needle directly, and in duplicate, into evacuated exetainer tubes (Labco, High Wycombe, UK) for the determination of the CO_2 isotopic ratio of $^{13}C/^{12}C$. Two minutes prior to the start of exercise, a further blood sample was collected and the first bolus of CHO test solution was provided (240 mL). The infusion rate of the deuterated glucose tracer was then doubled at the start of exercise (to 0.64 $\mu mol \cdot kg^{-1} \cdot min^{-1}$) to accommodate for the increased turnover of glucose during exercise and to maintain plasma enrichment.

2.6. The 2 h Preload Ride and Performance Task

Participants then completed a 2 h submaximal ride at 95% LT (185 ± 25 W, 59 ± 7% of VO_{2max}). In the last 3 min of each 15 min time segment a breath by breath gas capture was obtained for the determination of VO_2 and VCO_2 (Oxycon Pro, Mannheim, Germany). Immediately following the expired gas collection participants removed the mouth piece and provided a single end-tidal breath sample into a breath sample bag (Quintron QT00892 GasSampler, Milwaukee, WI 53215, USA) for the determination of $^{13}C/^{12}C$ ratio as per the baseline sample. Following the breath sampling a 10 mL blood sample (10 mL) was drawn and stored on ice prior to centrifugation. Finally, participants were asked to rate their perceived exertion [11] before ingesting a volume of test drink (220 mL). This was repeated every 15 min throughout the 2 h ride. Following this a performance task lasting approximately 30 min was conducted and is reported elsewhere [7].

2.7. Carbohydrate Solutions

During the 2 h preload ride, each of four solutions were consumed in randomized double-blind fashion: 0% water (control); 2.0%; 3.9%; or 6.4% glucose (single source CHO) based commercially available solutions. All test solutions were maintained at 10 °C and were consumed at a rate of 1 $L \cdot h^{-1}$ providing 0, 20, 39 and 64 $g \cdot h^{-1}$ of CHO respectively. The 20 $g \cdot h^{-1}$ solution contained 37 mg of sodium per 100 mL with the 39 and 64 $g \cdot h^{-1}$ solutions both containing 50 mg per 100 mL. Each solution was initially provided two minutes prior to the start of exercise (240 mL) with subsequent volumes (220 mL) consumed every 15 min. The final drink was provided at 120 min of exercise. All solutions except for the 0% were enriched by adding 50 mg L^{-1} of U-$^{13}C_6$ glucose (Cambridge Isotopes, Cambridge, UK) during preparation by the laboratory technician. The trial day experimental protocol is shown in Figure 1.

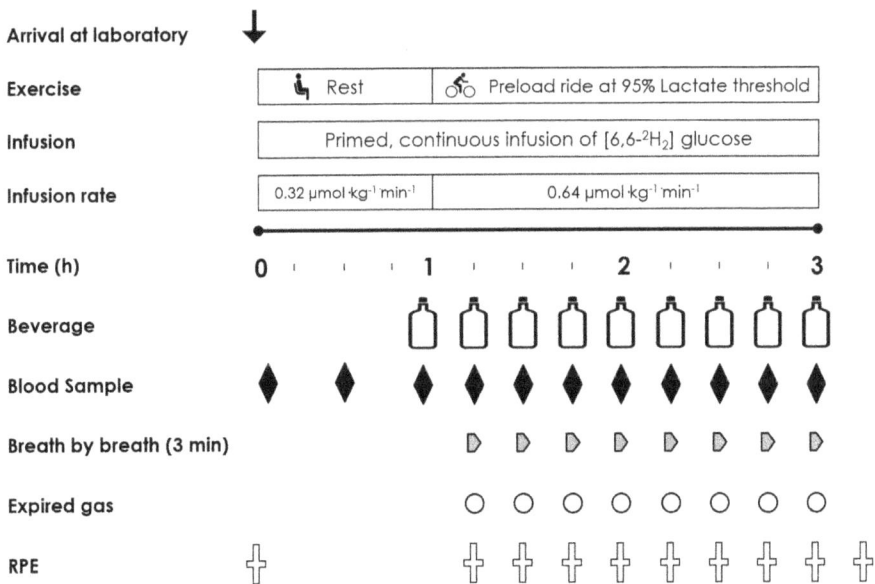

Figure 1. Trial visit time line indicating the infusion rate and time course and the frequency of measures taken throughout each experimental trial visit.

2.8. Analyses and Calculations

2.8.1. Blood

Blood samples were collected in EDTA-containing vacutainers and spun in a centrifuge at 3500 rpm for 10 min at 4 °C. Aliquots of plasma were then frozen and stored at −80 °C until further analysis. Plasma glucose, non-esterified fatty acids (NEFA), and lactate were analysed using enzymatic methods on an automated analyser (Ilab Aries, Instrumentation Laboratory, Warrington, UK). Plasma insulin and adrenaline concentrations were analysed using commercially available ELISA kits (Dimedic International, Hamburg, Germany and IBL International, Hamburg, Germany respectively). Both ELISAs were carried out following the manufacturer's instructions.

Plasma samples were derivatised for the analysis of $[^2H_2]$ glucose and $[^{13}C]$ glucose content. Briefly, 150 μL of plasma and 150 μL of distilled water with added hydrochloric acid (pH 2) was added to a glass vial and mixed vigorously for 10 s. 3 mL of methanol: chloroform (2.3:1) (500 mL = 348:152) was then added and mixed on a plate shaker (300 rpm) for 3 min. Samples were then centrifuged at 4 °C at 3500 rpm for 15 min. The supernatant was then pipetted into a new glass vial. Here 2 mL of chloroform and 1 mL of distilled water (pH 2) were added and mixed for 15 min on a plate shaker at 300 rpm. Samples were then centrifuged at 4 °C for 15 min at 3500 rpm. The supernatant was then pipetted into a new glass tube. The glass tubes were then transferred to a nitrogen drying rack and incubated at 40 °C for ~2 h until the vials were dry. Once dried 150 μL of butaneboronic acid (10 mg/1 mL pyridine) was added and mixed on a plate shaker for 15 min. Once mixed samples were then incubated at 95 °C for 30 min before 150 μL of acetic anhydride was added and mixed at 300 rpm for 90 min. Samples were then dried under nitrogen gas and incubated at 40 °C until dry. Samples were prepared for the GC-MS and GC-C-IRMS by adding 150 μL of ethylacetate and mixing for 10 min. $[6,6,^2H_2]$ enrichment was determined by gas chromatography mass spectroscopy (GCMS) using selected ion monitoring at molecular weights of 297 and 299 ($[^{12}C]$ and $[^2H_2]$ respectively). Plasma $[^{13}C]$ content was assessed using gas chromatography combustion isotope ratio mass spectroscopy (GC-C-IRMS). Plasma ^{13}C glucose enrichment was determined using the method of Pickert et al. [12], modified for use with gas chromatography-combustion-IRMS (GC-C-IRMS). The glucose derivative (1 μL) was injected into the GC (split ratio 1:15) and analysed by GC-C-IRMS (GC, Trace GC Ultra; C, GC Combustion III; IRMS, Delta Plus XP; all Thermo Finnigan, Herts, UK).

2.8.2. ^{13}C Breath Samples

Breath samples were analysed in duplicate for $^{13}C/^{12}C$ ratio by continuous-flow IRMS (GC, Trace GC Ultra; IRMS, Delta Plus XP; both Thermo Finnigan, Herts, UK).

2.8.3. Substrate Oxidation

Expired gas analysis was used to estimate rates of substrate oxidation from VO_2 and VCO_2 every 15 min. These breath measures were averaged every 4 breaths and the mean of these were taken from the last 60 s of a 3-min sampling period. Whole body substrate oxidation calculations were based on those proposed by Jeukendrup and Wallis [13]:

$$\text{CHO oxidation rate (g·min}^{-1}) = 4.210 \, VCO_2 - 2.962 \, VO_2 \qquad (2)$$

$$\text{Fat oxidation rate (g·min}^{-1}) = 1.695 \, VO_2 - 1.701 \, VCO_2 \qquad (3)$$

where VCO_2 and VO_2 are measured in litres per minute. Once the rate of substrate usage was determined during each 15-min breath by breath capture, the rates calculated in grams per minute were multiplied by 15 and summed from each time point to provide an estimate of the total substrate use during the whole exercise bout. Protein oxidation was considered as negligible.

2.8.4. Tracer Calculations

The isotopic enrichment in the expired breath samples was expressed as mean difference between the $^{13}C/^{12}C$ ratio of the sample and a known laboratory reference standard using the following formula to enable calculation of exogenous carbohydrate oxidation:

$$\text{Exogenous CHO oxidation (g·min}^{-1}) = VCO_2\ [(R_{exp} - R_{ref})/(R_{exo} - R_{ref})]/k \qquad (4)$$

where VCO_2 is in litres per minute, R_{exp} is the observed isotopic composition of expired CO_2, R_{ref} is the isotopic composition of expired CO_2 with the ingestion of the placebo, R_{exo} is the isotopic composition of exogenous glucose ingested in the drink and k (0.747 L·g^{-1}) is the volume of CO_2 produced by the complete oxidation of glucose.

2.8.5. Percentage Contribution of Substrates (Second Hour of Exercise)

Once the total amount of exogenous carbohydrate oxidation had been determined, this rate was extrapolated over the previous 15 min period to determine total grams of exogenous carbohydrate oxidised in each time period from 60 min of exercise onwards. Only the second hour of exercise was considered as exogenous carbohydrate oxidation rates are stable. The total exogenous carbohydrate oxidised was subtracted from the total carbohydrate oxidised over the same time period to give an estimate of endogenous carbohydrate oxidation. The endogenous and exogenous carbohydrate oxidised totals were then multiplied by 4.07 to provide total carbohydrate energy expenditure in kcal for each carbohydrate source. The total fat oxidised was multiplied by 9.75 to give total energy expenditure (kcal) for fat oxidation [13]. The total energy expenditure from all three substrates was then summed and each component was expressed as a percentage of the total energy expenditure over the second hour of exercise.

2.8.6. Glucose Kinetics

From the 6,6,$^2H_2$ tracer infusion the Ra and Rd of glucose were calculated with the single pool non-steady state equations of Steele, as modified for use with stable isotopes [14]. Total Ra represents the total splanchnic glucose from ingested CHO and liver derived glucose:

$$R_a\ total = F - (pV \times (C_1 + C_2)/2 \times (E_2 - E_1)/(t_2 - t_1))/(E_2 + E_1)/2) \qquad (5)$$

$$R_d\ total = R_a\ total - V \times (C_1 + C_2/t_2 - t_1) \qquad (6)$$

where F is the infusion rate (mg·kg^{-1}·min^{-1}); E_1 and E_2 are the [$^2H_2$] glucose enrichments (MPE) in plasma at time points t_1 and t_2 (min), respectively; C_1 and C_2 are glucose concentrations (mg·mL^{-1}) at t_1 and t_2, respectively; and pV is volume of distribution which was set at 40 mL·kg^{-1} to coincide with the findings of Wolfe et al. [15].

2.8.7. Estimation of Liver Glucose Contribution

Liver glucose contribution has been estimated from the following calculation:

$$\text{Whole body glucose Ra (Ra}_{body})\ g·min^{-1} = Ra \times body\ mass \times 1000 \qquad (7)$$

$$\text{Estimation of liver glucose contribution to glucose Ra (\%)} = 100 - ((EXO/Ra_{body}) \times 100) \qquad (8)$$

where Ra is the total Ra previously calculated (mg·kg^{-1}·min^{-1}), and the body mass is the pre-trial body mass measure taken before each trial (kg). The factor of 1000 is to convert from mg to grams. EXO is the exogenous CHO oxidation rate (g·min^{-1}) calculated previously. These calculations serve as an estimation of hepatic glucose contribution during the second hour of exercise [5].

2.9. Data Presentation and Statistical Analysis

All data are presented as mean (±SD) unless otherwise stated. Unfortunately, two participants had to be removed from all stable isotope and substrate use data due to measurement errors making these data $n = 18$. Specific reference to how many participants are included in each dataset is made for each variable considered. Three factor repeated measures analysis of variance was used to determine treatment, time, period (order) main effects and treatment x time interaction effects. Where a significant period effect was observed then period was used as a covariate and the analysis re-run. Significant main and interaction effects were explored using post hoc Tukey's comparisons to indicate where these differences occurred. Pearson correlation analysis was performed to examine associations between individual metabolic parameters and the performance outcome differences on 20, 39 and 64 g·h^{-1} trials compared to 0 g·h^{-1}. Stepwise linear regression analysis was used to find the best prediction model for performance outcome using multiple metabolic parameters. An alpha value of 0.15 was used for inclusion and exclusion of variables from the model at any given step. An alpha value of 0.15 was chosen to ensure variables were not included or excluded too easily from the model. In addition, a best subsets regression analysis was performed on all metabolic variables. In all cases statistical significance was accepted at $p < 0.05$.

3. Results

3.1. Participants

Twenty male competitive cyclists completed all trials in this study. All treatments were tolerated well by participants. Tremendous effort was made to ensure all data points were collected, though some data sets had to be removed due to technical problems. As such, all data for substrate oxidation are for $n = 18$ due to the absence of ^{13}C tracer on one CHO trial for one participant, and expired gas analysis analytical problems during exercise with one other participant. All other data are for $n = 20$.

3.2. Performance Task Outcomes

Performance task data ($n = 20$) has been reported elsewhere [7]. Briefly, endurance cycling performance was equally improved with carbohydrate provision when ingested at a rate of 39 or 64 g·h^{-1} in comparison to water placebo (0 g·h^{-1}). No significant difference in performance task time was noted between 20 g·h^{-1} and 0 g h^{-1} treatments, or between 39 and 64 g·h^{-1} treatments (Table 1).

Table 1. Performance task outcome data on trials where 0, 20, 39 and 64 g·h^{-1} of carbohydrate was ingested.

Variable	Performance Time (min)	Percentage Change from 0 g·h^{-1} (%)	Cohen's Size Effect from 0 g·h^{-1}
0 g·h^{-1}	37:01.9 ± 05:35.0	-	-
20 g·h^{-1}	35:17.6 ± 04:16.3	3.7 (95% CI −1.5–8.8; $p = 0.13$)	0.6 (95% CI −0.1–1.4)
39 g·h^{-1}	34:19.5 ± 03:07.1	6.1 (95% CI 1–11.3; $p = 0.02$)	1.0 (95% CI 0.2–1.7)
64 g·h^{-1}	34:11.3 ± 03:08.5	7.0 (95% CI 1–12, $p = 0.01$)	1.0 (95% CI 0.3–1.8)

3.3. Substrate Oxidation

3.3.1. Respiratory Exchange Ratio

RER data analysis ($n = 18$) revealed a significant main effect of time ($p < 0.01$), treatment ($p < 0.01$), and period ($p < 0.01$) but no interaction ($p = 0.39$). Period was subsequently treated as a covariate for all further analyses. Pairwise comparisons of time indicated that RER values declined with exercise duration. Comparisons of treatment indicated the 0 g·h^{-1} treatment RER was significantly ($p < 0.01$) lower in comparison to 20, 39 and 64 g·h^{-1} (Figure 2A). Mean (SD) RER on the trials was 0.90 (0.03), 0.91 (0.03), 0.92 (0.03) and 0.92 (0.03) for 0, 20, 39 and 64 g·h^{-1} respectively.

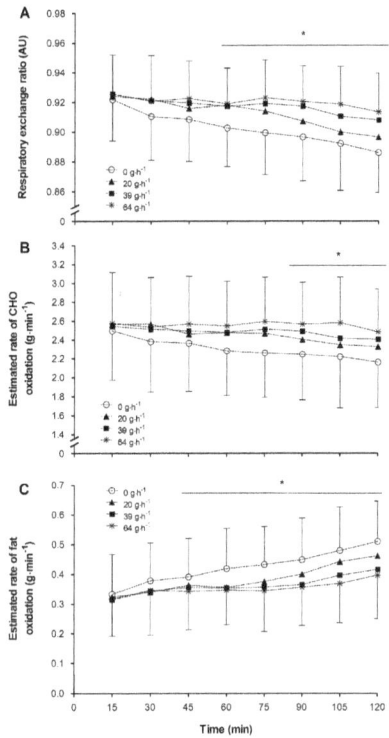

Figure 2. Mean (SD) (**A**) respiratory exchange ratio; (**B**) estimated rate of carbohydrate oxidation and (**C**) estimated rate of fat oxidation during submaximal exercise when consuming 0, 20, 39 and 64 $g \cdot h^{-1}$ of CHO. RER values are significantly (* $p < 0.01$) lower from time point 60 min onwards in comparison to 15 min values, with the 0 $g \cdot h^{-1}$ treatment eliciting a significantly ($p < 0.01$) lower mean RER over the two hours in comparison to 20, 39 and 64 $g \cdot h^{-1}$. A comparison of time indicated CHO oxidation rates at 90 min onwards were significantly lower than 15 min with the 0 $g \cdot h^{-1}$ treatment being significantly lower than 20, 39 and 64 $g \cdot h^{-1}$. Post hoc comparisons indicated the mean rate of fat oxidation was significantly lower when consuming 39 and 64 $g \cdot h^{-1}$ of CHO compared to 0 $g \cdot h^{-1}$. Additionally, time comparisons indicated an increase in fat oxidation from 45 min onwards in comparison to rates at 15 min.

3.3.2. Whole Body Substrate Oxidation

Analysis of the carbohydrate oxidation data ($n = 18$) indicated a significant effect of time ($p < 0.01$), treatment ($p < 0.01$), and period ($p < 0.01$) but no interaction effect ($p < 0.58$). Period was treated as a covariate for all subsequent analysis. Pairwise comparisons over time indicated that estimated rate of CHO oxidation was declining over time with measures from 90 min onwards being significantly lower than 15 min values. Treatment pairwise comparisons revealed that the lowest CHO oxidation rate occurred when consuming the 0 $g \cdot h^{-1}$ treatment with significant increases in oxidation when consuming 20 and further increases when consuming 39 and 64 $g \cdot h^{-1}$ in comparison to 20 $g \cdot h^{-1}$ (Figure 2B). Total CHO oxidation on the trials was 279 (58), 298 (52), 302 (47) and 312 (56) g for 0, 20, 39 and 64 $g \cdot h^{-1}$ respectively of which 0 was significantly different ($p < 0.01$) from 64 $g \cdot h^{-1}$.

Results for estimated rate of fat oxidation ($n = 18$) indicated a significant effect of time, treatment and period (all $p < 0.01$) but no interaction effect ($p = 0.82$). Period was treated as a covariate with all further comparisons. Pairwise comparisons of time indicated an increase in fat oxidation rates with increase in exercise duration. Fat oxidation was higher from 45 min onwards compared with 15 min values. Additional pairwise comparisons revealed that consuming the 0 g·h^{-1} treatment resulted in the significantly higher mean fat oxidation rates in comparison to 39 and 64 g·h^{-1} (Figure 2C). Estimated total fat oxidation was 51 (15), 45 (15), 43 (15) and 42 (15) g for 0, 20, 39 and 64 g·h^{-1} respectively of which no total oxidation was significantly different from another.

3.3.3. Exogenous Carbohydrate Oxidation and Whole-Body Substrate Contribution to Total Energy Expenditure

Data for exogenous carbohydrate oxidation ($n = 18$) indicated significant main effects of treatment ($p < 0.01$), time ($p < 0.01$), period ($p < 0.01$) and an interaction effect between treatment and time ($p < 0.01$). Period was included as a covariate for all further comparisons. Pairwise comparisons indicated that exogenous CHO oxidation rates were significantly different between all treatments from the 60-min time point until the end of exercise. Specifically, exogenous CHO oxidation rates were higher in comparison to the 20 g·h^{-1} treatment by 0.13 (95% CI: 0.10 to 0.15) and 0.29 (95% CI: 0.27–0.31) g·min^{-1} on the 39 and 64 g·h^{-1} treatments. Additionally, the 64 g·h^{-1} treatment increased exogenous oxidation rates at 75, 90, 105 and 120 min above the 60 min rates highlighting that exogenous CHO oxidation was still rising from 60 min onwards on this trial. Similarly, when consuming 39 g·h^{-1} the values at 90, 105 and 120 min also were significantly increased above the 60 min values, and at 105 and 120 min in comparison to 60 min for the 20 g·h^{-1} treatment (Figure 3A).

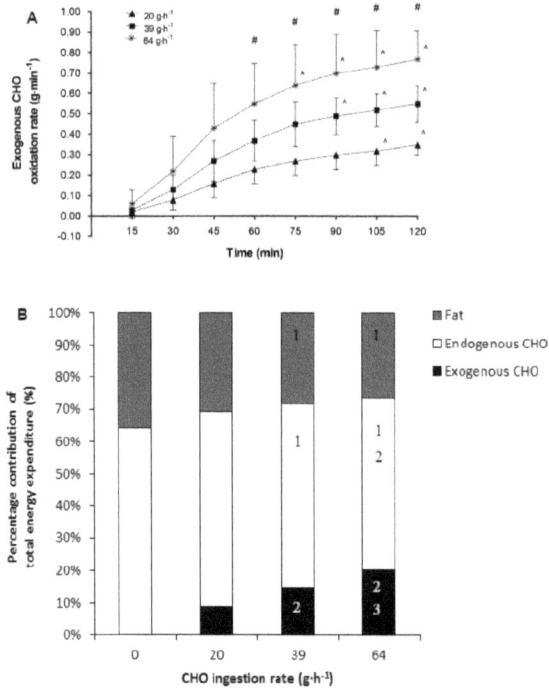

Figure 3. Mean (SD) exogenous carbohydrate (CHO) oxidation rates during submaximal exercise (**A**) while consuming 0, 20, 39 and 64 g·h^{-1} of CHO. ˆ indicates time point values significantly ($p < 0.01$) different in comparison to 60 min values; # indicates that all trials are significantly ($p < 0.01$) different from each other at the indicated time point; (**B**) Percentage contribution of total carbohydrate oxidation rates from endogenous and exogenous sources during the second hour of exercise; 1 Indicates significantly different from 0 g·h^{-1}; 2 indicates significantly different from 20 g·h^{-1}; and 3 indicates significantly different from 39 g·h^{-1}.

Percentage contribution of fat oxidised in the second hour of exercise ($n = 18$) revealed significant effects of treatment ($p < 0.01$) and period ($p < 0.01$). Following inclusion of period as a covariate, pairwise comparisons of treatment indicated the percentage contribution of fat oxidation was significantly lower when consuming 39 (-7.5, 95% CI: -1.6 to -13.4%) and 64 (-8.9, 95% CI: -3.1 to -14.8%) $g \cdot h^{-1}$ in comparisons to consuming 0 $g \cdot h^{-1}$. Endogenous carbohydrate percentage contribution highlighted significant effects of treatment ($p < 0.01$) but not period ($p = 0.39$). Pairwise comparisons of treatment indicated that endogenous carbohydrate percentage contribution was significantly suppressed in the 39 and 64 $g \cdot h^{-1}$ trials (-7.3, 95% CI: -1.6 to -13.1 and -11.2, 95% CI: -5.5 to -16.9 respectively) compared to the 0 $g \cdot h^{-1}$ treatment, respectively. Additionally, consuming carbohydrate at 64 $g \cdot h^{-1}$ suppressed endogenous carbohydrate percentage contribution by -7.2% (95% CI: -1.5–13.0%) in comparison to the 20 $g \cdot h^{-1}$ treatment. Exogenous carbohydrate oxidation percentage contribution demonstrated a significant effect of treatment ($p < 0.01$). Pairwise comparisons indicated that for exogenous carbohydrate oxidation all treatments were significantly different from one another (Figure 3B).

Glucose Ra values ($n = 18$) were mirrored by that of the Rd values, and as such statistical analysis outcomes for both data sets were almost identical. Analysis of the glucose Ra and Rd indicated significant effects of treatment ($p < 0.01$), time ($p < 0.01$), period ($p < 0.01$) and an interaction of treatment by time ($p < 0.01$). Period was treated as a covariate for all subsequent analysis. Post hoc comparisons revealed that consuming CHO resulted in a significantly higher glucose Ra of 1.98 (95% CI: 1.37–2.58), 2.12 (95% CI: 1.52–2.72), and 3.65 (95% CI: 3.05–4.25) $mg \cdot kg^{-1} \cdot min^{-1}$ for the 20, 39 and 64 $g \cdot h^{-1}$ treatments respectively, when compared to the 0 $g \cdot h^{-1}$ condition. Post hoc interaction comparisons indicated a significant increase in glucose Ra when consuming 39 and 64 $g \cdot h^{-1}$ compared to 0 $g \cdot h^{-1}$ at time points from 75 min onwards. Additionally, during the 20 $g \cdot h^{-1}$ trial glucose Ra was significantly increased over the control condition at time points from 90 min onwards. Glucose Ra values were significantly increased over the 60 min time point value from 75 to 120 min in the 64 $g \cdot h^{-1}$, 105 to 120 min with 39 $g \cdot h^{-1}$, and only at 120 min in the 20 $g \cdot h^{-1}$ Trial (Figure 4A,B).

In the analysis of the contribution of liver glucose to total Ra ($n = 18$), significant effects of treatment ($p < 0.01$) but not time ($p = 0.13$) or interaction ($p = 0.89$) were observed. Pairwise comparisons of treatment indicated that the percentage contribution of liver glucose to total Ra was significantly reduced between 20 and 39 $g \cdot h^{-1}$ feeding rates (-17.8%, 95% CI: -22.8 to -12.8%), and further significantly reduced when comparing 39 to 64 $g \cdot h^{-1}$ (-11.6%, 95% CI: -16.6 to -6.6%). These reductions represent a mean percentage suppression of liver glucose output of 43 (14%), 61 (14%) and 72 (23%) for the 20, 39 and 64 $g \cdot h^{-1}$ treatments, respectively, in comparison to the 0 $g \cdot h^{-1}$ (Figure 4C).

3.4. Blood Plasma Measures

3.4.1. Glucose

There were main effects of time ($p < 0.01$), treatment ($p < 0.01$), period ($p < 0.01$) and an interaction effect ($p < 0.01$) of treatment by time observed for plasma glucose response ($n = 20$). Period was then used as a covariate for all further analyses. Mean glucose concentration was higher when consuming 39 $g \cdot h^{-1}$ and 64 $g \cdot h^{-1}$ (0.41 $mmol \cdot L^{-1}$ (95% CI: 0.31–0.51) and 0.46 $mmol \cdot L^{-1}$ (95% CI 0.36–0.56), respectively) when pairwise comparisons to the 0 $g \cdot h^{-1}$ treatment were made. Consuming 39 and 64 $g \cdot h^{-1}$ also resulted in increased mean plasma glucose concentration by 0.23 (95% CI: 0.13–0.33) and 0.28 (95% CI: 0.18–0.38) $mmol \cdot L^{-1}$, respectively, in comparison to consuming 20 $g \cdot h^{-1}$. There was no evidence of a difference between 39 and 64 $g \cdot h^{-1}$ treatments. Treatment by time interaction analysis revealed that plasma glucose concentration was significantly increased above 0 min in the 64 $g \cdot h^{-1}$ treatment from 15 min until the end of the exercise period. Additionally, the 39 $g \cdot h^{-1}$ treatment significantly increased plasma glucose concentration from the 0 min value at 15, 30, 45 and 60 min, as did the 20 $g \cdot h^{-1}$ treatment at 30 and 45 min (Figure 5).

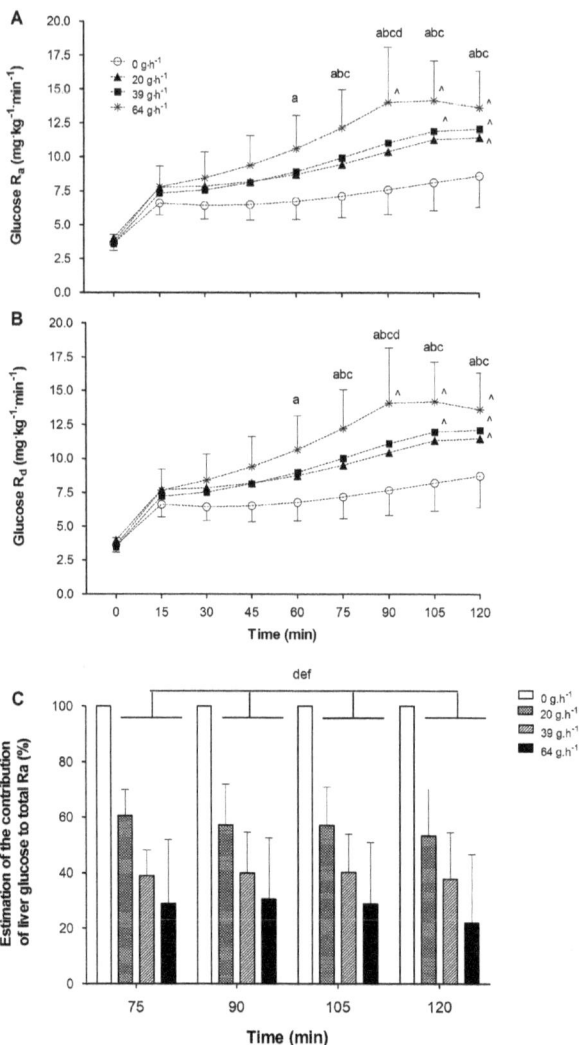

Figure 4. Mean (SD) glucose rate of appearance (**A**) and rate of disappearance (**B**) during submaximal exercise while consuming 0, 20, 39 and 64 $g \cdot h^{-1}$ of CHO. a, b and c indicates 64, 39 and 20 $g \cdot h^{-1}$ value is significantly different from 0 $g \cdot h^{-1}$ at the marked time point; d indicates 64 $g \cdot h^{-1}$ is significantly different from 20 $g \cdot h^{-1}$ at marked time point. ^ indicates time point values significantly ($p < 0.01$) different in comparison to 60 min values. Mean (SD) estimation of the contribution of liver glucose to total glucose Ra during submaximal exercise (**C**) while consuming 20, 39 and 64 $g \cdot h^{-1}$ of CHO. There was a significant treatment effect whereby 64 $g \cdot h^{-1}$ was significantly different from 20 $g \cdot h^{-1}$ (d) and 39 $g \cdot h^{-1}$ (e); and 39 $g \cdot h^{-1}$ was significantly ($p < 0.01$) different from 20 $g \cdot h^{-1}$ (f).

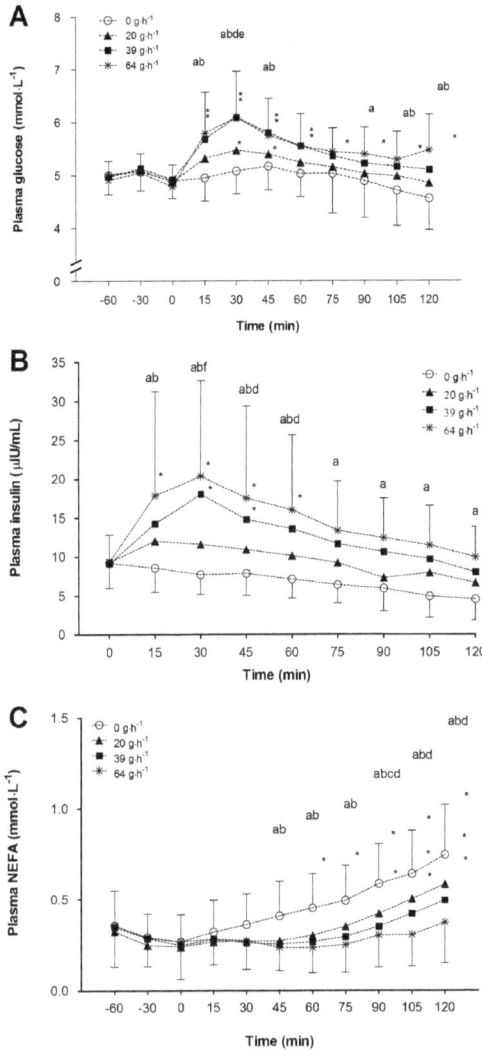

Figure 5. Mean (SD) plasma glucose (**A**); insulin (**B**); and non-esterified fatty acids (**C**) concentration during rest (−60, −30 and 0 min), and during submaximal exercise, (15–120 min), while consuming 0, 20, 39 and 64 g·h^{-1} of CHO. * Values significantly different from 0 min time point. a, b and c indicates 64, 39 and 20 g·h^{-1} value is significantly different from 0 g·h^{-1} at the marked time point; d indicates 64 g·h^{-1} is significantly different from 20 g·h^{-1} at marked time point; e indicates 39 g·h^{-1} is significantly different from 20 g·h^{-1} at marked time point; f indicates 39 and 64 g·h^{-1} are both different from the 20 g·h^{-1} at the marked time point.

3.4.2. Insulin

There were main effects of time ($p < 0.01$), treatment ($p < 0.01$), and an interaction effect between treatment and time ($p < 0.01$) on plasma insulin response ($n = 20$). There was no effect of period ($p = 0.14$). On average insulin concentration increased by 2.5 (95% CI: 1.3–3.7), 5.2 (95% CI: 4.0–6.4) and 7.3 (95% CI: 6.1–8.5) µIU·mL when consuming 20, 39 and 64 g·h^{-1} CHO, respectively, compared to the 0 g·h^{-1} trial. Insulin concentration significantly increased from pre ingestion (0 min) values at 15–60 min time points for 64 g·h^{-1}, and at 30 and 45 min for 39 g·h^{-1}. Further pairwise comparisons revealed that insulin concentration was significantly elevated in the 39 and 64 g·h^{-1} treatments when compared to the 0 g·h^{-1} at time points between 15 and 45 min. At 30 min, consuming 39 and 64 g·h^{-1} also significantly elevated insulin concentration over that of consuming 20 g·h^{-1}. The 64 g·h^{-1} treatment also significantly increased insulin concentration at time points 45 and 60 min when compared to the 20 g·h^{-1} treatment (Figure 5).

3.4.3. Non-Esterified Fatty Acids

There were main effects of time ($p < 0.01$), treatment ($p < 0.01$), period ($p < 0.01$) and an interaction of treatment by time ($p < 0.01$) on plasma NEFA ($n = 20$). Period was included as a covariate for all further analyses. Pairwise comparisons between treatments revealed that on the 0 g·h^{-1} treatment mean NEFA concentration was 0.10 (95% CI: 0.07–0.13), 0.12 (95% CI: 0.10–0.16) and 0.16 (95% CI: 0.13–0.19) mmol·L^{-1} higher than when consuming the 20, 39 and 64 g·h^{-1} treatments, respectively. Additionally, the NEFA concentration throughout exercise on 20 g·h^{-1} was significantly higher (0.06 mmol·L^{-1}, 95% CI: 0.03–0.09) than when consuming 64 g·h^{-1}. When consuming 0 g·h^{-1} all NEFA concentrations were significantly increased above the 0 min time point from 60 min onwards. On the 20 g·h^{-1} treatment plasma NEFA concentration was elevated compared to the 0 min time point at 90, 105 and 120 min. Additionally, on the 39 g·h^{-1} treatment NEFA concentration increased at time points 105 and 120 min compared to 0 min. No increase was observed on 64 g·h^{-1} treatment. Post hoc interaction comparisons revealed that mean NEFA concentration in the 0 g·h^{-1} treatment was significantly elevated compared to the 39 and 64 g·h^{-1} treatments from the 45 min time point until the end of exercise. Additionally, the 20 g·h^{-1} treatment was significantly different from 64 g·h^{-1} at 90, 105 and 120 min. Finally, the 20 g·h^{-1} treatment significantly elevated plasma NEFA concentration in comparison to the 0 g·h^{-1} at time point 90 min (Figure 5).

3.4.4. Lactate

Plasma lactate concentration ($n = 20$) revealed a significant effect of time ($p < 0.01$) and an effect of treatment ($p = 0.02$) but no interaction ($p = 0.84$), and no period effect ($p = 0.57$). Post hoc comparisons of time indicated that all exercising lactate concentrations were elevated above resting values, though there was no significant difference between trials. Mean plasma lactate concentration was 1.06 (0.38), 1.09 (0.35), 1.04 (0.29) and 1.10 (0.36) mmol·L^{-1} for 0, 20, 39 and 64 g·h^{-1} trials, respectively.

3.4.5. Adrenaline

Analysis of adrenaline concentration ($n = 20$) revealed there was a main effect of period ($p < 0.01$) time ($p < 0.01$), treatment ($p < 0.01$), but no interaction ($p = 0.10$). Period was treated as a covariate for all subsequent analysis. Pairwise comparisons of time indicated adrenaline concentrations were increasing over the duration of the exercise bout. Additionally, comparisons between treatments indicated that adrenaline concentration was highest on the 0 g·h^{-1} trial ($0.99 \pm 0.69 \text{ ng mL}^{-1}$) in comparison to the 39 g·h^{-1} ($0.78 \pm 0.38 \text{ ng mL}^{-1}$) and 64 g·h^{-1} ($0.78 \pm 0.41 \text{ ng mL}^{-1}$) trials. Mean adrenaline concentration on the 20 g·h^{-1} trial was 0.87 (0.48) ng mL^{-1}.

3.5. Associations between Metabolic Responses and Prediction of Subsequent Performance Outcomes

Association analysis between a number of key metabolic parameters during 2 h of exercise and subsequent change in performance task outcome are shown in Table 2. This analysis revealed a moderate positive association between increases in total exogenous CHO oxidized in the second hour of exercise, and total glucose rate of disappearance in the second hour of exercise, with an improvement in performance outcome. In addition, there was a tendency towards a moderate negative association between change in circulating NEFA concentration and subsequent performance outcome.

Factors such as liver glucose output suppression, mean plasma glucose, and mean plasma insulin concentrations were not associated with the changes in performance task outcome.

Table 2. Association analysis for selected metabolic parameters, obtained on the 20, 39 and 64 $g \cdot h^{-1}$ ingestion trials, with change in performance outcome compared to 0 $g \cdot h^{-1}$.

Variable	Pearson's Correlation	*p*-Value
Liver glucose suppression (%)	−0.055	0.691
Total exogenous CHO oxidation in 2nd hour (g)	0.269	0.049 *
Total glucose Rd in 2nd hour (g)	0.291	0.033 *
Mean glucose concentration	−0.045	0.747
Mean insulin concentration	0.158	0.253
Δ NEFA concentration from 0 $g \cdot h^{-1}$	−0.262	0.056 [†]

* Significant association between parameter and change in performance outcome from 0 $g \cdot h^{-1}$ trial; [†] tendency towards significant association.

Using these metabolic parameters in a stepwise linear regression analysis revealed that the combination of exogenous CHO oxidation in the second hour of exercise ($p = 0.011$) and the suppression of circulating NEFA ($p = 0.010$) provided the best model to predict subsequent performance outcomes as shown in Table 3. However, this model only explained ~19% of the variance in performance outcome observed. The regression equation that resulted was:

$$\text{Performance change from 0 (\%)} = 1.13 + 0.405\,(\text{Exo CHO}) - 34.6\,(\Delta\,\text{NEFA}) \qquad (9)$$

where Exo CHO is the total exogenous CHO oxidation in 2nd hour in grams and Δ NEFA is the difference in NEFA concentration ($mmol \cdot L^{-1}$) in comparison to consuming the 0 $g \cdot h^{-1}$. Finally, a best subsets analysis was conducted on all variables included in the regression. Including all variables reported in Table 2, except for mean glucose concentration, explained 23% of the variance in performance.

Table 3. Stepwise regression analysis of metabolic variables, obtained on the 20, 39 and 64 $g \cdot h^{-1}$ ingestion trials, with change in performance outcome compared to 0 $g \cdot h^{-1}$.

Model	Variable (ID)	R^2	*p* (Variable ID)
1	(1) Total glucose Rd in 2nd hour (g)	0.087	0.033 (1) *
2	(2) +Δ NEFA concentration from 0 $g \cdot h^{-1}$	0.126	0.129 (2), 0.074 (1) *
3	(3) +Total exogenous CHO oxidation in 2nd hour (g)	0.192	0.048 (3) *, 0.032 (2) *, 0.48 (1)
4	(4) −Total glucose Rd in 2nd hour (g)	0.184	0.011 (2) *, 0.010 (3) *

Alpha-to-enter = 0.15; Alpha-to-remove = 0.15; * indicates a significant ($p < 0.05$) component in the model.

3.6. Heart Rate and RPE

There were significant effects of time, treatment and period ($p < 0.01$) but no interaction effect ($p > 0.99$) for heart rate. Period was treated as a covariate for all subsequent analysis. Pairwise comparisons indicated that heart rate tended to increase with increasing exercise duration. In addition, post hoc comparisons of treatment indicated a significant difference between 0 and 64 $g \cdot h^{-1}$ with a mean difference of 4 (95% CI: 2, 6) beats per minute between the two trials. The mean heart rates for each treatment were 135, 137, 136 and 139 for 0, 20, 39 and 64 $g \cdot h^{-1}$ respectively.

There was a significant effect of time ($p < 0.01$) and period ($p = 0.02$), but not treatment ($p = 0.83$) or interaction ($p = 0.94$) effects on RPE responses to exercise. Period was treated as a covariate for all further comparisons. Post hoc comparisons indicated that mean RPE scores increased from 13 to 14 from minute 60 to minute 120.

4. Discussion

During this investigation, we aimed primarily to characterize the metabolic response of trained cyclists to the ingestion of graded amounts of CHO during a two-hour submaximal ride to explore the dose-response. Our secondary aim was to determine the strength of association between selected metabolic parameters and prediction of the performance task outcomes, previously reported elsewhere [7]. We observed that increasing rates of CHO ingestion (particularly at 39 and 64 g·h^{-1}) during non-exhaustive submaximal exercise resulted in: (1) a reduction in the contribution of endogenous carbohydrate and fat stores to total energy provision; (2) a decrease in hepatic glucose output in a dose response manner; (3) an increase in the contribution of exogenous CHO oxidation to total energy contribution in a dose response manner; (4) an increase in rate of total carbohydrate oxidation and plasma glucose turnover; (5) increased circulating blood glucose and insulin concentration; and (6) a blunting of the circulating NEFA response to exercise. While individually these responses to increasing doses of CHO feeding are not unforeseen the novel aspect of the present study is the examination of dose-response effects in all of these responses. Moreover, the correlation and regression analyses indicate that the rate of exogenous CHO oxidation and suppression of NEFA are the two most closely linked to a significant prediction of subsequent performance task outcome.

The significant alterations in fuel selection observed with the ingestion of 39 and 64 g·h^{-1} of CHO in comparison to 0 or 20 g·h^{-1} closely compliment the performance task outcome data previously reported [7]. In addition, the ingestion of 64 g·h^{-1} had no added effect, over 39 g·h^{-1}, on many of the key metabolic responses to exercise, but it did result in an increased rate of exogenous substrate oxidation, and a further blunting of hepatic glucose output, when compared with the 39 g·h^{-1} trial. A difference in exogenous CHO oxidation rate and hepatic glucose output, two key metabolic parameters, between the 39 and 64 g·h^{-1} trials might be expected to impact subsequent performance, but no impact was observed. Interestingly, the best subsets regression analysis indicated that some other metabolic parameters were also associated with change in performance in relation to graded doses of CHO ingestion. These other predictors were mean insulin concentration, and suppression in circulating plasma NEFA. The lack of difference in insulin or NEFA response between the 39 and 64 g·h^{-1} trials suggests that, with ingestion of a dose of single source CHO up to 64 g·h^{-1} over a two-hour exercise bout, there seems little metabolic advantage of going beyond ~40 g·h^{-1}.

Many investigators have observed a significant difference in plasma insulin and NEFA concentration with the ingestion of CHO during submaximal exercise, and a subsequent alteration of fuel utilization [5,16]. However, the only previous dose-response study [4] did not observe differences in insulin or NEFA response between the two highest CHO doses ingested (30 and 60 g·h^{-1}). The present study data corroborate these observations and, with a more suitably powered design, suggest that moderate amounts of CHO in the region of only 40 g·h^{-1} are sufficient to modulate metabolic responses enough to impact upon subsequent performance task outcome. By utilizing stable isotopes researchers have been able to quantify the movement of glucose into and out of the plasma pool during exercise when carbohydrate is consumed [15]. During exercise, blood glucose can be maintained or increased by the augmented release of glucose from the liver. In the present study, the 20 g·h^{-1} treatment reduced hepatic glucose output by 43% but the performance outcome (3.7% improvement) previously reported [7] was too variable for it to be considered a significant performance enhancement. Hepatic glucose output calculations in the current investigation reveal that all CHO ingestion rates resulted in a reduction in hepatic glucose output, and that the magnitude of reduction essentially followed a dose-response pattern. Higher CHO ingestion rates of 39 and 64 g·h^{-1} both suppressed hepatic glucose output to a greater extent than the 20 g·h^{-1} treatment. Interestingly, the magnitude of suppression in the 39 and 64 g·h^{-1} trials was similar or greater than that observed by McConnell et al. [8] when they fed CHO at 100 g·h^{-1}. This observation could suggest that even low doses of CHO at 39 and 64 g·h^{-1} are resulting in a near maximal suppression of hepatic glucose output, which is not exceeded unless very high doses of CHO are ingested (i.e., 180 g·h^{-1}; Jeukendrup [5]). The lack of any association between hepatic glucose output suppression and performance outcome suggests that even modest reductions in liver glucose output, induced by feeding only 39 g·h^{-1} of CHO during two hours of exercise, are sufficient to impact upon subsequent endurance task performance.

Exogenous CHO oxidation rates increase when CHO is ingested, but when a single source of CHO is ingested these typically only reach rates of ~1 g·min^{-1} [5,17,18]. In the present study, rates of exogenous CHO oxidation followed a dose-response pattern with the highest rates of around 0.75 g·min^{-1} achieved on the 64 g·h^{-1} trial. On the 39 g·h^{-1} trial exogenous CHO oxidation rates reached 0.55 g·min^{-1}. These data are slightly higher than those obtained by Smith et al. [4] in their dose-response study. Smith et al. [4] noted rates of ~0.3 and ~0.5 g·min^{-1} for their 30 and 60 g·h^{-1} CHO ingestion trials, respectively. The lack of performance task improvement with increasing rate of oxidation of exogenous CHO in the present study, and in the Smith et al. [4] study, suggests that capacity to oxidize exogenous CHO at a high rate is not for the key factor driving performance improvement. In addition, there was only a weak, albeit significant, association observed between exogenous CHO oxidation rate and subsequent performance outcome, as well as a modest contribution from exogenous CHO oxidation rate to prediction of performance outcome in the regression analyses. Overall, these data suggest that higher exogenous CHO delivery and higher subsequent oxidation likely contribute to endogenous (hepatic) glycogen sparing during two hours of endurance cycling, and can have some impact upon subsequent performance task outcome. However, as a note of caution, these observations may be particular to the exercise model investigated. For example, in longer exercise task durations exceeding three hours of total activity it may well is that higher feeding rates and higher exogenous CHO oxidation would translate to improved performance.

The blunting of fat oxidation observed only on the two highest CHO doses (39 and 64 g·h^{-1}) subsequently would drive fuel utilization towards a CHO dependent state. The suppression of fat oxidation and circulating NEFA concentration was similar on both 39 and 64 g·h^{-1} feeding rates. Thus, it seems that feeding of only 39 g·h^{-1} is sufficient to sustain exercise. Van Loon et al. [19] reported that a suppression in adipose tissue lipolysis increases glycogen utilization in exercising humans. While a greater dependence upon CHO oxidation was observed between 39 and 64 g·h^{-1} feeding rates compared with 0 g·h^{-1}, there was no difference in CHO usage between the two highest feeding rates in the present study. These observations suggest that near optimal substrate metabolism changes occurred with a feeding rate of close to 40 g·h^{-1}.

Our results highlight that CHO provision leading to an increased oxidation of exogenous CHO, increased total glucose disposal, and reduction in circulating NEFA, have the closest associations with subsequent performance task outcome. Higher CHO feeding rates that reach a threshold level to blunt circulating NEFA concentration, increase reliance on CHO oxidation, and enhance exogenous CHO oxidation, will have the biggest impact upon subsequent performance task outcome. The threshold required for these outcomes appears to be around 40 g·h^{-1} in the present study. However, these associations are low to moderate and the threshold of CHO ingestion rate could well be influenced by the total task duration and/or training status of participants. With longer task durations (>3 h) an increased reliance on exogenous CHO oxidation later in exercise could enhance liver glycogen sparing and could improve subsequent performance outcomes. Furthermore, with improved training status comes improved capacity to oxidize substrates [20,21] which might drive the CHO provision threshold beyond 40–60 g·h^{-1}.

Prediction of performance task outcome from metabolic parameters was not particularly strong, with only 19–23% of the variance in subsequent performance task outcome explained by the key metabolic parameters in the model. Interestingly, the prediction model containing only the two variables of exogenous CHO oxidation rate and suppression of plasma NEFA response provided almost all of the predictive power of the model. Given that these two parameters are most closely aligned to the actual CHO dose administered, it would seem plausible to suggest that higher doses of CHO should result in better performance outcomes. However, further investigation into higher rates of CHO provision and performance outcome are required before this can be categorically stated.

Of particular interest would be studies in which comparisons are made between ingestion rates within the 40–60 g·h^{-1} range using single source CHO, with those in the 90–120 g·h^{-1} range using multiple transportable CHO. To date, only one study, by Baur et al. [22], has compared a single source trial with a practically relevant feeding rate of glucose, against a multiple transportable CHO trial designed to maximize rate of exogenous CHO oxidation. Their study compared feeding rates of glucose at 62 and 93 g·h^{-1} with a 2:1 glucose: fructose beverage ingested at 93 g·h^{-1}, during three hours of endurance cycling. Their data revealed that when compared to the 62 g·h^{-1} glucose trial, there was no clear evidence of a benefit to performance compared with ingestion of 93 g·h^{-1} of the glucose: fructose beverage. These data indicate that aiming to increase exogenous CHO oxidation through consumption of composite CHO drinks at high feeding rates will not necessarily lead to meaningful performance gains. Thus, it seems that there is a need for further investigation of CHO dose and performance outcome. So, at high ingestion rates the use of multiple transportable CHO will likely reduce gastrointestinal symptoms, but this does not necessarily translate into enhanced endurance exercise performance. These previous findings may explain the lack of a strong association between exogenous CHO oxidation rate and subsequent performance outcome in the present dataset. Thus, it seems that in endurance tasks lasting <3 h, a feeding rate of 40 g·h^{-1} of single source CHO could be considered near optimal to provide sufficient metabolic advantages to maximize performance gains.

5. Conclusions

Researchers have been aiming to identify the optimal ingestion rate of CHO to elicit the greatest improvements in endurance performance for many years. We have reported that the ingestion of 39 and 64 g·h^{-1} of single source CHO were equally effective at improving endurance exercise performance in comparison to a control condition (0 g·h^{-1}). The data presented in the current manuscript further demonstrate that the ingestion of 39 g·h^{-1} of CHO appears sufficient to alter substrate utilization during a two-hour submaximal exercise bout, and lead to performance gains. These performance gains partly come from preservation of endogenous glycogen stores, most likely hepatic stores, and maintaining high rates of CHO oxidation through suppression of circulating NEFA concentration. Ingestion of CHO at a lower rate (20 g·h^{-1}) is insufficient for these particular metabolic changes to occur, while increasing the rate of ingestion to 64 g·h^{-1} does not appear to have any additional benefit. The lack of any additional change in many metabolic parameters when consuming 64 g·h^{-1} could be partly responsible for a lack of any additional improvement in subsequent performance task outcome. From these present observations, we conclude that an ingestion rate of 39 g·h^{-1} is a dose beyond which there a no further performance or metabolic benefits, suggesting that it could be an optimal ingestion rate, to elicit a sufficient alteration in fuel provision during submaximal exercise. While a 39 g·h^{-1} dose appears effective in this investigation, the observations should be confined to the particular task duration and participant group studied. Further work is required to explore the metabolic advantages and potential performance enhancement from higher feeding rates in more well-trained/elite competitors, in female participants, and in tasks lasting longer than three hours.

Acknowledgments: Funding for the present work was provided by Lucozade Ribena Suntory Ltd., Uxbridge, U.K. to University of Stirling. We would like to thank Gillian Dreczkowski for her assistance with the analytical components of the trials.

Author Contributions: S.D.R.G., K.D.T. and A.M.H. conceived and designed the experiments; M.L.N., A.M.H. and S.D.R.G. performed the experiments; M.L.N., S.D.R.G. and G.A.W. analysed the data; G.A.W. contributed reagents/materials/analysis tools; M.L.N. and S.D.R.G. wrote the paper. All other authors contributed to editing of the final manuscript.

Conflicts of Interest: The authors declare no conflict of interest. The founding sponsors had some input into the design of the study, but were not involved in the collection, analyses, or interpretation of data; or in the writing of the manuscript, and in the decision to publish the results.

References

1. Coyle, E.F.; Coggan, A.R.; Hemmert, M.K.; Ivy, J.L. Muscle glycogen utilization during prolonged strenuous exercise when fed carbohydrate. *J. Appl. Physiol. (1985)* **1986**, *61*, 165–172. [CrossRef] [PubMed]

2. Coggan, A.R.; Coyle, E.F. Reversal of fatigue during prolonged exercise by carbohydrate infusion or ingestion. *J. Appl. Physiol. (1985)* **1987**, *63*, 2388–2395. [CrossRef] [PubMed]

3. Nybo, L. CNS Fatigue and prolonged exercise: Effect of glucose supplementation. *Med. Sci. Sports Exerc.* **2003**, *35*, 589–594. [CrossRef] [PubMed]

4. Smith, J.W.; Zachwieja, J.J.; Peronnet, F.; Passe, D.H.; Massicotte, D.; Lavoie, C.; Pascoe, D.D. Fuel selection and cycling endurance performance with ingestion of [^{13}C] glucose: Evidence for a carbohydrate dose response. *J. Appl. Physiol. (1985)* **2010**, *108*, 1520–1529. [CrossRef] [PubMed]

5. Jeukendrup, A.E.; Wagenmakers, A.J.; Stegen, J.H.; Gijsen, A.P.; Brouns, F.; Saris, W.H. Carbohydrate ingestion can completely suppress endogenous glucose production during exercise. *Am. J. Physiol.* **1999**, *276*, E672–E683. [CrossRef] [PubMed]

6. Rowlands, D.S.; Houltham, S.; Musa-Veloso, K.; Brown, F.; Paulionis, L.; Bailey, D. Fructose–glucose composite carbohydrates and endurance performance: Critical review and future perspectives. *Sports Med.* **2015**, *45*, 1561–1576. [CrossRef] [PubMed]

7. Newell, M.L.; Hunter, A.M.; Lawrence, C.; Tipton, K.D.; Galloway, S.D. The ingestion of 39 or 64 g·h^{-1} of carbohydrate is equally effective at improving endurance exercise performance in cyclists. *Int. J. Sport Nutr. Exerc. Metab.* **2015**, *25*, 285–292. [CrossRef] [PubMed]

8. McConell, G.; Fabris, S.; Proietto, J.; Hargreaves, M. Effect of carbohydrate ingestion on glucose kinetics during exercise. *J. Appl. Physiol. (1985)* **1994**, *77*, 1537–1541. [CrossRef] [PubMed]

9. Neal, C.M.; Hunter, A.M.; Brennan, L.; O'Sullivan, A.; Hamilton, D.L.; De Vito, G.; Galloway, S.D. Six weeks of a polarized training-intensity distribution leads to greater physiological and performance adaptations than a threshold model in trained cyclists. *J. Appl. Physiol. (1985)* **2013**, *114*, 461–471. [CrossRef] [PubMed]

10. Kuipers, H.; Verstappen, F.; Keizer, H.; Geurten, P.; Van Kranenburg, G. Variability of aerobic performance in the laboratory and its physiologic correlates. *Int. J. Sports Med.* **1985**, *6*, 197–201. [CrossRef] [PubMed]

11. Borg, G.A. Psychophysical bases of perceived exertion. *Med. Sci. Sports Exerc.* **1982**, *14*, 377–381. [CrossRef] [PubMed]

12. Pickert, A.; Overkamp, D.; Renn, W.; Liebich, H.; Eggstein, M. Selected ion monitoring Gas Chromatography/Mass Spectrometry using uniformly labelled (^{13}C) glucose for determination of glucose turnover in man. *Biol. Mass Spectrom.* **1991**, *20*, 203–209. [CrossRef] [PubMed]

13. Jeukendrup, A.; Wallis, G. Measurement of substrate oxidation during exercise by means of gas exchange measurements. *Int. J. SPORTS Med.* **2005**, *26*, S28–S37. [CrossRef] [PubMed]

14. Proietto, J. Estimation of glucose kinetics following an oral glucose load. Methods and applications. *Horm. Metab. Res. Suppl.* **1990**, *24*, 25–30. [PubMed]

15. Wolfe, R.R.; Chinkes, D.L. *Isotope Tracers in Metabolic Research: Principles and Practice of Kinetic Analysis*; John Wiley & Sons: Hoboken, NJ, USA, 2005.

16. Wallis, G.A.; Yeo, S.E.; Blannin, A.K.; Jeukendrup, A.E. Dose-response effects of ingested carbohydrate on exercise metabolism in women. *Med. Sci. Sports Exerc.* **2007**, *39*, 131–138. [CrossRef] [PubMed]

17. Jentjens, R.L.; Shaw, C.; Birtles, T.; Waring, R.H.; Harding, L.K.; Jeukendrup, A.E. Oxidation of combined ingestion of glucose and sucrose during exercise. *Metab. Clin. Exp.* **2005**, *54*, 610–618. [CrossRef] [PubMed]

18. Harvey, C.R.; Frew, R.; Massicotte, D.; Peronnet, F.; Rehrer, N.J. Muscle glycogen oxidation during prolonged exercise measured with oral [^{13}C] glucose: Comparison with changes in muscle glycogen content. *J. Appl. Physiol. (1985)* **2007**, *102*, 1773–1779. [CrossRef] [PubMed]

19. Van Loon, L.J.; Thomason-Hughes, M.; Constantin-Teodosiu, D.; Koopman, R.; Greenhaff, P.L.; Hardie, D.G.; Keizer, H.A.; Saris, W.H.; Wagenmakers, A.J. Inhibition of adipose tissue lipolysis increases intramuscular lipid and glycogen use in vivo in humans. *Am. J. Physiol. Endocrinol. Metab.* **2005**, *289*, E482–E493. [CrossRef] [PubMed]

20. Lund, J.; Tangen, D.S.; Wiig, H.; Stadheim, H.K.; Helle, S.A.; Birk, J.B.; Ingemann-Hansen, T.; Rustan, A.C.; Thoresen, G.H.; Wojtaszewski, J.F. Glucose metabolism and metabolic flexibility in cultured skeletal muscle cells is related to exercise status in young male subjects. *Arch. Physiol. Biochem.* **2017**, 1–12. [CrossRef] [PubMed]

21. Hargreaves, M.; Spriet, L.L. Exercise metabolism: Fuels for the fire. *Cold Spring Harb Perspect. Med.* **2017**. [CrossRef] [PubMed]

22. Baur, D.A.; Schroer, A.B.; Luden, N.D.; Womack, C.J.; Smyth, S.A.; Saunders, M.J. Glucose-fructose enhances performance versus isocaloric, but not moderate, glucose. *Med. Sci. Sports Exerc.* **2014**, *46*, 1778–1786. [CrossRef] [PubMed]

nutrients

MDPI

Article

Chronic Ketogenic Low Carbohydrate High Fat Diet Has Minimal Effects on Acid–Base Status in Elite Athletes

Amelia J. Carr [1,*], Avish P. Sharma [2,3], Megan L. Ross [4,5], Marijke Welvaert [3,6], Gary J. Slater [7] and Louise M. Burke [4,5]

[1]   Centre for Sport Research, Deakin University, Burwood, VIC 3125, Australia
[2]   Physiology, Australian Institute of Sport, Bruce, ACT 2617, Australia; avish.sharma@ausport.gov.au
[3]   Research Institute for Sport and Exercise, University of Canberra, Belconnen, ACT 2616, Australia;
      marijke.welvaert@ausport.gov.au
[4]   Sports Nutrition, Australian Institute of Sport, Bruce, ACT 2617, Australia;
      meg.ross@ausport.gov.au (M.L.R.); louise.burke@ausport.gov.au (L.M.B.)
[5]   Mary MacKillop Institute for Health Research, Australian Catholic University,
      Melbourne, VIC 3000, Australia
[6]   Innovation, Research and Development, Australian Institute of Sport, Bruce, ACT 2617, Australia
[7]   School of Health and Sport Sciences, University of the Sunshine Coast, Maroochydore, QLD 4558, Australia;
      gslater@usc.edu.au
*   Correspondence: amelia.carr@deakin.edu.au; Tel.: +61-3-9251-7309

Received: 20 January 2018; Accepted: 13 February 2018; Published: 18 February 2018

Abstract: Although short (up to 3 days) exposure to major shifts in macronutrient intake appears to alter acid–base status, the effects of sustained (>1 week) interventions in elite athletes has not been determined. Using a non-randomized, parallel design, we examined the effect of adaptations to 21 days of a ketogenic low carbohydrate high fat (LCHF) or periodized carbohydrate (PCHO) diet on pre- and post-exercise blood pH, and concentrations of bicarbonate [HCO_3^-] and lactate [La^-] in comparison to a high carbohydrate (HCHO) control. Twenty-four (17 male and 7 female) elite-level race walkers completed 21 days of either LCHF ($n = 9$), PCHO ($n = 7$), or HCHO ($n = 8$) under controlled diet and training conditions. At baseline and post-intervention, blood pH, blood [HCO_3^-], and blood [La^-] were measured before and after a graded exercise test. Net endogenous acid production (NEAP) over the previous 48–72 h was also calculated from monitored dietary intake. LCHF was not associated with significant differences in blood pH, [HCO_3^-], or [La^-], compared with the HCHO diet pre- or post-exercise, despite a significantly higher NEAP (mEq·day^{-1}) (95% CI = (10.44; 36.04)). Our results indicate that chronic dietary interventions are unlikely to influence acid–base status in elite athletes, which may be due to pre-existing training adaptations, such as an enhanced buffering capacity, or the actions of respiratory and renal pathways, which have a greater influence on regulation of acid–base status than nutritional intake.

Keywords: dietary interventions; periodized carbohydrate diet; fat adaptation; keto-adaptation

1. Introduction

Low carbohydrate high fat (LCHF) diets have previously been implemented in the context of epilepsy treatment [1,2] and as a weight loss strategy [3,4]. More recently, however, there has been a re-emergence of interest in their potential role in sports nutrition, with claims that adaptation to restricted carbohydrate (CHO) intake and high levels of circulating ketone bodies by trained individuals achieves significant changes in substrate utilization during sub-maximal exercise, to shift reliance from glycogen utilization to the relatively unlimited stores of body fat [5]. Indeed, both early [6] and more recent [7–9] studies have shown that sustained (3 weeks to several years) exposure to such a diet causes these robust shifts in exercise fuel use. However, beneficial effects on performance remain unsubstantiated, with reports of maintained capacity for submaximal cycling under fasted conditions in well-trained cyclists [6], but a reduction in exercise economy and failure to improve 10,000 m race performance following a block of intensified training in elite race walkers [9], in comparison to a more traditional diet providing high CHO availability [10]. A third approach to nutrition support for endurance sport is the periodized CHO diet, which integrates strategies to achieve high CHO availability to support key training sessions with protocols for low CHO availability to enhance adaptive responses to selected lower intensity sessions [11]. This redistribution of CHO intake to target the individualized goals of each training session has been shown to alter substrate utilization during submaximal exercise [12], and to produce performance benefits in sub-elite [11,13], but not elite [9,14] athletes. Thus, it appears that sports performance is determined by factors other than a simple change in substrate utilization.

One of the less well-studied effects of sustained manipulations of the macronutrient composition of the diet is the alteration in acid–base status. Diets high in fat can increase blood acidity, attributed to the stimulation of lipolysis, and the release of acidic ketone bodies [4]. A small number of studies have investigated the effects of short-term dietary modifications on acid–base status and exercise capacity in healthy but essentially untrained individuals [15–18]. For example, measurement of pre-exercise blood pH and bicarbonate concentrations after an overnight fast showed that three days of a low (<10% energy intake) CHO diet was associated with a reduction in blood alkalinity, compared with a high (>65% energy) CHO diet [15]. Other studies by the same group confirmed the effect of various macronutrient concentrations. After a three-day high-protein (24% total energy intake), high fat diet (73% total energy intake) [17], and a three-day low-carbohydrate diet (4% total energy intake) [16], pre- and post-exercise blood pH and blood bicarbonate concentrations decreased, compared with a high carbohydrate diet. While these results indicate modifications of acid–base status after acute alterations in dietary macronutrient intake, and potentially a separate and additional contribution to changes in exercise capacity, the effects have yet to be determined for sustained dietary interventions, particularly in athletic populations.

Acid–base status has primarily been investigated in terms of the deleterious health implications of metabolic acidosis [19–22]. Damage to bone and muscle can occur when additional calcium is excreted in response to the excess of protons, and adverse effects, such as kidney stones, can be caused by a reduced urine pH, a further compensatory mechanism against acidosis [23,24]. Small changes to acid–base status place substantial stress on the body's buffering mechanisms [23,25], and one contributing factor is the composition of the diet [26]. It is acknowledged that normal metabolism incorporates many reactions that produce and consume acids and bases, and that acid–base status is routinely corrected by a tightly controlled acid–base regulatory system. The discrepancy between the acid and base forming reactions due to dietary intake is, however, the primary contributing factor toward net endogenous acid production (NEAP) [25].

NEAP can be calculated via validated equations [26,27]. One widely used method focuses on the consumption of protein and potassium, quantified over a 24 h period [26]; since protein is a sulfuric acid precursor, and potassium ingestion results in bicarbonate formation, the two dietary components have a substantial influence on acid production [26]. Another more comprehensive estimation of acid–base status involves more variables, including the intake of protein and specified micronutrients (potassium, phosphate, magnesium, and calcium) over 24 h, body surface area (which affects acid excretion rates), and established intestinal absorption rates of different dietary components [27]. Several previous studies have estimated acid production due to short-term (7 days or less) dietary interventions [15–18,28,29] using a similar principle to the two approaches described, but NEAP has not been calculated for sustained dietary interventions. NEAP calculations could potentially provide an indication of the effect of different dietary regimes on acid–base status, particularly when combined with direct measures of acid–base status, such as blood pH. There are potential implications for athletes' performance, given that an increased alkalosis or buffering capacity can improve performance, and a reduction in blood or muscle pH can be detrimental to high-intensity exercise performance [30,31]. Furthermore, differences in acid–base status between athletes and untrained individuals have been reported, associated with differences in buffering capacity [32,33].

The goal of the current study was to fill gaps in our current knowledge on the longer-term changes in acid–base status associated with diets that have been shown to alter blood pH and bicarbonate concentrations when followed for short periods. In particular, we wanted to investigate the effects of adaptation to a sustained ketogenic diet or periodized carbohydrate availability on such markers before and after exercise in elite athletes, and whether NEAP calculations of endogenous acid production were able to predict or explain any changes. Therefore, our aims were to determine the effect of adaptations to two sustained dietary interventions of current interest to endurance athletes (a LCHF diet and a periodized carbohydrate availability diet [9]) on blood bicarbonate concentration, blood pH, and blood lactate concentration (pre and post exercise), and net endogenous acid production (NEAP).

2. Materials and Methods

A non-randomized parallel groups study design was used to determine the effect of three sustained (3 week) dietary interventions (a control high CHO diet—HCHO; low carbohydrate high fat diet—LCHF, and periodized carbohydrate availability diet—PCHO) on acid base balance, at rest, and following a graded exercise test in national and international level race walkers (Figure 1). Acid–base status was measured via capillary blood samples, which were analyzed for blood pH, blood bicarbonate $[HCO_3^-]$ concentration, and blood lactate $[La^-]$ concentration. Furthermore, calculations were undertaken to determine net endogenous acid production (NEAP) of the diets from dietary records kept over the 48 h prior to each exercise test. These measurements were conducted at two timepoints: baseline (prior to any intervention and with participants following a self-chosen diet), and post-testing (after a three-week dietary and training intervention, during which all food was provided to participants and consumed under supervision). The study was part of a larger research project, conducted during a residential training camp held throughout January and February 2017 at the Australian Institute of Sport (AIS; Canberra, Australian Capital Territory, Australia). All participants gave their written informed consent for inclusion, prior to their participation in the study. The study was conducted in accordance with the Declaration of Helsinki, and the protocol was approved by the AIS Ethics Committee (Project Approval Code: 20161201).

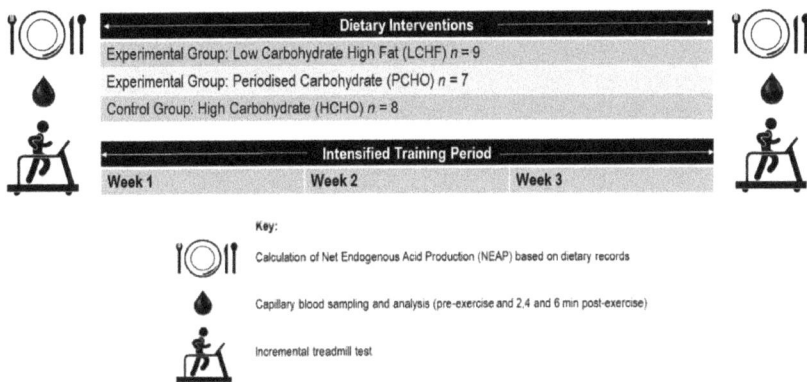

Figure 1. Overview of testing conducted in the study. All testing was conducted at baseline (prior to any intervention), and post-testing (after a supervised three-week training and dietary intervention).

2.1. Participants

Twenty-eight participants volunteered for this study, with twenty-four participants completing all testing (Table 1). Two males and one female were unable to complete the requirements due to injury, and one male was unable to provide a full data set for the baseline testing due to their delayed arrival at the training camp. The cohort consisted of elite race walkers, competitive at international or national level (mean $\pm$ SD International Association of Athletics Federations points = 1130 $\pm$ 52) [34] who participated in a large dietary intervention study conducted at the Australian Institute of Sport with the support of Athletics Australia (further details are provided by Burke et al., submitted for publication).The study included participants from Australia, New Zealand, Poland, United Kingdom, South Africa, Lithuania, Canada, United States of America, Chile, Hungary, Japan, and Spain. A sample size estimation performed for the larger study, based upon outcome measures associated with performance and substrate utilization from recent research from our group with a similar cohort, indicated that eight participants per group would be required to detect physiological differences [9]. Participants were assigned to one of the two dietary intervention groups (LCHF or PCHO), or the HCHO control group, according to their preferred dietary intervention.

Table 1. Participant characteristics (mean $\pm$ SD). Nutritional intake is presented as daily intake relative to body mass (BM), and as a percentage of total energy intake.

	Males (n = 18)	Females (n = 7)
BASELINE TEST RESULTS		
Height (cm)	178.4 $\pm$. 6.3	165.9 $\pm$ 5.8
Mass (kg)	67.7 $\pm$ 5.4	54.1 $\pm$ 5.1
VO$_2$max (mL·kg·min^{-1})	60.5 $\pm$ 4.5	56.2 $\pm$ 3.6
BASELINE NUTRITIONAL INTAKE		
Carbohydrate (g·kg^{-1} BM)	6.6 $\pm$ 1.3 (53 $\pm$ 7%)	7.1 $\pm$ 1.5 (53 $\pm$ 9%)
Protein (g·kg^{-1} BM)	2.5 $\pm$ 0.6 (20 $\pm$ 4%)	2.5 $\pm$ 0.6 (18 $\pm$ 4%)
Fat (g·kg^{-1} BM)	1.5$\pm$ 0.5 (26 $\pm$ 5%)	1.6 $\pm$ 0.4 (26 $\pm$ 5%)

2.2. Training Intervention

Athletes followed a supervised and monitored training program across a three-week period of intensified training (Table 2). Of the ~12 weekly training sessions, six involved mandatory and supervised group sessions of race walking. There were two long sessions of $\geq$20 km, a track session involving sustained high-intensity 1-km repetitions on a 6-min cycle, a 14-km hill training session (change in elevation ~285 m), and two low–moderate intensity "recovery" sessions, while two resistance training and three hydrotherapy sessions were completed. The athletes could modify their remaining sessions, undertaking them as additional race walking sessions, or substituting with cross training, such as swimming or cycling. All athletes recorded their daily training in standardized training logs.

Table 2. Template for weekly training program. Shading indicates mandatory sessions; remaining sessions could be modified by individual athletes.

DAY	Monday	Tuesday	Wednesday	Thursday	Friday	Saturday	Sunday
AM	10 km walk	10–15 km walk [#] Resistance training	>20 km long walk Hydrotherapy strategies	10 km walk Resistance training	Hill session [†]	>20 km long walk Hydrotherapy strategies	10 km walk or rest
PM	1 km reps [†] Hydrotherapy strategies	10 km walk	10 km walk or rest	10–15 km walk	10–15 km walk [#]	10 km walk or rest	

[#] Indicates training session with low carbohydrate availability prior to the session (PCHO group only). [†] Indicates training session with low carbohydrate availability after the session (PCHO group only).

2.3. Dietary Intervention

All meals were prepared by chefs, according to standardized recipes, and were consumed by participants in a group environment, under the supervision of registered sports dietitians, according to a previously established practice [32]. For baseline testing, participants were able to choose freely from items provided as a buffet-style menu at the AIS Dining Hall, and were assisted to weigh and record all of their food intake during this period, using calibrated food scales (SJ-5001HS, A&D Weighing, Australia). Once the dietary intervention commenced, participants were provided with all meals and snacks according to their intervention group (PCHO or LCHF, or HCHO) with menus being individualized to BM and training load (to provide an energy availability of ~40 kcal·kg^{-1} lean BM; LBM), and specific dietary requirements including food allergies, intolerances, and dietary preferences. Each athlete could request an increase or decrease in the quantities of foods or drinks provided according to hunger, changes in training load, or fluctuations in BM. Such variations were accommodated by adhering to the macronutrient composition of that individual's treatment group. During this period, all food intake was weighed prior to consumption, with allowances made for unfinished portions or additional snacks chosen from a menu-specific list, as per food diary logs. Energy and nutrient intake provided by the diets was calculated from a food analysis database specific to Australian foods (Foodworks Version 9, Highgate Hill, Australia) by the same registered dietitian. Full details of the methodology for creating, providing, and recording food intake can be found elsewhere [35].

The three dietary interventions were chosen to represent different approaches to the support of high volume training programs [9].

- HCHO: traditional sports nutrition guidelines promoting high carbohydrate availability for all training sessions: CHO: ~8 g·kg^{-1} BM, 60–65% energy intake; protein: ~1.8 g·kg^{-1}·day^{-1} 15–20% energy; fat: ~20% energy intake [10].
- PCHO: contemporary approach to sports nutrition, with same energy and macronutrient composition as HCHO, but manipulated across and between days to provide high CHO availability for key training sessions and low CHO availability for other sessions [11,36].
- LCHF: popular ketogenic low CHO high fat diet: CHO: <50 g·day^{-1}, protein: ~1.8 g·kg^{-1}·day^{-1} 15–20% protein, and 75–80% fat [6,34].

2.4. Exercise Testing

All participants completed a graded maximal exercise test at the baseline and post-testing timepoints, under controlled laboratory conditions. All participants conducted each test in a fasted state. The test was performed on a custom-built, motorized treadmill (Australian Institute of Sport, Canberra, Australia). The test was comprised of four submaximal stages (for determination of submaximal VO$_2$ and walking economy). Each stage, three minutes in length, was immediately followed by an incremental ramp to exhaustion for determination of VO$_2$peak. As such, the total test duration was approximately 13 to 20 min, depending on a participant's time to exhaustion (TTE). The treadmill velocity for the first stage was dependent on each participants' most recent 10 km race time (9–12 km·h^{-1}), at 0% gradient, with the velocity being increased by 1 km·h^{-1} with each subsequent stage. After each stage, a small (5 microlitres; µL) capillary blood sample was taken from the fingertip, to measure blood [La$^-$] (Lactate Pro, Arkray, Kyoto, Japan). Immediately following the completion of the fourth submaximal stage (approximately equivalent to 20 km race-walk speed), the gradient of the treadmill was increased by 0.5 degrees every 30 s, until the participant reached volitional exhaustion. Heart rate (Polar heart rate monitor, Polar Electro, Kempele, Finland) was measured throughout the test. Expired ventilation samples were collected continuously throughout the test, using a custom-built open-circuit indirect calorimetry system (Australian Institute of Sport, Canberra, Australia). The system was calibrated prior to each test [37].

Prior to starting the maximal exercise test, 100 μL of capillary blood was collected from the fingertip. The hand was first immersed in warm water for ~1 min to increase blood flow to the area, and then the fingertip was pierced using a sterile retractable lancet (Accu-Check, Roche, Sydney, Australia). Blood samples were immediately analyzed for blood pH, blood [HCO_3^-], and blood [La^-] using a portable blood-gas analyzer (i-STAT, Abbott, Chicago, IL, USA), which was calibrated prior to testing, in accordance with manufacturers guidelines. The blood sampling and analysis procedure was repeated 2, 4, and 6 min after the completion of the maximal exercise test.

2.5. NEAP Calculations

The NEAP of dietary intake preceding the baseline and post-testing timepoints was calculated from the participant's dietary records. Mean values were used to account for any fluctuation in dietary intake across days, and incorporated as much dietary information as was possible. Therefore, for baseline testing, values were determined from the mean intake recorded for the previous 48 h (the time between participants arriving at the training camp and the commencement of their dietary intervention), and post-testing calculations were taken from mean intake over the previous 72 h (the final three days of their supervised dietary intervention). Several published algorithms were used to estimate NEAP. These included two equations established by Frassetto et al. [26] which use daily protein and potassium intake, and one further equation developed by Remer and Manz [27], which incorporates daily protein, phosphorus, potassium, magnesium, and calcium intake, as well as anthropometric measurements. The three equations used are listed below.

1. Estimated NEAP ($mEq \cdot day^{-1}$) = ($0.91 \times$ protein ($g \cdot day^{-1}$)) − ($0.57 \times$ potassium ($mEq \cdot day^{-1}$)) + 21 [26] ($NEAP_{F1}$).
2. Estimated NEAP ($mEq \cdot day^{-1}$) = ($54.5 \times$ protein ($g \cdot day^{-1}$)/potassium ($mEq \cdot day^{-1}$)) − 10.2 [26] ($NEAP_{F2}$).
3. Estimated NEAP ($mEq \cdot day^{-1}$) = Potential Renal Acid Load (PRAL) ($mEq \cdot day^{-1}$) + Estimated Urinary Organic Anions (OA_{est}) [38]. Within Equation (3), PRAL is calculated as: ($mEq \cdot day^{-1}$) = $0.488 \times$ protein ($g \cdot day^{-1}$) + $0.037 \times$ phosphorous ($mg \cdot day^{-1}$) − $0.021 \times$ potassium ($mg \cdot day^{-1}$) − $0.026 \times$ magnesium ($mg \cdot day^{-1}$) − $0.013 \times$ calcium ($mg \cdot day^{-1}$), whereby: OA_{est} ($mEq \cdot day^{-1}$) = ($0.007184 \times$ height$^{0.725} \times$ mass$^{0.425}$) $\times$ (41/1.73) ($NEAP_R$) [27].

2.6. Statistical Analysis

The data were analyzed with a general linear mixed model using the R package lme4 (R Core Team, Vienna, Austria) [39]. A random intercept for participants included interindividual homogeneity. All models were estimated using restricted maximum likelihood. Visual inspection of residual plots did not reveal any obvious deviations from homoscedasticity or normality. *p*-Values were obtained using Type II Wald F tests with Kenward-Roger degrees of freedom as implemented in the R package car [40]. Results for all variables (blood pH, blood [HCO_3^-], blood [La^-], NEAP) are reported as mean estimates and 95% confidence intervals. Initial models included all possible interactions, but non-significant interaction terms were dropped from the models for ease of interpretation.

3. Results

3.1. Participants

Participants' pre-intervention VO_2max was 57.6 ± 4.6 $mL \cdot kg \cdot min^{-1}$ BM for HCHO, 58.1 ± 3.3 mL for PCHO and 61.1 ± 5.3 for LCHF. Post-intervention, VO_2max was 58.3 ± 4.1, 60.2 ± 3.8, and 63.4 ± 4.1 $mL \cdot kg \cdot min^{-1}$ BM, respectively. Within each of the three groups, there was a significant increase in VO_2max compared with baseline ($p < 0.05$).

3.2. Blood pH

At baseline, there were no differences in blood pH between HCHO, PCHO, and LCHF, at any timepoint ($F_{(2,20.97)} = 0.86$; $p = 0.44$), and as expected, there was a significant decrease in blood pH from pre-exercise to all post-exercise timepoints for all groups ($F_{(1,144.10)} = 8.64$; $p = 0.003$). Post-intervention, there were no significant differences between groups at the pre-exercise collection (95% CI = (-0.03; 0.05) and (-0.01; 0.07) for HCHO and PCHO, compared to LCHF, respectively), but at the 2 min post-exercise timepoint, pH was significantly lower for PCHO compared with LCHF (95% CI = (-0.16; -0.04)), with no significant difference between HCHO and LCHF (95% CI = (-0.11; 0.01)). Similarly, at 4 min and 6 min post-exercise, pH was significantly lower for PCHO compared to LCHF (95% CI = (-0.15; -0.04) and (-0.15; -0.03), respectively), with no significant difference between HCHO and LCHF (95% CI = (-0.10; 0.02) and (-0.08; 0.04), respectively) (Figure 2).

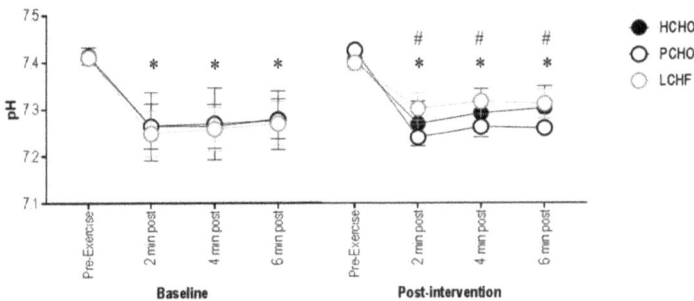

Figure 2. Blood pH for the high carbohydrate (HCHO), periodized carbohydrate (PCHO), and low carbohydrate high fat (LCHF) groups for baseline and post-intervention, at pre-exercise, plus 2, 4, and 6 min post-test. * Significantly different to pre-exercise ($p < 0.05$). # PCHO significantly different to LCHF ($p < 0.05$).

3.3. Blood Bicarbonate Concentration

There were no significant differences in pre-exercise or post-exercise blood [HCO_3^-] between groups at baseline ($F_{(2,20.99)} = 3.31$; $p = 0.06$), or post-training intervention ($F_{(1,161.15)} = 0.14$; $p = 0.71$). As with blood pH, there was a significant decrease in blood [HCO_3^-] (95% CI = (-10.74; -9.53)) between pre-exercise and all post-exercise timepoints for HCHO, PCHO, and LCHF ($F_{(1,161.16)} = 511.24$; $p < 0.001$) (Figure 3).

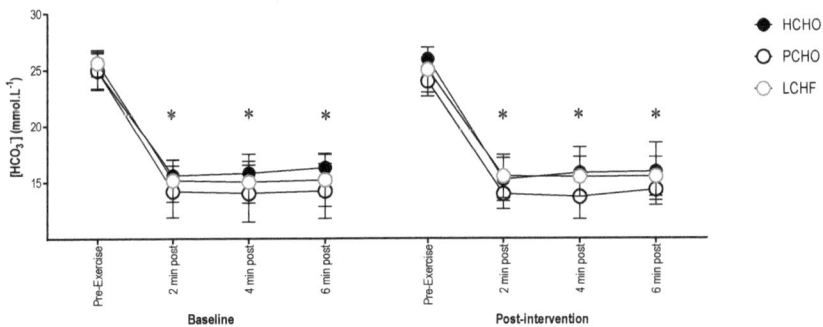

Figure 3. Blood [HCO^{3-}] for the high carbohydrate (HCHO), periodized carbohydrate (PCHO), and low carbohydrate high fat (LCHF) groups for baseline and post-testing at pre-exercise, plus 2, 4, and 6 min post-test. * Significantly different to pre-exercise ($p < 0.001$).

3.4. Blood Lactate Concentration

There were no significant differences in blood [La$^-$] between groups at baseline for any timepoint ($F(2,20.98) = 0.96$, $p = 0.40$). Post-intervention, there were no significant differences between groups ($F(14,311.07) = 1.28$; $p = 0.22$) for pre-exercise blood [La$^-$] (95% CI = (-2.15; 2.07) and (-2.14; 2.19) for LCHF and HCHO, compared to PCHO, respectively), or at any stage during the economy test (95% CI = (-2.22; 3.53) and (-2.13; 4.00) for LCHF and PCHO, compared to HCHO, respectively), or 2 min post-test (95% CI = (-5.13; 0.62) and (-1.27; 4.86) for LCHF and PCHO, compared to HCHO, respectively). At 4 min post-test, blood [La$^-$] was significantly higher for PCHO than LCHF (95% CI = (1.52; 7.57)), with no difference between PCHO and HCHO control (95% CI = (-5.12; 1.07)). Similarly, blood [La$^-$] was significantly higher 6 min post-exercise with PCHO compared with LCHF (95% CI = (1.77; 7.74)), and there was no difference compared with HCHO (95% CI = (-5.27; 0.86)) (Figure 4).

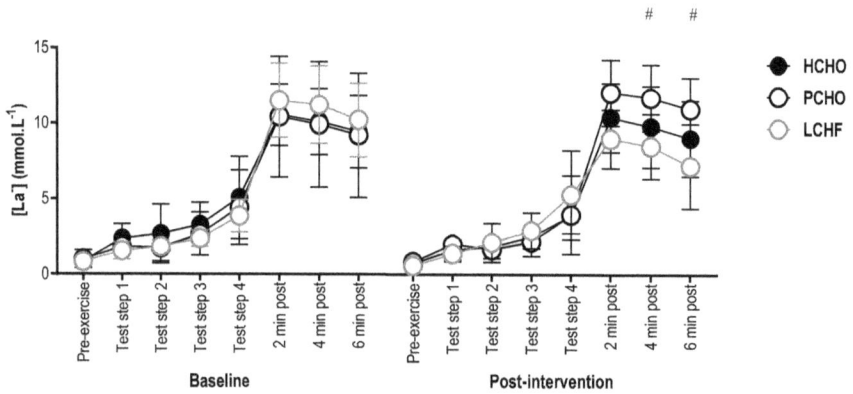

Figure 4. Blood [La$^-$] for the high carbohydrate (HCHO), periodized carbohydrate (PCHO) and low carbohydrate high fat (LCHF) groups for baseline and post-testing at pre-exercise, steps 1, 2, 3, and 4 of the economy test, and 2, 4, and 6 min post-test. # PCHO significantly different to LCHF ($p < 0.05$).

3.5. Net Endogenous Acid Production

There were significant differences between NEAP$_{F1}$, NEAP$_{F2}$, and NEAP$_R$ ($F(2,115) = 39.02$, $p < 0.001$). For all groups and at both baseline and post-intervention, NEAP$_{F1}$ values were significantly higher compared to NEAP$_R$ (95% CI = (17.40; 30.44)), with no significant difference between NEAP$_{F2}$ and NEAP$_R$ (95% CI = (-9.90; 3.13)) (Table 3). At baseline, NEAP was lower for HCHO compared to LCHF (95% CI = (-28.22; -3.54)), and there was no difference between PCHO and LCHF (95% CI = (-20.82; 4.77)). Post-intervention, NEAP values remained unchanged for LCHF (95% CI = (-6.91; 10.47)), but there was a significant decrease for the HCHO (95% CI = (-20.70; -227)) and PCHO (95% CI = (-23.29; -3.58)). Post-intervention, NEAP was significantly higher in LCHF compared with HCHO control (95% CI = (10.44; 36.04)) but there was no difference for PCHO compared with HCHO (95% CI = (-19.06; 7.23)).

Table 3. Net endogenous acid production (NEAP) estimations for high carbohydrate (HCHO), periodized carbohydrate (PCHO) and low carbohydrate, high fat (LCHF) groups using two equations by Frassetto et al. (NEAP$_{F1}$, NEAP$_{F2}$) [25], and one equation by Remer and Manz (NEAP$_R$) [26]. NEAP estimations are provided for baseline and post-testing.

	HCHO	PCHO	LCHF
BASELINE			
NEAP$_{F1}$ (mEq·day^{-1})	75 ± 19	84 ± 18	104 ± 35
NEAP$_{F2}$ (mEq·day^{-1})	49 ± 9	62 ± 16	61 ± 15
NEAP$_R$ (mEq·day^{-1})	62 ± 19 *	63 ± 14 *	68 ± 31 *
POST-INTERVENTION			
NEAP$_{F1}$ (mEq·day^{-1})	65 ± 12	71 ± 16	98 ± 15
NEAP$_{F2}$ (mEq·day^{-1})	45 ± 4	48 ± 4	70 ± 7
NEAP$_R$ (mEq·day^{-1})	40 ± 10 *	58 ± 35 *	69 ± 17 *

* significantly different to NEAP$_{F1}$ ($p < 0.05$).

4. Discussion

This is the first study to investigate the effects of sustained manipulation of macronutrient intake on acid–base status in elite athletes. The main finding of this investigation was that substantial restriction of dietary carbohydrate intake with a ketogenic low carbohydrate high fat diet over a three-week period had no influence on resting or exercise-associated blood indices of acid–base status, despite changes in the calculated net endogenous acid production of the diet. While it is plausible that this may be an anomaly that is specific to the training status and its effect on muscle buffering capacity of elite athletes in this study, it may also reflect the neutralization or expulsion of dietary acids through the respiratory or renal pathways.

The findings of this study contrast with earlier investigations, where low carbohydrate high fat diets were associated with decreased resting blood pH and blood [HCO$_3$$^-$] [15–17]. However, there are several important differences between our study and those conducted previously. Our study involved elite athletes, most of whom had represented their countries at World Championships, Olympic Games, and International Association of Athletics Federations (IAAF) World Cup events, whereas the participants in previous studies were healthy but untrained [15–17]. Furthermore, the dietary intervention was applied for 21 days in the present investigation, in comparison to acute (three or four-day) interventions [15–17], potentially affording sufficient time for acids to be neutralized by blood and tissue buffer systems, or eliminated through the respiratory or renal systems [25]. Finally, unlike previous studies, a high level of control was imposed in the current investigation, with dietary intake being individually prescribed and consumed under supervision of qualified dietitians [32].

Race walking competitions require sustained efforts at high velocities [41], and athletes maintain a high percentage of maximal heart rate throughout races [42]. High-intensity training sessions are an important component of race walkers' training programs [42,43], and include 1 km (~4 min) and 2 km (~8.5 min) repetitions performed close to race pace, plus hill sessions [43]. Indeed, evidence of the intensity of several of the weekly training sessions in the current study was provided by observations of post-exercise blood lactate concentrations as high as 19.2 mmol·L^{-1} after the prescribed session of 1 km repetitions. Such training sessions are likely to contribute to highly adapted buffering capacities, particularly within skeletal muscle, among the race walkers in this study [44,45]. Indeed, the buffering capacity in athletes is typically more adapted than in untrained individuals [32,33], which may explain why we did not observe any post-intervention differences in resting acid–base status between participants in the LCHF and PCHO or HCHO groups, despite elevations in ketone body concentration (typically sustained at ~1 mmol·L^{-1}) in the LCHF group (Burke et al.; submitted for publication). Furthermore, it is likely that the well-developed buffering capacity of the athletes was evident prior to the start of the study, rather than specifically improving over the course of the study as a result of the intensified training that was performed, given that no differences between baseline and post-intervention sub-maximal exercise blood lactate concentrations were observed.

There is some evidence from other approaches to modifying acid–base status in elite athletes that buffering capacity is already robust, and perhaps already able to cope with their regular exposure to perturbations associated with high-intensity exercise. Indeed, among the few studies specifically examining responses to supplementation with extra-cellular buffering agents, such as bicarbonate or citrate, in elite athletes across different exercise modalities [46,47], the improvements in exercise capacity or performance normally observed in modestly trained or lower caliber athletes [48] are not detected.

While some differences in post-exercise blood pH were evident as a consequence of sustained macronutrient manipulation, this may merely reflect differences in work capacity. Specifically, post-exercise blood pH was lower, while blood [La$^-$] was higher following PCHO, compared with LCHF. However, the PCHO group in this study performed more work during the post-intervention graded exercise test, with a mean test duration of 17.6 min for PCHO, compared with 15.8 min for LCHF, and race performance was impaired in the LCHF group (Burke et al., submitted for publication). Similar suppression of blood [La$^-$] associated with impaired work capacity has been reported in elite athletes of a similar caliber and training status to the current study [9].

In this study, we observed a higher post-intervention NEAP with LCHF, compared with both PCHO and HCHO. The fact that a similar trend was not observed in either post-intervention blood bicarbonate concentration or blood pH, despite evidence of increased ketone body concentration (Burke et al.; submitted for publication), which can contribute to acidosis [4] in the LCHF group, may provide indirect evidence of well-developed blood buffering capacity of the elite athletes in the present study. This observation suggests that NEAP calculations may have little applicability to highly trained athletes.

One limitation of this study is that the results are specific to our participant population, where the sample size was limited both by the resources needed to undertake the investigation and, by the definition of elite, to a small sub-group with special characteristics. Future research is warranted, whereby NEAP and direct measures of acid–base status are monitored after dietary interventions in different participant populations, with a similar level of rigor applied to the dietary control to that implemented in the current investigation. A controlled environment, such as that involved in the current investigation may lead to an alteration in dietary intake from usual choices [49], and an increase in the day-to-day variability of NEAP. Indeed, the self-chosen diets at baseline in the present study showed a greater range than the prescribed diets, even when consumed from the same menu options. Nevertheless in terms of adequately testing hypotheses around this topic, dietary standardization is important [49]. A further limitation of the current investigation was that 24 h urinary net acid excretion (NAE) was not included in the measures taken. It is acknowledged that the collection of NAE allows more accurate estimations of NEAP than equations based on dietary constituents [50], however, it was not feasible to collect 24 h urine samples in the current study. Future investigations may incorporate the direct quantification of NEAP via the measurement of NAE, or if resources are constrained, 24 h pH may be investigated since it reasonably correlates with, and explains the variability in, 24 h NAE [27,51].

5. Conclusions

Our results indicate that sustained manipulation of macronutrient intake is unlikely to influence acid–base status in elite athletes. Thus, any performance implications of a ketogenic low carbohydrate high fat diet are unlikely to result from perturbations in acid–base status, a system which is tightly regulated via the interaction of the blood and tissues, plus respiratory and renal systems. Finally, the implications of sustained manipulation of dietary carbohydrate on NEAP may be best assessed via the criterion 24 h urinary net acid excretion (NAE) method, over indirect NEAP calculations, in highly trained athletes.

Supplementary Materials: The Supplementary Materials are available online at http://www.mdpi.com/2072-6643/10/2/236/s1.

Acknowledgments: A Program Grant from Australian Catholic University (2013000800) provided funding for this study. The authors would like to thank all participants for their time and effort in completing this study.

Author Contributions: A.J.C., G.J.S. and L.M.B. conceived and designed the experiments; the experiments were performed by L.M.B., M.L.R., A.P.S. and A.J.C.; M.W. analyzed the data; A.J.C., L.M.B., M.W. and G.J.S. wrote the paper; L.M.B., M.W., G.J.S., M.L.R. and A.P.S. provided feedback on the manuscript.

Conflicts of Interest: The authors declare no conflict of interest.

References

1. Freeman, J.; Veggiotti, P.; Lanzi, G.; Tagliabue, A.; Perucca, E. The ketogenic diet: From molecular mechanisms to clinical effects. *Epilepsy Res.* **2006**, *68*, 145–180. [PubMed]
2. Lefevre, F.; Aronson, N. Ketogenic diet for the treatment of refractory epilepsy in children: A systematic review of efficacy. *Pediatrics* **2000**, *105*, e46. [CrossRef] [PubMed]
3. Johnstone, A.; Horgan, G.; Murison, S.; Bremner, D.; Lobley, G. Effects of a high-protein ketogenic diet on hunger, appetite, and weight loss in obese men feeding ad libitum. *Am. J. Clin. Nutr.* **2008**, *87*, 44–55. [CrossRef] [PubMed]
4. Yancy, W.; Olsen, M.; Dudley, T.; Westman, E. Acid-base analysis of individuals following two weight loss diets. *Eur. J. Clin. Nutr.* **2007**, *61*, 1416–1422. [CrossRef] [PubMed]
5. Volek, J.; Noakes, T.; Phinney, S. Rethinking fat as a fuel for endurance exercise. *Eur. J. Sport Sci.* **2015**, *15*, 13–20. [CrossRef] [PubMed]
6. Phinney, S.; Bistrian, B.; Evans, W.; Gervino, E.; Blackburn, G. The human metabolic response to chronic ketosis without caloric restriction: Preservation of submaximal exercise capability with reduced carbohydrate oxidation. *Metabolism* **1983**, *8*, 769–776. [CrossRef]
7. Volek, J.; Freidenreich, J.; Saenz, C.; Kunces, L.; Creighton, B.; Bartley, J.; Davitt, P.; Munoz, C.; Anderson, J.; Maresh, C.; et al. Metabolic characteristics of keto-adapted ultra-endurance runners. *Metabolism* **2016**, *65*, 100–110. [CrossRef] [PubMed]
8. Webster, C.; Noakes, T.; Chacko, S.; Swart, J.; Kohn, T.; Smith, J. Gluconeogenesis during endurance exercise in cyclists habituated to a long-term low carbohydrate high-fat diet. *J. Physiol.* **2016**, *15*, 4389–4405. [CrossRef] [PubMed]
9. Burke, L.; Ross, M.; Garvican-Lewis, L.; Welvaert, M.; Heikura, I.; Forbes, S.; Mirtschin, J.; Cato, L.; Strobel, N.; Sharma, A.; et al. Low carbohydrate, high fat diet impairs exercise economy and negates the performance effect from intensified training in elite race walkers. *J. Physiol.* **2017**, *595*, 2785–2807. [CrossRef] [PubMed]
10. Coyle, E. Timing and method of increased carbohydrate intake to cope with heavy training, competition and recovery. *J. Sports Sci.* **1991**, *9*, 29–51. [CrossRef] [PubMed]
11. Marquet, L.; Brisswalter, J.; Louis, J.; Tiollier, E.; Burke, L.; Hawley, J.; Hausswirth, C. Enhanced endurance performance by periodization of carbohydrate intake: "Sleep low" strategy. *Med. Sci. Sports Exerc.* **2016**, *48*, 663–672. [CrossRef] [PubMed]
12. Yeo, W.; Lessard, A.; Chen, Z.; Garnham, A.; Burke, L.; Rivas, D.; Kemp, B.; Hawley, J. Fat adaptation followed by carbohydrate restoration increases AMPK activity in skeletal muscle from trained humans. *J. Appl. Physiol.* **2008**, *105*, 1519–1526. [CrossRef] [PubMed]
13. Marquet, L.; Hausswirth, C.; Molle, O.; Hawley, J.; Burke, L.; Tiollier, E.; Brisswalter, J. Periodization of carbohydrate intake: Short-term effect on performance. *Nutrients* **2016**, *8*, 755. [CrossRef] [PubMed]
14. Gejl, K.; Ørtenblad, N.; Andersson, E.; Plomgaard, P.; Holmberg, H.-C.; Nielsen, J. Local depletion of glycogen with supramaximal exercise in human skeleton fibres. *J. Physiol.* **2017**, *595*, 2809–2821. [CrossRef] [PubMed]
15. Greenhaff, P.; Gleeson, M.; Whiting, P.; Maughan, R. Dietary composition and acid-base status: Limiting factors in the performance of maximal exercise in man? *Eur. J. Appl. Physiol.* **1987**, *56*, 444–450. [CrossRef]
16. Greenhaff, P.; Gleeson, M.; Maughan, R. The effects of a glycogen loading regimen on acid-base status and blood lactate concentration before and after a fixed period of high intensity exercise in man. *Eur. J. Appl. Physiol.* **1988**, *57*, 254–259. [CrossRef]
17. Greenhaff, P.; Gleeson, M.; Maughan, R. The effects of diet on muscle pH and metabolism during high intensity exercise. *Eur. J. Appl. Physiol.* **1988**, *57*, 531–539. [CrossRef]

18. Greenhaff, P.; Gleeson, M.; Maughan, R. The effects of dietary manipulation on blood acid-base status and the performance of high intensity exercise. *Eur. J. Appl. Physiol.* **1987**, *56*, 331–337. [CrossRef]

19. Lutz, J. Calcium balance and acid-base status of women as affected by increased protein intake and by sodium bicarbonate ingestion. *Am. J. Clin. Nutr.* **1984**, *39*, 281–288. [CrossRef] [PubMed]

20. Nowson, C.; Wattanapenpaiboon, N.; Pachett, A. Low-sodium dietary approaches to stop hypertension. *Nutr. Res.* **2009**, *29*, 8–18. [CrossRef] [PubMed]

21. Luis, D.; Huang, X.; Riserus, U.; Sjögren, P.; Lindholm, B.; Arnlöv, J.; Cederholm, T.; Carrero, J. Estimated dietary acid load is not associated with blood pressure or hypertension incidence in men who are approximately 70 years old. *J. Nutr.* **2015**, *145*, 315–321. [CrossRef] [PubMed]

22. Frassetto, L.; Hardcastle, A.; Sebastian, A.; Aucott, L.; Fraser, W.; Reid, D.; Macdonald, H. No evidence that the skeletal non-response to potassium alkali supplements in healthy postmenopausal women depends on blood pressure or sodium chloride intake. *Eur. J. Clin. Nutr.* **2012**, *66*, 1315–1322. [CrossRef] [PubMed]

23. Pizzorno, J.; Frassetto, L.; Katzinger, J. Diet-indicued acidosis: Is it real and clinically relevant? *Br. J. Nutr.* **2010**, *103*, 1185–1194. [PubMed]

24. Adeva, M.; Souto, G. Diet-induced metabolic acidosis. *Clin. Nutr.* **2011**, *30*, 416–421. [CrossRef] [PubMed]

25. Poupin, N.; Calvez, J.; Lassale, C.; Chesneau, C.; Tomé, D. Impact of diet on net endogenous acid production and acid-base balance. *Clin. Nutr.* **2012**, *31*, 313–321. [CrossRef] [PubMed]

26. Frassetto, L.; Todd, K.; Morris, R.; Sebastian, A. Estimation of net endogenous noncarbonic acid production in humans from diet potassium and protein contents. *Am. J. Clin. Nutr.* **1998**, *68*, 576–583. [CrossRef] [PubMed]

27. Remer, T.; Manz, F. Potential renal acid loads of foods and its influence on urine pH. *J. Am. Diet. Assoc.* **1995**, *95*, 791–797. [CrossRef]

28. Ball, D.; Greenhaff, P.; Maughan, R. The acute reversal of a diet-induced metabolic acidosis does not restore endurance capacity during high-intensity exercise in man. *Eur. J. Appl. Physiol.* **1996**, *73*, 105–112. [CrossRef]

29. Ball, D.; Maughan, R. Blood and urine acid-base status of premenopausal omnivorous and vegetarian women. *Br. J. Nutr.* **1997**, *78*, 683–693. [CrossRef] [PubMed]

30. Cairns, S. Lactic acid and exercise performance: Culprit or friend? *Sports Med.* **2006**, *36*, 279–291. [CrossRef] [PubMed]

31. Messionnier, L.; Kristensen, M.; Juel, C.; Denis, C. Importance of pH regulation and lactate/H+ transport capacity for work production during supramaximal exercise in humans. *J. Appl. Physiol.* **2007**, *102*, 1936–1944. [CrossRef] [PubMed]

32. Edge, J.; Bishop, D.; Hill-Hass, S.; Dawson, B.; Goodman, C. Comparison of muscle buffer capacity and repeated-sprint ability of untrained, endurance-trained and team-sport athletes. *Eur. J. Appl. Physiol.* **2006**, *96*, 225–234. [CrossRef] [PubMed]

33. Bishop, D.; Spencer, M. Determinants of repeated-sprint ability in well-trained team-sport athletes and endurance-trained athletes. *J. Sports Med. Phys. Fit.* **2004**, *44*, 1–7.

34. Spiriev, B. IAAFscoring Tables of Athletics. Available online: https://www.iaaf.org/news/iaaf-news/scoring-tables-2017 (accessed on 20 January 2018).

35. Mirtschin, J.; Forbes, S.; Cato, L.; Heikura, I.; Strobel, N.; Hall, R.; Burke, L. Organisation of dietary control for training-nutrition intervention involving low carbohydrate high fat (LCHF) diet. *Int. J. Sport Nutr. Exerc. Metab.* **2017**. [CrossRef]

36. Bartlett, J.; Hawley, J.; Morton, J. Carbohydrate availability and exercise training adaptation: Too much of a good thing? *Eur. J. Sport Sci.* **2015**, *15*, 3–12. [CrossRef] [PubMed]

37. Saunders, P.; Telford, R.; Pyne, D.; Cunningham, R.; Gore, C.; Hahn, A.; Hawley, J. Improved running economy in elite runners after 20 days of simulated moderate-altitude exposure. *J. Appl. Physiol.* **2004**, *96*, 931–937. [CrossRef] [PubMed]

38. Wang, Y.; Moss, J. Predictors of body surface area. *J. Clin. Anaesth.* **1992**, *4*, 4–10. [CrossRef]

39. Bates, D.; Maechler, M.; Bolker, B.; Walker, S. Fitting linear mixed models using lme4. *J. Stat. Softw.* **2015**, *67*, 1–48. [CrossRef]

40. Fox, J.; Wrisberg, S. *An R Companion to Applied Regression*, 2nd ed.; Sage Publications: Thousand Oaks, CA, USA, 2011; Available online: http://socserv.socsi.mcmaster.ca/jfox/Books/Companion (accessed on 8 December 2017).

Nutrients **2018**, *10*, 236

41. Hanley, B. An analysis of racing profiles of world-class racewalkers. *Int. J. Sports Physiol. Perform.* **2013**, *8*, 435–441. [CrossRef] [PubMed]
42. Vernillo, G.; Agnello, L.; Drake, A.; Padulo, J.; Piacentini, M.; La Torre, A. An observational study on the perceptive and physiological variables during a 10,000-m race walking competition. *J. Strength Cond. Res.* **2012**, *26*, 2741–2747. [CrossRef] [PubMed]
43. Pugliese, L.; Serpiello, F.; Millet, G.; La Torre, A. Training diaries during altitude training camp in two olympic champions: An observational case study. *J. Sports Sci. Med.* **2014**, *13*, 666–672. [PubMed]
44. Weston, A.; Myburgh, K.; Lindsay, F.; Dennis, S.; Noakes, T.; Hawley, J. Skeletal muscle buffering capacity and endurance performance after high-intensity interval training by well-trained cyclists. *Eur. J. Appl. Physiol.* **1997**, *75*, 7–13. [CrossRef]
45. Edge, J.; Bishop, D.; Goodman, C. The effects of training intensity on muscle buffer capacity in females. *Eur. J. Appl. Physiol.* **2006**, *96*, 97–105. [CrossRef] [PubMed]
46. Cameron, S.; McLay-Cooke, R.; Brown, R.; Gray, A.; Fairbairn, K. Increased blood pH but not performance with sodium bicarbonate supplementation in elite rugby union players. *Int. J. Sport Nutr. Exerc. Metab.* **2010**, *20*, 307–321. [CrossRef] [PubMed]
47. Tan, F.; Polglaze, T.; Cox, G.; Dawson, B.; Mujika, I.; Clark, S. Effects of induced alkalosis on simulated match performance in elite female water polo players. *Int. J. Sport Nutr. Exerc. Metab.* **2010**, *20*, 198–205. [CrossRef] [PubMed]
48. Carr, A.; Hopkins, W.; Gore, C. Effects of acute alkalosis and acidosis on performance: A meta-analysis. *Sports Med.* **2011**, *41*, 801–814. [CrossRef] [PubMed]
49. Reilly, T.; Waterhouse, J.; Burke, L.; Alonso, J. Nutrition for travel. *J. Sports Sci.* **2007**, *25*, S125–S134. [CrossRef] [PubMed]
50. Frassetto, L.; Lanham-New, S.; Macdonald, H.; Remer, T.; Sebastian, A.; Tucker, K.; Tylavsky, F. Standardizing terminology for estimating the diet-dependent net acid load to the metabolic system. *J. Nutr.* **2007**, *137*, 1491–1492. [CrossRef] [PubMed]
51. Parmenter, B.; Slater, G.; Frassetto, L. Spot-testing urine pH, a novel dietary biomarker? A randomised cross-over trial. *Nutr. Diet.* **2017**, *74*, 313–319. [CrossRef] [PubMed]

nutrients

MDPI

Review

Interstitial Glucose and Physical Exercise in Type 1 Diabetes: Integrative Physiology, Technology, and the Gap In-Between

Othmar Moser [1,2], Jane E. Yardley [3] and Richard M. Bracken [1,2,*]

1   Diabetes Research Group, Medical School, Swansea University, Swansea SA2 8PP, UK;
    othmar.moser@swansea.ac.uk
2   Applied Sport, Technology, Exercise and Medicine Research Centre (A-STEM), College of Engineering,
    Swansea University, Swansea SA1 8EN, UK
3   Kinesiology, Sport and Recreation, Social Sciences, University of Alberta, Edmonton, AB T4V 2R3, Canada;
    jane.yardley@ualberta.ca
*   Correspondence: r.m.bracken@swansea.ac.uk; Tel.: +44-179-251-3059

Received: 27 November 2017; Accepted: 12 January 2018; Published: 15 January 2018

Abstract: Continuous and flash glucose monitoring systems measure interstitial fluid glucose concentrations within a body compartment that is dramatically altered by posture and is responsive to the physiological and metabolic changes that enable exercise performance in individuals with type 1 diabetes. Body fluid redistribution within the interstitial compartment, alterations in interstitial fluid volume, changes in rate and direction of fluid flow between the vasculature, interstitium and lymphatics, as well as alterations in the rate of glucose production and uptake by exercising tissues, make for caution when interpreting device read-outs in a rapidly changing internal environment during acute exercise. We present an understanding of the physiological and metabolic changes taking place with acute exercise and detail the blood and interstitial glucose responses with different forms of exercise, namely sustained endurance, high-intensity, and strength exercises in individuals with type 1 diabetes. Further, we detail novel technical information on currently available patient devices. As more health services and insurance companies advocate their use, understanding continuous and flash glucose monitoring for its strengths and limitations may offer more confidence for patients aiming to manage glycemia around exercise.

Keywords: continuous glucose monitoring; flash glucose monitoring; exercise; interstitium

1. Introduction

Type 1 diabetes is characterized by the autoimmune destruction of pancreatic β-cells within the islets of Langerhans and loss of endogenous insulin production. Most diagnoses are made early in life and result in a life-long dependency on pharmacological insulin, carbohydrate counting, and frequent blood glucose measurements to manage blood glucose levels [1]. In the United Kingdom, over a quarter of a million people have type 1 diabetes and this number is rising. The cost to the National Health Service (NHS) for treatment and care is approximately £1.8+ billion a year [2]. Low levels of physical activity are evident in people with type 1 diabetes [3] and large cohort studies demonstrate an increased risk of cardiovascular disease [4,5]; however, regular physical activity is advocated [6,7] and conveys important health benefits [8]. Nevertheless, the increased risk of developing hypoglycemia and the loss of glucose control around exercise are major concerns for people with type 1 diabetes, contributing to the failure to initiate or maintain regular physical activity [9].

Frequent finger prick blood sampling and adjustments in exogenous basal and/or bolus insulin and carbohydrate intake are necessary to minimize glycemic disturbances with physical exercise [6,7].

However, mimicking the natural secretory pattern of endogenous insulin in individuals without diabetes whilst maintaining glycemia within a normal physiological range is challenging around exercise [6,10]. In the Diabetes Control and Complications Trial, intensive treatment to maintain euglycemia between 4 and 7 mmol/L with either an external insulin pump or three or more daily insulin injections and frequent blood glucose measurements delayed the onset and slowed the progression of retinopathy, nephropathy, and neuropathy compared with conventional therapy of one or two daily insulin injections over a 6.5-year period [11–14]. However, an increased occurrence of severe hypoglycemia was evident with intensive treatment compared to the conventional arm [15]. Thus, more frequent finger prick sampling is inconvenient for patients, increases cost to healthcare providers, and might still not stabilize glycemic control.

In recent years, devices that monitor glucose levels continuously offer opportunities for painless observations of glucose that may guide better glucose management in the daily life of people with type 1 diabetes [16]. Emerging work now demonstrates longitudinal improvements in glucose management with the use of continuous glucose monitoring systems in the daily life of people with type 1 diabetes [17]. Flash glucose monitoring systems, similar to continuous glucose monitoring systems, are now available on prescription in the UK NHS for patients with type 1 diabetes. Current devices monitor dynamic interstitial glucose information for up to two weeks without the need for confirmatory finger prick blood glucose measures. However, whilst mean absolute relative differences demonstrate strong similarity to reference blood values [18], these data are generally reported during rest and not during dynamic exercise.

Performing an acute bout of exercise evokes rapid changes in many physiological systems and demands constant adjustment to facilitate glucose provision to the exercising tissues in both healthy individuals and people with type 1 diabetes [10,19]. The net exercise-induced glucose response in the circulation results from changes in carbohydrate ingestion, hepatic release and uptake, skeletal muscle and adipose tissue uptake (amongst other organs), or hepatic gluconeogenesis, and is dependent on the exercise characteristics: type, mode, intensity, and duration [20–23]. Importantly, (and often overlooked) acute exercise also causes significant redistribution in body fluid between the interstitium, lymphatics, and bloodstream [24]. Changes in posture, sweating, muscle contraction, and hydration dramatically impact interstitial fluid volume and result in altered concentrations of glucose due to fluid redistribution within this compartment.

This review does not seek to revise the historical development of interstitial glucose monitoring technology or its inclusion around physical activity *per se* but rather seeks to focus on our current understanding of acute physiological changes around exercise involving the interstitium and device performance around acute physical exercise in individuals with type 1 diabetes. A firm understanding of the complex physiological adaptations to acute exercise may inform the reader of the strengths and limitations in tracking glucose using continuous and flash monitoring systems in a dynamic metabolic environment such as exercise and offer more confidence for people with type 1 diabetes to better manage glycemia around exercise.

2. Physiological Adjustments to Increased Blood Flow and Glucose Provision during Physical Exercise

2.1. Initiation of Physical Exercise

Complex physiological adaptations dictate the ability of skeletal muscle to generate and maintain power in healthy individuals and people with type 1 diabetes [25,26]. The cardiorespiratory system plays a critical role in facilitating increased provision of glucose and oxygen (alongside the removal of carbon dioxide and other metabolic by-products) to exercising muscle. Yet, physical exercise begins with mental activity! In the brain, increased nerve impulses from the cerebral cortex act on the medulla oblongata and suppress vagal tone to stimulate an increase in cardiac output (i.e., heart rate × stroke volume) from ~5 L/min at rest to over 25 L/min in individuals with type 1 diabetes [27]. Central brain areas also raise the baroreflex level to allow blood pressure regulation at a higher set-point, which facilitates higher sympathoadrenal activity and allows further increases in cardiac output. Skeletal muscle arterioles dilate via the actions of increased circulating catecholamines and sympathetic nerve activity on vascular β-adrenoceptors [20]. In non-active splanchnic and renal beds, a greater vascular constriction occurs to allow more blood to flow to exercising areas [27].

2.2. Continuation of Physical Exercise

As exercise continues, cardiovascular and respiratory systems continuously adjust to the exercise workload to ensure glucose- and oxygen (O_2)-rich blood reaches working muscles at the required rate. Elevated cerebral motor center activity and skeletal muscle afferent fiber mechanical (tension) and metabolic (e.g., H^+, K^+, and local ischemia) signals stimulate sympathetic nervous system activity and causes increases in circulating catecholamines that act on adrenoceptors on vessels and skeletal muscle cells and consequently vasodilation in healthy individuals [28]. Local to the exercising muscle and heart, metabolites cause vasodilation of skeletal muscle and coronary arterioles. This exercise-induced hyperemia or 'exercise pressor reflex' decreases vascular resistance and increases blood flow concomitant with increased capillary recruitment [29]. In high-intensity activities such as sprinting or strength exercises that involve a Valsalva-type maneuver, vascular compression and decreased blood flow induce the buildup of metabolites that activate chemoreceptors, resulting in a greater exercise pressor reflex than dynamic activities, and cause increased cardiac output, total peripheral resistance, and tachycardia that alters blood flow rate in healthy individuals [29]. Increased sympathoadrenal activity constricts arterioles in the non-exercising areas. Also, exercise-induced sympathoadrenal activity increases suppression of vagal activity proportional to the exercise intensity and vasoconstriction of the smooth muscle arterioles. Cutaneous blood flow is initially reduced via vasoconstriction but then increases as vasodilatation occurs due to a rise in body temperature to promote cooling [20]. However, with long duration high-intensity exercise, cutaneous perfusion may again fall as vasoconstriction re-diverts blood flow to muscle. Brain blood flow tends to remain largely constant. During inspiration, pressure in the thoracic cavity falls. With diaphragm contraction during inspiration, there is an increased abdominal pressure that compresses veins and venous blood flow in the abdomen and drives blood back to the heart [30]. The thoracoabdominal pump is also appreciable during heavy exercise as tidal volume and breath frequency increase ventilation rates to over 150 L/min, which was observed in healthy and diseased individuals.

2.3. Blood Flow Distribution to Skeletal Muscle during Exercise

At rest, around two-thirds of the blood volume is present in veins and most capillaries are closed; blood flow rates to skeletal muscle is approximately 20–30 mL·min^{-1}·kg muscle^{-1} at an open capillary density of ~100 capillaries per mm^2 in healthy individuals [31]. During exercise, open capillaries increase five-fold due to vasodilation from sympathoadrenal activity and locally from factors emanating from skeletal muscle metabolism (decrease in pO_2, rise in pCO_2, decreased pH, temperature, adenosine, NO, P_i, K^+). Thus, increased cardiac output and localized vasodilation together account for up to a 100-fold increase in blood flow to working tissues (2–3 L·min^{-1}·kg muscle^{-1} during heavy exercise). Further, to facilitate this increased rate of blood flow, a rapid shunting of venous blood to the arterial circuit occurs by vasoconstriction of major veins.

Notwithstanding rapid blood redistribution within the circulation, plasma volume may fall ~20% (1 L) with intense exercise over 15 min due to fluid movement towards active muscle [28]. Alterations in the osmolarity of skeletal myocyte cytoplasm and interstitial fluid due to local release of metabolic factors from skeletal muscle increase solutes in the interstitial fluid [28]. Also, an increased perfusion of capillary beds in active muscle increases capillary hydrostatic pressure, which drives more fluid from the plasma to the interstitium [20]. The extent of plasma volume reduction can be magnified by the ambient temperature and humidity and offset by oral consumption of fluids. The overall effect of a net loss in plasma volume is an apparent increase in red blood cell count and hemoglobin concentration, which increases the O_2-carrying capacity per liter of blood, but at the expense of a reduction in total blood volume and an increase in viscosity. The movement of fluid from the plasma to the interstitium is offset to a certain extent by fluid movement back into the circulation from tissues that undergo vasoconstriction (gut, kidneys) [32].

3. The Interstitial Compartment and Factors Influencing Movement of Interstitial Fluid during Exercise

The interstitial space lies between the circulatory system and organs such as skeletal muscle, myocardium, and liver and contains approximately three-quarters of the body's total extracellular fluid (e.g., ~10–12 L in a 70-kg person) [33]. Plasma and lymph fluid represent most of the remaining quarter (~3 L) of extracellular fluid. Interstitial fluid represents a quarter of total body water mixing with plasma, lymph fluid, and sweat [33].

The factors that determine the speed and magnitude of movement of fluid between capillaries and interstitium are determined by Starling forces [34]. The net filtration pressure (NFP) is determined by two pressure differentials that determine the net drive for directional fluid movement between the capillary and interstitial space, namely:

(i) Colloid osmotic pressure between capillary (P_c) and interstitial space (P_{if})

Capillary colloidal pressure is exerted by proteins such as albumin, which causes fluid to be pulled back into the capillary at the venous end due to a higher solute concentration [35]. The remaining fluid left in the interstitial space is drained by the lymphatic capillaries and returned to the circulation. The interstitial colloidal pressure is regulated by the interstitial fluid concentration and the permeability of the capillary to different proteins. As plasma contains more protein than the interstitial fluid, capillary colloidal pressure tends to dominate all the way along the capillary wall [35].

(ii) Hydrostatic pressures from capillary (π_c) to interstitial space (π_{if})

Colloidal hydrostatic pressure is exerted by the plasma fluid against the capillary wall, moving the fluid into the interstitium through filtration and is greater at the arterial end of a capillary due to the higher blood pressure [36]. The interstitial hydrostatic pressure is determined by the interstitial fluid volume and the surrounding tissues' compliance with changes in volume and pressure; skin and skeletal muscle have higher compliance to changes in volume and pressure with more soft tissue surrounding them than the brain [37].

Thus, net filtration pressure (NFP) = $P_c + \pi_{if} - P_{if} - \pi_c$ and the transendothelial fluid filtration rate (cm^3/s) is represented by Equation (1):

$$J_v = L_p S \left[(P_c - P_{if}) - \sigma(\pi_c - \pi_{if}) \right] \tag{1}$$

where

- L_p is the hydraulic conductivity (which provides a measure of water permeability; m^2·s·kg^{-1} or m·s^{-1}·mmHg^{-1});
- S is the surface area available for fluid exchange in m^2;
- σ is the osmotic reflection coefficient (dimensionless unit) where σ close to 1.0 indicates the capillary is fully effective in allowing fluid and smaller solutes to filter to the interstitial space while larger protein molecules such as albumin are retained. Where σ is <1.0, capillary filter function is reduced.

Acute physical exercise does not alter the characteristics of the microvascular exchange vessels [38–40]. Thus, S (the surface area available for fluid exchange) contributes most to the enhanced filtration rates with acute exercise but alterations in hydraulic conductivity (L_p) and/or the osmotic reflection coefficient (σ) do not.

Although most pressure fluctuations are absorbed at the level of precapillary resistance vessels, even modest increases in capillary filtration drive lead to fluid accumulation in the interstitial compartment and can increase from 0.008–0.015 mL/min/mmHg for resting skeletal muscle to greater than 4-fold during exercise in healthy individuals [24]. Thus, the rate of fluid filtration from the blood to the tissue interstitial space can reach 1.5 mL·min^{-1}·100 g muscle^{-1} during intense rhythmic contraction [39–42]. This rapid rate of fluid movement with exercise initially exceeds the drainage capacity of the lymphatic system and, converse to the loss of plasma fluid, is a gain in interstitial fluid volume of ~15–20% in exercising tissue [43]. This is due to several factors: (i) exercise-induced capillary hydrostatic pressure increasing with arteriolar vasodilation [43]; (ii) relaxation of vascular smooth muscle in arterioles increasing capillary surface area by recruiting capillaries that were previously closed [41]; and (iii) a muscular contraction-induced increase in interstitial fluid osmolarity due to released metabolites (e.g., K^+, lactate, H^+, adenosine, ammonia) from active skeletal muscle cells that transiently raises interstitial fluid osmolarity by ~20–30 mmol/L or 7–10% [41]. Also, an expanded interstitial fluid volume increases the diffusion distance for glucose molecules and may hinder cellular metabolism in swollen tissues. Interstitial (or cellular) edema can also impair glucose tissue perfusion by collapsing capillaries in swollen tissue [44].

The initial rate of hypervolemia in the interstitium slows as exercise continues due to (i) a compensatory absorption of interstitial fluid by vasoconstricted vascular beds where capillary pressure is lower in non-exercising tissues; and (ii) the subsequent reduction in flow rate allows better lymphatic drainage [43] with one-way flow [44] and flow-induced lymphatic dilation. Muscle contraction increases interstitial fluid pressure and reduces the transcapillary hydrostatic gradient, further reducing fluid accumulation with sustained exercise [45].

4. Interstitial Glucose Metabolism during Exercise

The small amount of glucose in the interstitium of an adult individual with type 1 diabetes (e.g., 10 mmol/L = 1.8 g/L = 18 g with an interstitial fluid volume of 10 L) represents the point measurement of glucose changes at that measured site. Patients in a fasted, resting state produce small rate changes in the disappearance and/or appearance of glucose. However, acute exercise dramatically increases carbohydrate combustion and the metabolic turnover [46] of glucose increases 40 to 50-fold in exercising tissues (Table 1) [33,47]. The magnitude of change in interstitial glucose with acute physical exercise in people with type 1 diabetes is dependent on the type, mode, intensity, and duration of exercise, as well as the time of day, circulating insulin levels, and whether the individual is fasted or fed.

Table 1. Glucose transport from blood to skeletal muscle (per 110/g) [33,47].

	Rest	Heavy Exercise	Fractional Δ (Exercise/Rest)
Skeletal muscle glucose consumption (J_s), μmol/min	1.4	60	×43
Arterial glucose concentration (C_a), mM	5.0	5.0	-
Venous concentration (C_v), mM	4.4	4.0	×0.9
Extraction E, %	11.2	20	×1.8
Blood flow, mL/min	2.5	60	×24
Perfused capillary density, per mm^2	250	1000	×4
Diffusion capacity (PS), cm^3/min	5	20	×4
Mean concentration difference across capillary wall ($\Delta C = J_s/PS$), mM	0.3	3.0	×10
Mean pericapillary concentration (Ci), mM	4.7	2.0	×0.4
Krogh Cylinder radius, μm	36	18	×0.5

4.1. Glucose Transport to Exercising Muscle

Capillary recruitment shortens the diffusion distance and increases the surface area, with the diffusional capacity increased further by an effect of blood flow on permeability. A fall in tissue glucose concentrations raises the gradient across the capillary wall and the increase in diffusional flux raises the fractional extraction and arteriovenous concentration difference (C_a-C_v) [48]. An increased blood flow delivers glucose faster to the capillary and prevents a major fall in the mean intra-capillary plasma concentration, thereby avoiding a flow-limitation of exchange. The new trans-capillary glucose flux is, thus, determined from the Fick principle equaling blood flow × increased arteriovenous concentration difference (Table 1) [48].

4.2. Exercising Skeletal Muscle Tissue Is a Glucose Consumer; Inactive Tissue Acts as a Glucose Store

Glucose entry into muscle occurs across the cell membrane via the glucose transporter type-4 (GLUT-4) [49]. Once inside the muscle, glucose is irreversibly converted to glucose 6-phosphate via the enzyme hexokinase to prevent loss of this valuable nutrient from muscle and is either utilized via glycolysis or stored as glycogen until needed [49]. Skeletal muscle glycogen stores are about 300–350 g and can be manipulated with low carbohydrate diets to reduce estimated whole-body stores from 300 to 50 mmol·glucosyl units per kg^{-1} wet weight in healthy individuals [50]. Conversely, short-term high carbohydrate diets markedly increase muscle glycogen stores to more than 500 mmol·glucosyl units per kg^{-1} wet weight. Carbohydrate combustion in the absence of oxygen is important for releasing energy during high-intensity exercises such as sprinting or strength exercises. For example, a 6 second cycle sprint reduces quadriceps muscle glycogen concentration by 14% from 316 ± 75 to 273 ± 80 mmol·glucosyl units per kg^{-1} wet weight [51] and by 32% after a 30 second sprint [21]. Consequently, blood lactate and H+ concentrations increase from ~1.0 to over 20 mmol/L and 7.4 to <7 pH units and alters osmolarity in the interstitial space [21,51]. The net glucose uptake into tissues during exercise depends on the exercise characteristics; moderate intensity exercise causes similar whole-body rates of glucose appearance and disappearance. In contrast, high-intensity exercise causes a disproportionate rate of increase in glucose appearance compared to disappearance [52].

4.3. Hepatocytes Take Up and Release Glucose

The liver glycogen content is approximately 80–110 g for a 70-kg person and similar to muscle is easily manipulated with low carbohydrate diets, reducing hepatic stores from 232 to 24–55 mmol·glucosyl units per kg^{-1} wet weight within 24 h in healthy people [53]. Conversely, high carbohydrate diets markedly increase hepatic glycogen stores to supernormal values of 424–624 mmol·glucosyl units per kg^{-1} wet weight [53], especially if preceded by glycogen-depleting exercise, a concept known as glycogen supercompensation [50]. In contrast to skeletal myocytes, hepatocytes contain a phosphatase enzyme that catalyzes stored glucose, allowing the glucose to leave the liver cell and enter the interstitial space prior to appearing in the circulation. A strong counter-regulatory hormone response elicited by high-intensity exercise acts on hepatocytes to promote hepatic glucose release and consequent exercise-induced hyperglycemia in people with type 1 diabetes [54]. Carbohydrate combustion in the presence of oxygen can increase from resting values of ~0.25 g/min to over 3.0 g/min and the liver plays an important role in the later stages of endurance exercise to support exercise metabolism beyond that of lipid [20]. On the other hand, one day of starvation and/or prolonged exercise might almost deplete endogenous carbohydrate stores, excepting for the body's ability to minimize this loss through the manufacture of new glucose from amino acids, ketoacids, or lactate in the liver; the contribution of gluconeogenesis to exercise is important [20,22].

4.4. Exogenous 'On-Board' Insulin Concentrations Differ When Compared with Endogenous Insulin

In people that do not have diabetes, plasma insulin falls with exercise duration and/or intensity due to exercise-induced catecholamine inhibition on pancreatic ß-cell function [55]. However, no such reductions on exogenous insulin are seen in individuals with type 1 diabetes. Indeed, there is evidence of increased circulating insulin with exercise [56]. Insulin stimulates vasodilation around subcutaneous depots where injected insulin may be flushed into the extracellular space and increase glucose uptake and hepatic glucose inhibition [57].

4.5. Ingested Carbohydrate from the Gastrointestinal Tract

Ingested carbohydrates (e.g., a mixture of glucose and fructose) can be assimilated from the small intestine at an upper limit of ~1.8 g/min [58]. Interestingly, glucose alone can be absorbed and utilized at rates of ~1.0–1.1 g/min, due to rate limits in the capacities of the glucose transporter type-2 (GLUT-2), glucose transporter type-5 (GLUT-5), and/or sodium–glucose linked transporter-1 (SGLT-1) carriers in the intestinal membrane [59]. In people with type 1 diabetes, carbohydrate ingestion in readiness for exercise increases blood glucose with the overall magnitude dependent on factors such as amount, digestibility, and glycemic index [6]. This new glucose appearance into body fluid compartments alters body fluid osmolarity.

5. Interstitial Glucose Responses to Different Forms of Exercise

Continuous glucose monitoring systems use an enzymatic technique that reacts with circulating glucose in the interstitium [60]. This reaction releases one electron for each glucose molecule and transmits it to an electrode. Within this electrode, an electric current is generated that is passed from a transmitter (attached or incorporated) to a reader (e.g., mobile phone) to reflect the interstitial glucose levels [60].

Data on head-to-head comparisons between continuous and flash glucose monitoring systems remain rare. Aberer et al. [61] have shown that the Abbott FreeStyle Libre (Abbott, Alameda, CA, USA) flash glucose monitoring system was the most accurate (median absolute relative differences (MARD) 8.7 ± 5.9%, $n = 13$), followed by Dexcom G4 Platinum (Dexcom, San Diego, CA, USA) (MARD 15.7 ± 14.6%, $n = 24$), and Medtronic MiniMed 640 G (Medtronic MiniMed, Inc., Northridge, CA, USA) (MARD 19.4 ± 13.5%, $n = 22$) during 30 min (2 × 15 with a 5 min rest) of low-intensity continuous exercise. However, it must be mentioned that only thirteen comparison points between blood glucose and interstitial glucose concentrations were available for the flash glucose monitoring system. Unfortunately, in this trial, glucose concentrations were not classified for hypo-, eu-, and hyperglycemia during exercise. Disagreements in absolute glucose values during exercise between different compartments were found in healthy individuals [62] and in patients with type 1 diabetes [63–65]. Interstitial glucose concentrations measured via the SEVEN® PLUS CGM (Dexcom, San Diego, CA, USA) system during exercise under euglycemic conditions are higher compared to capillary (finger sticks) and venous blood, and the capillary blood levels seem to be higher than the venous blood levels [62]. In patients with type 1 diabetes performing high-intensity interval exercise and continuous exercise, no clear trend was found in comparing the capillary blood glucose concentration (earlobe) and interstitial glucose concentration when using the Guardian REAL-time system and the Enlite sensor (Medtronic MiniMed, Inc., Northridge, CA, USA) (no classification for glycemic ranges was performed) [63]. However, in a different study comparing the interstitial glucose (G4 Platinum, Dexcom, San Diego, CA, USA) and venous blood glucose concentration, a trend towards lower levels were found in venous blood when exercising within a euglycemic range (blood glucose during high-intensity interval exercise: 7.56 ± 0.21; blood glucose during continuous exercise: 6.71 ± 0.23) [64]. Interestingly, significantly lower glucose concentrations were found in the interstitium compared to the venous system for continuous exercise and for resistance exercise under a euglycemic range when using the CGMS® System Gold® (Medtronic MiniMed, Inc., Northridge, CA, USA) [65].

The duration of exercise seems not to influence the differences in glucose levels in different compartments [63–65]. However, it might be that longer-duration exercise and the potential for dehydration could decrease the glucose supply within the interstitium, resulting in lower glucose levels compared to the capillary and venous systems. Furthermore, the lag time in interstitium could be influenced by the intensity of exercise and the amount of circulating insulin. Low insulin reductions combined with high exercise intensities could increase the lag time between the blood and the interstitium, resulting in severe impairments for continuous and flash glucose monitoring systems. The type and mode of exercise seem not to influence continuous glucose monitoring accuracy [63–65]. During high-intensity interval exercise and continuous exercise, continuous glucose monitoring technology shows sufficient potential to trace changes in glucose concentrations [63,64]. During resistance exercise, a mean difference of −0.71 mmol/L was found [65]. However, it should be considered that during resistance exercise the site of sensor placement might directly influence its performance. Therefore, it is prudent to place the sensor away from working muscles.

Only a single study assessed continuous glucose monitoring accuracy during exercise under hypoglycemic conditions [66]. When a hypoglycemic alarm was set at 5.5 mmol/L, the system overestimated the interstitial glucose by 1.6 mmol/L. Therefore, a higher alarm level setting is recommended to avoid exercise-induced hypoglycemia.

Future research should clearly evaluate sensor performance under hypo-, eu-, and hyperglycemic ranges during exercise. Additionally, reported sensor performance during exercise needs to be adjusted for any treatments (e.g., carbohydrate ingestion, bolus insulin correction).

6. Continuous and Flash Glucose Monitor Performance

Twenty years of intensive research reveal marked improvements in continuous glucose monitoring performance in the reflection of accuracy, precision, and reliability. Since the start of the millennium, MARDs have been halved from 20% to 10% [67,68]. The Dexcom G5™ Mobile continuous glucose monitoring system, which is approved for patients aged two years and older, shows a MARD of 9% for adults and 10% for pediatric patients [68]. Sensor lifetime is 7 days accompanied with 2 calibrations to capillary blood glucose concentration per day. Its predecessor model, the Dexcom G4® system (Dexcom, San Diego, CA, USA), was found to be less accurate with MARDs of 13% for adults and 15% for pediatric patients [69–72]. Sensor lifetime is also 7 days, accompanied with 2 calibrations to capillary blood glucose concentration per day.

Medtronic's (Medtronic MiniMed, Inc., Northridge, CA, USA) flagships are the MiniMed 670 G hybrid closed-loop system and 640 G with SmartGuard® (predictive low glucose), as well as the Medtronic Paradigm Minimed Veo® (530 G) with low glucose suspend. Recently, the MARD for the Medtronic 640 G system was found at 9.6%, using the Guardian™ sensor 3 [73]. Predecessor models using Enlite™ sensor technologies resulted in MARDs of 13.6–14.2% for adults and of about 21% for children from 4–14 years for blood glucose levels below 6.7 mmol/L [72,74]. Sensors need to be changed after 6 days, requiring 2 calibrations per day.

A slightly different technology, Abbotts FreeStyle® Libre FGM system (Abbott, Alameda, CA, USA), was also found to track changes in interstitial glucose accurately with a MARD of 10–11.4% [18,72]. This device shows a sensor lifetime of 14 days and is factory-calibrated. The FreeStyle® Navigator II CGM system has a MARD of 12.3% with a sensor lifetime of up to 5 days and needs to be calibrated 3–4 times per day [72].

Notable, a novel analytical framework that enables spectroscopy-based longitudinal tracking of blood glucose without extensive a priori information was investigated by Spegazzini et al. [75]. Using blood glucose monitoring by Raman spectroscopy as an example, it was found that the efficacy of this approach was comparable to conventional calibration methods (35% reduction in error over partial least squares regression when applied to glucose tolerance tests).

7. Continuous and Flash Glucose Monitoring Systems' (Dis) Advantages and Exercise

These commonly used devices reveal different advantages and disadvantages for daily life; furthermore, accuracy during exercise might be divergent to general MARDs (Table 2).

Table 2. Advantages, disadvantages, and exercise performance in continuous and flash glucose monitoring systems.

	Advantage	Disadvantage	Exercise Performance
Dexcom G5™	Hypo- and hyperglycemia alerts; rise and fall (rate of change) alerts; compatible with mobile devices; online live monitoring with different mobile devices (e.g., for parents), CE [1] mark (European Union). Approved for non-adjuvant use; compatible with Apple iPhone 4S and subsequent iOS models	Requires calibration to blood glucose; no integrated bolus wizard	N/A for Dexcom G5™
Dexcom G4™	Hypo- and hyperglycemia alerts; rise and fall (rate of change) alerts; available integrated with the Animas Vibe pump	Requires calibration to blood glucose; no integrated bolus wizard	Continuous exercise: MARD [2] 13.6–18.6%; interval exercise: MARD 13.3–17.7% [61,64,76] (interstitial glucose compared to venous plasma glucose)
Medtronic 670 G with SmartGuard®	Hybrid closed-loop system when combined with insulin pump (automatic insulin delivery when glucose is high); predictive low glucose-suspend when combined with insulin pump; predictive low glucose alert; hypo- and hyperglycemia alerts; rise and fall (rate of change) alerts; Bluetooth connected to glucometer (CONTOUR, NEXT LINK 2.4); bolus wizard	Requires calibration to blood glucose	N/A [3] for this specific system
Medtronic 640 G with SmartGuard®	Predictive low glucose-suspend when combined with insulin pump; predictive low glucose alert; hypo- and hyperglycemia alerts; rise and fall (rate of change) alerts; Bluetooth connected to glucometer (CONTOUR, NEXT LINK 2.4); bolus wizard	Requires calibration to blood glucose	Continuous exercise: MARD 19.4% [61] (interstitial glucose compared to venous plasma glucose)
Medtronic Paradigm Minimed® Veo (530 G)	Low-glucose-suspend when combined with insulin pump; hypo- and hyperglycemia alerts; rise and fall (rate of change) alerts; Bluetooth connected to glucometer (CONTOUR, NEXT LINK 2.4); bolus wizard;	Requires calibration to blood glucose	Continuous exercise: MARD 12.8–23.7%; interval exercise: MARD 15.5–26.5% [63] (interstitial glucose compared to capillary blood glucose)
FreeStyle® Libre Flash glucose monitoring	Factory-calibrated; long sensor lifetime (14 days); integrated glucometer; integrated blood ketone measurement; cheap sensor costs; integrated bolus wizard	No automatic hypo- or hyperglycemia alerts; not combinable with pump	Continuous exercise: MARD 8.7% [61] (interstitial glucose compared to venous plasma glucose)
FreeStyle® Navigator II CGM system	30 m transmission range; new result every minute; hypo- and hyperglycemia alerts; early warning alarms; integrated glucometer	Fixed time points for calibration: 1, 2, 10, 24, and 72 h after sensor insertion	N/A for the second generation

[1] CE (Conformité Européenne) marking is a certification mark that indicates conformity with health, safety, and environmental protection standards for products sold within the European Economic Area. [2] MARD: Median Absolute Relative Difference. [3] N/A: Not Applicable.

8. Continuous and Flash Glucose Monitor Algorithms

As continuous and flash glucose monitoring systems measure glucose within the interstitial fluid, a certain lag time is present for the glucose to diffuse from the blood into the interstitium. To overcome this lag time and to reduce the risk of inadequate therapy decisions (e.g., overdose in bolus insulin correction), several algorithms are incorporated in continuous and flash glucose monitoring systems. Generally, so-called algorithmically "smart sensors" incorporate three software modules that need to work in real time [77,78]:

- denoising the random noise component
- enhancing the accuracy
- predicting the future glucose concentration

As used by Facchinetti et al. [77], an adaptive self-tunable Bayesian smoother is able to automatically estimate the signal-to-noise ratio present in the continuous glucose monitoringdata [79]. For the enhancement of accuracy, a stochastic deconvolution-based re-calibration algorithm is applied, which re-scales the interstitial glucose data using a simple linear regressor whose parameters are re-calculated for each self-measured blood glucose value [78,80]. Especially this or a similar denoising and enhancement algorithm is inwrought in the Dexcom (Dexcom, San Diego, CA, USA) G4 and G5 sensor technology (505 algorithm).

Predicting future glucose concentrations is based on an autoregressive model of order, one based on the findings of Sparacino et al. [81]. For example, the Medtronic 640 G with SmartGuard® (Medtronic MiniMed, Inc., Northridge, CA, USA) uses an algorithm that predicts the decrease of 20 mg/dL within the next 30 min above the pre-defined "low-threshold". When combined with an insulin pump, it suspends the infusion of insulin before reaching this value and automatically restarts when it predicts that glucose levels will increase by 20 $mg \cdot dL^{-1}$ above the pre-defined low glucose threshold [78].

9. Future Directions

9.1. Artificial Pancreas

In line with a recent review, a threshold low-glucose suspend of insulin infusion might be the first step towards an artificial pancreas [72]. The future of artificial pancreas devices promises a more accurate mimicking of endogenous responses to glucose fluctuations; achieved via dual hormone closed-loop systems using insulin and glucagon [82,83]. This system is able to reduce significantly hypoglycemic episodes. Intriguingly, the beneficial effects of dual- compared to single hormone closed-loop systems were observed during continuous and interval exercises [84]. In this study, glucagon was given a microboluses to avoid drastic increases in blood glucose concentrations. Glucagon delivery was based on logical rules considering glucose concentration estimates and their trends. Importantly, the insulin delivery algorithm took account of the injected glucagon onboard via the dual-hormone artificial pancreas. During continuous exercise, a total glucagon dose of 0.126 ± 0.057 mg was administered and during interval exercise 0.093 ± 0.068 mg. The split total glucagon dose was delivered every 10 min during the 60 min exercise sessions. Mini-doses of glucagon were found to be efficacious and safe to treat mild hypoglycemia in adults with type 1 diabetes [85]. Successful hypoglycemia-treatment criteria were met for 94% when using a mini-dose of glucagon in comparison to 95% when giving oral glucose tablets.

Instead of glucagon selective antagonism of somatostatin receptor type 2 (SSTR2) might also be reasonable (within the artificial pancreas) [86]. SSTR2 antagonism after recurrent hypoglycemia ameliorates the glucagon and corticosterone responses, as well as decreases the risk of insulin-induced hypoglycemia in rats with type 1 diabetes. However, a safety profile of SSTR2 antagonism will need to be established. Potential adverse effects might be increased gastric acid secretion and effects on pituitary and adrenal hormone secretion [87].

In real life, the only commercially available system (FDA approved), which is similar to an artificial pancreas, is the hybrid closed-loop system from Medtronic (670 G with SmartGuard®) (Medtronic MiniMed, Inc., Northridge, CA, USA). However, from the authors' point of view, the fundamental problem in the development of a "real" artificial pancreas might be the sensor inaccuracy and sensor/interstitial glucose lag time to blood glucose concentration. For example, the Abbott FreeStyle Libre (Abbott, Alameda, CA, USA) flash glucose monitoring system shows an average lag time of 4.5–4.8 min (for both clinic and home-phases) [18]. In a rapidly changing internal environment during acute exercise combined with low doses of circulating bolus insulin this lag time could potentially rise by 200%, as observed in our lab in two ongoing clinical trials (U1111-1174-6676; DRKS00013477). Furthermore, if the artificial pancreas is designated to work autonomously, the use of detectors of physical activity (e.g., heart rate monitor, accelerometer, small spirometry devices) needs to be incorporated from the very early stages of development.

9.2. Implantable Sensors

In a recent pivotal study, the MARD of the Eversense implantable continuous glucose monitoring sensor (Senseonics Inc., Germantown, MD, USA) was found at 11.1% against reference glucose values above 4.2 mmol/L [88]. Implantable continuous glucose monitoring systems may provide advantages since frequent sensor insertions are not needed and the transmitter can be removed without sensor replacement. However, the implantation and removal are a minor surgical procedure and accompanied by some discomfort.

10. Conclusions

As more health services and insurance companies advocate their use, understanding continuous and flash glucose monitoring systems for its strengths and limitations around exercise may offer more confidence for patients aiming to better manage glycemia. This review detailed the complexities of acute physical exercise and offers an integrated understanding of the efficacy of glucose monitoring with current technology around a heightened metabolic state such as exercise.

Author Contributions: O.M., J.E.Y. and R.M.B. wrote the paper.

Conflicts of Interest: O.M. has received lecture fees from Medtronic (Medtronic MiniMed, Inc., Northridge, CA, USA), travel fees from Novo Nordisk A/S and received a grant from Sêr Cymru II COFUND fellowship/European Union. No conflict of interest for J.E.Y. R.M.B. reports having received honoraria, travel, and educational grant support from Boehringer-Ingelheim (Ingelheim am Rhein, Germany), Eli Lily and Company (Indianapolis, IN, USA), Novo Nordisk (Bagsværd, Denmark), and Sanofi-Aventis (Paris, France).

References

1. Chiang, J.L.; Kirkman, M.S.; Laffel, L.M.B.; Peters, A.L. Type 1 diabetes through the life span: A position statement of the American Diabetes Association. *Diabetes Care* **2014**, *37*, 2034–2054. [CrossRef] [PubMed]
2. The Health and Social Care Information Centre National Diabetes Audit 2012–2013, Report 2. Available online: http://content.digital.nhs.uk/catalogue/PUB16496/nati-diab-audi-12-13-rep2.pdf (accessed on 27 November 2017).
3. Plotnikoff, R.C.; Taylor, L.M.; Wilson, P.M.; Courneya, K.S.; Sigal, R.J.; Birkett, N.; Raine, K.; Svenson, L.W. Factors associated with physical activity in Canadian adults with diabetes. *Med. Sci. Sports Exerc.* **2006**, *38*, 1526–1534. [CrossRef] [PubMed]
4. Laing, S.P.; Swerdlow, A.J.; Slater, S.D.; Burden, A.C.; Morris, A.; Waugh, N.R.; Gatling, W.; Bingley, P.J.; Patterson, C.C. Mortality from heart disease in a cohort of 23,000 patients with insulin-treated diabetes. *Diabetologia* **2003**, *46*, 760–765. [CrossRef] [PubMed]
5. Dorman, J.S.; Laporte, R.E.; Kuller, L.H.; Cruickshanks, K.J.; Orchard, T.J.; Wagener, D.K.; Becker, D.J.; Cavender, D.E.; Drash, A.L. The Pittsburgh insulin-dependent diabetes mellitus (IDDM) morbidity and mortality study: Mortality results. *Diabetes* **1984**, *33*, 271–276. [CrossRef] [PubMed]
6. Riddell, M.C.; Gallen, I.W.; Smart, C.E.; Taplin, C.E.; Adolfsson, P.; Lumb, A.N.; Kowalski, A.; Rabasa-Lhoret, R.; McCrimmon, R.J.; Hume, C.; et al. Exercise management in type 1 diabetes: A consensus statement. *Lancet Diabetes Endocrinol.* **2017**, *8587*, 1–14. [CrossRef]
7. Colberg, S.R.; Sigal, R.J.; Yardley, J.E.; Riddell, M.C.; Dunstan, D.W.; Dempsey, P.C.; Horton, E.S.; Castorino, K.; Tate, D.F. Physical activity/exercise and diabetes: A position statement of the American Diabetes Association. *Diabetes Care* **2016**, *39*, 2065–2079. [CrossRef] [PubMed]
8. Chimen, M.; Kennedy, A.; Nirantharakumar, K.; Pang, T.T.; Andrews, R.; Narendran, P. What are the health benefits of physical activity in type 1 diabetes mellitus? A literature review. *Diabetologia* **2012**, *55*, 542–551. [CrossRef] [PubMed]

9. Brazeau, A.-S.; Rabasa-Lhoret, R.; Strychar, I.; Mircescu, H. Barriers to physical activity among patients with type 1 diabetes. *Diabetes Care* **2008**, *31*, 2108–2109. [CrossRef] [PubMed]
10. Galassetti, P.; Riddell, M.C. Exercise and type 1 diabetes (T1DM). *Compr. Physiol.* **2013**, *3*, 1309–1336. [PubMed]
11. The Diabetes Control and Complications Trial Research Group. The effect of intensive treatment of diabetes on the development and progression of long-term complications in insulin-dependent diabetes mellitus. *N. Engl. J. Med.* **1993**, *329*, 977–986.
12. The Diabetes Control and Complications Trial Research Group. The effect of intensive diabetes treatment on the development and progression of long-term complications in adolescents with insulin-dependent diabetes mellitus: Diabetes control and complications trial. *J. Pediatr.* **1994**, *125*, 177–188.
13. The Epidemiology of Diabetes Interventions and Complications Research Group. Intensive diabetes therapy and carotid intima–media thickness in type 1 diabetes mellitus. *N. Engl. J. Med.* **2003**, *348*, 2294–2303.
14. Diabetes Control and Complications Trial/Epidemiology of Diabetes Interventions and Complications (DCCT/EDIC) Study Research Group. Intensive diabetes treatment and cardiovascular disease in patients with type 1 diabetes. *N. Engl. J. Med.* **2005**, *353*, 2643–2653.
15. The DCCT Research Group. Epidemiology of severe hypoglycemia in the diabetes control and complications trial. *Am. J. Med.* **1991**, *90*, 450–459.
16. Rodbard, D. Continuous glucose monitoring: A review of successes, challenges, and opportunities. *Diabetes Technol. Ther.* **2016**, *18*, 3–13. [CrossRef] [PubMed]
17. Parkin, C.G.; Graham, C.; Smolskis, J. Continuous glucose monitoring use in type 1 diabetes: Longitudinal analysis demonstrates meaningful improvements in HbA1c and reductions in health care utilization. *J. Diabetes Sci. Technol.* **2017**, *11*, 522–528. [CrossRef] [PubMed]
18. Bailey, T.; Bode, B.W.; Christiansen, M.P.; Klaff, L.J.; Alva, S. The performance and usability of a factory-calibrated flash glucose monitoring system. *Diabetes Technol. Ther.* **2015**, *17*, 787–794. [CrossRef] [PubMed]
19. Richter, E.A.; Derave, W.; Wojtaszewski, J.F.P. Glucose, exercise and insulin: Emerging concepts. *J. Physiol.* **2001**, *535*, 313–322. [CrossRef] [PubMed]
20. Brooks, G.A.; Fahey, T.D.; Baldwin, K.M. *Exercise Physiology: Human Bioenergetics and Its Applications*, 4th ed.; McGraw-Hill: New York, NY, USA, 2005.
21. Nevill, M.E.; Boobis, L.H.; Brooks, S.; Williams, C. Effect of training on muscle metabolism during treadmill sprinting. *J. Appl. Physiol.* **1989**, *67*, 2376–2382. [CrossRef] [PubMed]
22. Emhoff, C.-A.W.; Messonnier, L.A.; Horning, M.A.; Fattor, J.A.; Carlson, T.J.; Brooks, G.A. Gluconeogenesis and hepatic glycogenolysis during exercise at the lactate threshold. *J. Appl. Physiol.* **2013**, *114*, 297–306. [CrossRef] [PubMed]
23. Bergman, B.C.; Horning, M.A.; Casazza, G.A.; Wolfel, E.E.; Butterfield, G.E.; Brooks, G.A. Endurance training increases gluconeogenesis during rest and exercise in men. *Am. J. Physiol. Endocrinol. Metab.* **2000**, *278*, 244E–251E. [CrossRef] [PubMed]
24. Korthuis, R.J. *Skeletal Muscle Circulation*; Morgan & Claypool Life Sciences: Williston, ND, USA, 2011.
25. Spriet, L.L. New insights into the interaction of carbohydrate and fat metabolism during exercise. *Sports Med.* **2014**, *44*, 87–96. [CrossRef] [PubMed]
26. Jenni, S.; Oetliker, C.; Allemann, S.; Ith, M.; Tappy, L.; Wuerth, S.; Egger, A.; Boesch, C.; Schneiter, P.; Diem, P.; et al. Fuel metabolism during exercise in euglycaemia and hyperglycaemia in patients with type 1 diabetes mellitus—A prospective single-blinded randomised crossover trial. *Diabetologia* **2008**, *51*, 1457–1465. [CrossRef] [PubMed]
27. Rissanen, A.P.E.; Tikkanen, H.O.; Koponen, A.S.; Aho, J.M.; Peltonen, J.E. Central and peripheral cardiovascular impairments limit V O2peak in Type 1 diabetes. *Med. Sci. Sports Exerc.* **2015**, *47*, 223–230. [CrossRef] [PubMed]
28. Astrand, P.-O.; Rodahl, K.; Dahl, H.A.; Stromme, S.B. *Textbook of Work Physiology: Physiological Bases of Exercise*, 3rd ed.; McGraw Hill: New York, NY, USA, 1986.
29. Secher, N.H.; Amann, M. Human investigations into the exercise pressor reflex. *Exp. Physiol.* **2012**, *97*, 59–69. [CrossRef] [PubMed]
30. Saltin, B.; Boushel, R.; Niels, S.; Jere, M. *Exercise and Circulation in Health and Disease*; Human Kinetics: Champaign, IL, USA, 1999.

31. Saltin, B.; Radegran, G.; Koskolou, M.D.; Roach, R.C. Skeletal muscle blood flow in humans and its regulation during exercise. *Acta Physiol. Scand.* **1998**, *1*, 421–436. [CrossRef] [PubMed]

32. Middlekauff, H.R.; Nitzsche, E.U.; Nguyen, A.H.; Hoh, C.K.; Gibbs, G.G. Modulation of renal cortical blood flow during static exercise in humans. *Circ. Res.* **1997**, *80*, 62–68. [CrossRef] [PubMed]

33. Levick, J.R. *An Introduction to Cardiovascular Physiology*, 5th ed.; Taylor & Francis Ltd.: London, UK, 2010.

34. Kvietys, P.R. *The Gastrointestinal Circulation*; Morgan & Claypool Life Sciences: Williston, ND, USA, 2010.

35. Metheny, N.M.; Metheny, N.M. *Fluid and Electrolyte Balance: Nursing Considerations*, 5th ed.; Jones & Bartlett Learning: Burlington, ON, Canada, 2012.

36. Martini, F.H.; Nath, J.L. *Fundamentals of Anatomy and Physiology*; Wiley-Blackwell: Hoboken, NJ, USA, 2009.

37. Klabunde, R.E. *Cardiovascular Physiology Concepts*, 2nd ed.; Lippincott Williams & Wilkins: Philadelphia, PA, USA, 2004.

38. Arturson, G.; Kjellmer, I. Capillary permeability in skeletal muscle during rest and activity. *Acta Physiol. Scand.* **1964**, *62*, 41–45. [CrossRef] [PubMed]

39. Jacobsson, S.; Kjellmer, I. Flow and protein content of lymph in resting and exercising skeletal muscle. *Acta Physiol. Scand.* **1964**, *60*, 278–285. [CrossRef] [PubMed]

40. Jacobsson, S.; Kjellmer, I. Accumulation of fluid in exercising skeletal muscle. *Acta Physiol. Scand.* **1964**, *60*, 286–292. [CrossRef] [PubMed]

41. Lundvall, J. Tissue hyperosmolality as a mediator of vasodilatation and transcapillary fluid flux in exercising skeletal muscle. *Acta Physiol. Scand. Suppl.* **1972**, *379*, 1–142. [PubMed]

42. Lundvall, J.; Mellander, S.; Westling, H.; White, T. Fluid transfer between blood and tissues during exercise. *Acta Physiol. Scand.* **1972**, *85*, 258–269. [CrossRef] [PubMed]

43. Kjellmer, I. Studies on exercise hyperemia. *Acta Physiol. Scand. Suppl.* **1964**, *244*, 1–27.

44. Scallan, J.; Huxley, V.H.; Korthuis, R.J. *Capillary Fluid Exchange: Regulation, Functions, and Pathology*; Morgan & Claypool Life Sciences: Williston, ND, USA, 2010.

45. Stick, C.; Jaeger, H.; Witzleb, E. Measurements of volume changes and venous pressure in the human lower leg during walking and running. *J. Appl. Physiol.* **1992**, *72*, 2063–2068. [CrossRef] [PubMed]

46. Richter, E.A.; Ploug, T.; Galbo, H. Increased muscle glucose uptake after exercise. No need for insulin during exercise. *Diabetes* **1985**, *34*, 1041–1048. [CrossRef] [PubMed]

47. Renkin, E.M.; Michel, C.C. Capillary Permeability to Small Solutes. In *Handbook of Physiology: The Cardiovascular System, Microcirculation*; Crone, C., Levitt, D.G., Eds.; American Physiological Society: Bethesda, MD, USA, 1984.

48. Blomstrand, E.; Saltin, B. Effect of muscle glycogen on glucose, lactate and amino acid metabolism during exercise and recovery in human subjects. *J. Physiol.* **1999**, *514*, 293–302. [CrossRef] [PubMed]

49. Rose, A.J.; Richter, E.A. Skeletal muscle glucose uptake during exercise: How is it regulated? *Physiology (Bethesda)* **2005**, *20*, 260–270. [CrossRef] [PubMed]

50. Hultman, E. Studies on muscle metabolism of glycogen and active phosphate in man with special reference to exercise and diet. *Scand. J. Clin. Lab. Investig. Suppl.* **1967**, *94*, 1–63.

51. Gaitanos, G.; Willians, L.; Boobis, L.; Brooks, S. Human muscle metabolism maximal exercise. *Appl. Physiol.* **1993**, *75*, 712–719. [CrossRef] [PubMed]

52. Marliss, E.B.; Vranic, M. Intense exercise has unique effects on both insulin release and its roles in glucoregulation: Implications for diabetes. *Diabetes* **2002**, *51*, 271S–283S. [CrossRef]

53. Nilsson, L.H.; Hultman, E. Liver glycogen in man—The effect of total starvation or a carbohydrate—Poor diet followed by carbohydrate refeeding. *Scand. J. Clin. Lab. Investig.* **1973**, *32*, 325–330. [CrossRef]

54. Fahey, A.J.; Paramalingam, N.; Davey, R.J.; Davis, E.A.; Jones, T.W.; Fournier, P.A. The effect of a short sprint on postexercise whole-body glucose production and utilization rates in individuals with type 1 diabetes mellitus. *J. Clin. Endocrinol. Metab.* **2012**, *97*, 4193–4200. [CrossRef] [PubMed]

55. Gyntelberg, F.; Rennie, M.J.; Hickson, R.C.; Holloszy, J.O. Effect of training on the response of plasma glucagon to exercise. *J. Appl. Physiol.* **1977**, *43*, 302–305. [CrossRef] [PubMed]

56. West, D.J.; Morton, R.D.; Bain, S.C.; Stephens, J.W.; Bracken, R.M. Blood glucose responses to reductions in pre-exercise rapid-acting insulin for 24 h after running in individuals with type 1 diabetes. *J. Sports Sci.* **2010**, *28*, 781–788. [CrossRef] [PubMed]

57. Clark, M.G.; Wallis, M.G.; Barrett, E.J.; Vincent, M.A.; Richards, S.M.; Clerk, L.H.; Rattigan, S. Blood flow and muscle metabolism: A focus on insulin action. *Am. J. Physiol. Endocrinol. Metab.* **2003**, *284*, 241E–258E. [CrossRef] [PubMed]

58. Gonzalez, J.T.; Fuchs, C.J.; Betts, J.A.; van Loon, L.J. Glucose plus fructose ingestion for post—Exercise recovery—Greater than the sum of its parts? *Nutrients* **2017**, *9*, E344. [CrossRef] [PubMed]

59. Jeukendrup, A.E. Training the gut for athletes. *Sports Med.* **2017**, *47*, 101–110. [CrossRef] [PubMed]

60. Klonoff, D.C.; Ahn, D.; Drincic, A. Continuous glucose monitoring: A review of the technology and clinical use. *Diabetes Res. Clin. Pract.* **2017**, *133*, 178–192. [CrossRef] [PubMed]

61. Aberer, F.; Hajnsek, M.; Rumpler, M.; Zenz, S.; Baumann, P.M.; Elsayed, H.; Puffing, A.; Treiber, G.; Pieber, T.R.; Sourij, H.; et al. Evaluation of subcutaneous glucose monitoring systems under routine environmental conditions in patients with type 1 diabetes. *Diabetes Obes. Metab.* **2017**, *19*, 1051–1055. [CrossRef] [PubMed]

62. Herrington, S.J.; Gee, D.L.; Dow, S.D.; Monosky, K.A.; And, E.D.; Pritchett, K.L. Comparison of glucose monitoring methods during steady-state exercise in women. *Nutrients* **2012**, *4*, 1282–1292. [CrossRef] [PubMed]

63. Moser, O.; Mader, J.; Tschakert, G.; Mueller, A.; Groeschl, W.; Pieber, T.; Koehler, G.; Messerschmidt, J.; Hofmann, P. Accuracy of continuous glucose monitoring (CGM) during continuous and high-Intensity interval exercise in patients with type 1 diabetes mellitus. *Nutrients* **2016**, *8*, 489. [CrossRef] [PubMed]

64. Bally, L.; Zueger, T.; Pasi, N.; Carlos, C.; Paganini, D.; Stettler, C. Accuracy of continuous glucose monitoring during differing exercise conditions. *Diabetes Res. Clin. Pract.* **2016**, *112*, 1–5. [CrossRef] [PubMed]

65. Yardley, J.E.; Sigal, R.J.; Kenny, G.P.; Riddell, M.C.; Lovblom, L.E.; Perkins, B. A Point accuracy of interstitial continuous glucose monitoring during exercise in type 1 diabetes. *Diabetes Technol. Ther.* **2013**, *15*, 46–49. [CrossRef] [PubMed]

66. Iscoe, K.E.; Davey, R.J.; Fournier, P.A. Increasing the low-glucose alarm of a continuous glucose monitoring system prevents exercise-induced hypoglycemia without triggering any false alarms. *Diabetes Care* **2011**, *34*, 109e. [CrossRef] [PubMed]

67. Gross, T.M.; Bode, B.W.; Einhorn, D.; Kayne, D.M.; Reed, J.H.; White, N.H.; Mastrototaro, J.J. Performance evaluation of the MiniMed continuous glucose monitoring system during patient home use. *Diabetes Technol. Ther.* **2000**, *2*, 49–56. [CrossRef] [PubMed]

68. FDA Advisory Panel. FDA advisory panel votes to recommend non-adjunctive use of Dexcom G5 mobile CGM. *Diabetes Technol. Ther.* **2016**, *18*, 512–516.

69. Matuleviciene, V.; Joseph, J.I.; Andelin, M.; Hirsch, I.B.; Attvall, S.; Pivodic, A.; Dahlqvist, S.; Klonoff, D.; Haraldsson, B.; Lind, M. A clinical trial of the accuracy and treatment experience of the Dexcom G4 sensor (Dexcom G4 system) and Enlite sensor (Guardian REAL-time system) tested simultaneously in ambulatory patients with type 1 diabetes. *Diabetes Technol. Ther.* **2014**, *16*, 759–767. [CrossRef] [PubMed]

70. Kropff, J.; Bruttomesso, D.; Doll, W.; Farret, A.; Galasso, S.; Luijf, Y.M.; Mader, J.K.; Place, J.; Boscari, F.; Pieber, T.R.; et al. Accuracy of two continuous glucose monitoring systems: A head-to-head comparison under clinical research centre and daily life conditions. *Diabetes Obes. Metab.* **2014**, *17*, 343–349. [CrossRef] [PubMed]

71. Luijf, Y.M.; Mader, J.K.; Doll, W.; Pieber, T.; Farret, A.; Place, J.; Renard, E.; Bruttomesso, D.; Filippi, A.; Avogaro, A.; et al. Accuracy and reliability of continuous glucose monitoring systems: A head-to-head comparison. *Diabetes Technol. Ther.* **2013**, *15*, 721 726. [CrossRef] [PubMed]

72. Rodbard, D. Continuous glucose monitoring: A review of recent studies demonstrating improved glycemic outcomes. *Diabetes Technol. Ther.* **2017**, *19* (Suppl. 3), 25S–37S. [CrossRef] [PubMed]

73. Christiansen, M.P.; Garg, S.K.; Brazg, R.; Bode, B.W.; Bailey, T.S.; Slover, R.H.; Sullivan, A.; Huang, S.; Shin, J.; Lee, S.W.; et al. Accuracy of a fourth-generation subcutaneous continuous glucose sensor. *Diabetes Technol. Ther.* **2017**, *19*, 446–456. [CrossRef] [PubMed]

74. Lawson, M.L.; Bradley, B.; McAssey, K.; Clarson, C.; Kirsch, S.E.; Mahmud, F.H.; Curtis, J.R.; Richardson, C.; Courtney, J.; Cooper, T.; et al. The JDRF CCTN CGM TIME Trial: Timing of initiation of continuous glucose monitoring in established pediatric type 1 diabetes: Study protocol, recruitment and baseline characteristics. *BMC Pediatr.* **2014**, *14*, 183. [CrossRef] [PubMed]

75. Spegazzini, N.; Barman, I.; Dingari, N.C.; Pandey, R.; Soares, J.S.; Ozaki, Y.; Dasari, R.R. Spectroscopic approach for dynamic bioanalyte tracking with minimal concentration information. *Sci. Rep.* **2014**, *4*, 7013. [CrossRef] [PubMed]

76. Taleb, N.; Emami, A.; Suppere, C.; Messier, V.; Legault, L.; Chiasson, J.-L.; Rabasa-Lhoret, R.; Haidar, A. Comparison of two continuous glucose monitoring systems, Dexcom G4 Platinum and Medtronic Paradigm Veo Enlite system, at rest and during exercise. *Diabetes Technol. Ther.* **2016**, *18*, 561–567. [CrossRef] [PubMed]

77. Facchinetti, A.; Sparacino, G.; Guerra, S.; Luijf, Y.M.; DeVries, J.H.; Mader, J.K.; Ellmerer, M.; Benesch, C.; Heinemann, L.; Bruttomesso, D.; et al. Real-time improvement of continuous glucose monitoring accuracy: The smart sensor concept. *Diabetes Care* **2013**, *36*, 793–800. [CrossRef] [PubMed]

78. Facchinetti, A. Continuous glucose monitoring sensors: Past, present and future algorithmic challenges. *Sensors* **2016**, *16*, 2093. [CrossRef] [PubMed]

79. Facchinetti, A.; Sparacino, G.; Cobelli, C. An online self-tunable method to denoise CGM sensor data. *IEEE Trans. Biomed. Eng.* **2010**, *57*, 634–641. [CrossRef] [PubMed]

80. Guerra, S.; Facchinetti, A.; Sparacino, G.; De Nicolao, G.; Cobelli, C. Enhancing the accuracy of subcutaneous glucose sensors: A real-time deconvolution-based approach. *IEEE Trans. Biomed. Eng.* **2012**, *59*, 1658–1669. [CrossRef] [PubMed]

81. Sparacino, G.; Zanderigo, F.; Corazza, S.; Maran, A.; Facchinetti, A.; Cobelli, C. Glucose Concentration can be Predicted Ahead in Time From Continuous Glucose Monitoring Sensor Time-Series. *IEEE Trans. Biomed. Eng.* **2007**, *54*, 931–937. [CrossRef] [PubMed]

82. Gingras, V.; Rabasa-Lhoret, R.; Messier, V.; Ladouceur, M.; Legault, L.; Haidar, A. Efficacy of dual-hormone artificial pancreas to alleviate the carbohydrate-counting burden of type 1 diabetes: A randomized crossover trial. *Diabetes Metab.* **2016**, *42*, 47–54. [CrossRef] [PubMed]

83. El-Khatib, F.H.; Balliro, C.; Hillard, M.A.; Magyar, K.L.; Ekhlaspour, L.; Sinha, M.; Mondesir, D.; Esmaeili, A.; Hartigan, C.; Thompson, M.J.; et al. Home use of a bihormonal bionic pancreas versus insulin pump therapy in adults with type 1 diabetes: A multicentre randomised crossover trial. *Lancet* **2017**, *389*, 369–380. [CrossRef]

84. Taleb, N.; Emami, A.; Suppere, C.; Messier, V.; Legault, L.; Ladouceur, M.; Chiasson, J.-L.; Haidar, A.; Rabasa-Lhoret, R. Efficacy of single-hormone and dual-hormone artificial pancreas during continuous and interval exercise in adult patients with type 1 diabetes: Randomised controlled crossover trial. *Diabetologia* **2016**, *59*, 2561–2571. [CrossRef] [PubMed]

85. Haymond, M.W.; DuBose, S.N.; Rickels, M.R.; Wolpert, H.; Shah, V.N.; Sherr, J.L.; Weinstock, R.S.; Agarwal, S.; Verdejo, A.S.; Cummins, M.J.; et al. Efficacy and safety of mini-dose glucagon for treatment of non-severe hypoglycemia in adults with type 1 diabetes. *J. Clin. Endocrinol. Metab.* **2017**, *102*, 2994–3001. [CrossRef] [PubMed]

86. Yue, J.T.Y.; Riddell, M.C.; Burdett, E.; Coy, D.H.; Efendic, S.; Vranic, M. Amelioration of hypoglycemia via somatostatin receptor type 2 antagonism in recurrently hypoglycemic diabetic rats. *Diabetes* **2013**, *62*, 2215–2222. [CrossRef] [PubMed]

87. Taleb, N.; Rabasa-Lhoret, R. Can somatostatin antagonism prevent hypoglycaemia during exercise in type 1 diabetes? *Diabetologia* **2016**, *59*, 1632–1635. [CrossRef] [PubMed]

88. Kropff, J.; Choudhary, P.; Neupane, S.; Barnard, K.; Bain, S.C.; Kapitza, C.; Forst, T.; Link, M.; Dehennis, A.; De Vries, J.H. Accuracy and longevity of an implantable continuous glucose sensor in the PRECISE study: A 180-day, prospective, multicenter, pivotal trial. *Diabetes Care* **2017**, *40*, 63–68. [CrossRef] [PubMed]

nutrients

MDPI

Review

Metabolic Effects of High Glycaemic Index Diets: A Systematic Review and Meta-Analysis of Feeding Studies in Mice and Rats

Grace J. Campbell [1],*, Alistair M. Senior [2] and Kim S. Bell-Anderson [1]

[1] Charles Perkins Centre, School of Life and Environmental Sciences, University of Sydney, Sydney, NSW 2006, Australia; kim.bell-anderson@sydney.edu.au
[2] Charles Perkins Centre, School of Mathematics and Statistics, University of Sydney, Sydney, NSW 2006, Australia; alistair.senior@sydney.edu.au
*   Correspondence: grace.campbell@sydney.edu.au; Tel.: +61-2-8627-1996

Received: 18 May 2017; Accepted: 17 June 2017; Published: 22 June 2017

Abstract: Low glycaemic index (LGI) diets are often reported to benefit metabolic health, but the mechanism(s) responsible are not clear. This review aimed to systematically identify studies investigating metabolic effects of high glycaemic index (HGI) versus LGI diets in mice and rats. A meta-analysis was conducted to calculate an overall effect size, Hedge's standardised mean differences (hereafter d), for each trait, with moderator variables considered in subsequent meta-regressions. Across 30 articles, a HGI diet increased five of the seven traits examined: body weight ($d = 0.55$; 95% confidence interval: 0.31, 0.79), fat mass ($d = 1.08$; 0.67, 1.49), fasting circulating insulin levels ($d = 0.40$; 0.09, 0.71), and glucose ($d = 0.80$; 0.35, 1.25) and insulin ($d = 1.14$; 0.50, 1.77) area under the curve during a glucose tolerance test. However, there was substantial heterogeneity among the effects for all traits and the small number of studies enabled only limited investigation of possible confounding factors. HGI diets favour body weight gain, increased adiposity and detrimentally affect parameters of glucose homeostasis in mice and rats, but these effects may not be a direct result of GI per se; rather they may be due to variation in other dietary constituents, such as dietary fibre, a factor which is known to reduce the GI of food and promote health via GI-independent mechanisms.

Keywords: glycaemic index; glycaemic index; mice; rats; metabolism; glucose homeostasis

1. Introduction

Obesity is a chronic lifestyle disease that is brought about by many factors, the easiest of which to identify are diet and behaviour. Increased consumption of "Western" diets combined with increased sedentary behaviours are thought to be major contributors to the rapid rise in the prevalence of obesity worldwide [1]. Obesity is defined as an excess accumulation of fat mass and is associated with insulin resistance, increasing risk for type 2 diabetes mellitus, cardiovascular disease and some cancers. Diet is key to addressing this urgent public health issue, and the delineation of mechanisms linking dietary patterns and metabolic health is needed to establish optimal nutrition for individuals and populations.

Dietary guidelines recommend that carbohydrate, including sugars, starch and fibre, contribute roughly half the calories of our daily energy intake [2,3]. The nutritional importance of the subtypes of carbohydrate and their distinct physiological effects on health are not fully known. However, it is generally recommended that intake of added sugars be limited and wholegrain and fibre content increased [4,5]. Changes in food-processing technology since the Industrial Revolution have resulted in an increased supply and consumption of refined starchy foods, which feature heavily in a "Western" diet and are also associated with increased energy density, reduced fibre density and micronutrient dilution.

Starch can be further categorised as rapidly digestible starch (RDS), slowly digestible starch (SDS), or resistant starch (RS) based on the rate of digestion in the body. The two types of digestible starch (DS) are often distinguished based on their effect on postprandial blood glucose levels, with RDS quickly increasing and then decreasing blood glucose levels, and SDS inducing a slight rise that is maintained over a longer duration [6]. RS is the portion of carbohydrate that escapes digestion in the small intestine, and passes to the large intestine where it undergoes microbial fermentation, resulting in production of short chain fatty acids. RS may also be considered a source of fibre, as opposed to a form of carbohydrate, according to relatively new definitions of fibre [7].

The glycaemic index (GI) is used to compare different carbohydrate containing foods based on their effects on postprandial blood glucose levels in humans [8,9]. High glycaemic index (HGI) carbohydrates have a dramatic effect on blood glucose, with a quick release of glucose into the blood stream and a similarly quick dispersal. Low glycaemic index (LGI) carbohydrates result in a more gradual release of glucose that is digested and absorbed over a greater period of time. Hence, HGI carbohydrates consist mainly of RDS and LGI less so. The association between LGI and health is commonly criticized because a number of other factors lower GI, such as increased fat, protein and fibre content. While increased fibre, such as RS, within a diet can lower the GI of food [10], the well-documented beneficial effect of fibre on glucose homeostasis may be independent of its effects on postprandial glycaemia, and may act via other mechanisms such as those due to interactions among diet, microbiome and host.

Perhaps due to the more gradual perturbation in blood glucose levels, LGI foods lead to better control of blood glucose levels compared to HGI foods, particularly in type 2 diabetic patients [11,12]. To better understand this mechanism, the effect of both HGI and LGI diets on blood glucose homeostasis have been studied in humans and animals. Animal studies are essential for deeper investigation of molecular mechanisms attributed to the GI, and help isolate these effects of the environment, or diet, from genetic variability. While there have been several studies investigating the effect of dietary GI on rodent metabolism, these studies report somewhat contradictory results and it is unclear if these heterogeneous results are due to differences in experimental design and methodology, such as diet, or simply sampling variance. In this review, studies in mice and rats were identified and, after subjecting them to pre-set inclusion and exclusion criteria, data relating to macronutrient composition, carbohydrate source and their effect on metabolic characteristics were extracted and effect sizes derived providing the first meta-analysis of the effect of high glycaemic index diets in mice and rats.

2. Methods

2.1. Eligibility Criteria and Literature Search

The method for this review was adapted from Ainge et al. 2011 [13] and, as such, complied with the Cochrane Guidelines, Version 5.0.1 Part 2-General Methods for Cochrane Reviews [14]. In order to minimise bias during the search, a set of specific inclusion and exclusion criteria was specified a priori (Table 1). These criteria were obtained based on the set question this review aimed to answer: "What is the effect of a high versus low glycaemic index diet on metabolic characteristics in mice and rats?" As the aim was to observe the effect of GI, carbohydrate was required to contribute the majority of the energy composition. Additionally, sufficient protein content was necessary in order for the mice and rats to remain healthy and produce their own protein [15–17].

Table 1. Inclusion and exclusion criteria. This criterion was used to assess all articles and determine their eligibility for inclusion in this review. HGI, high glycaemic index; LGI, low glycaemic index.

Inclusion Criteria	Exclusion Criteria
Mice or rat studies	Transgenic animal models
Experiments performed in live whole animals	Cell cultures
HGI diet	No control diet (controls include chow or LGI)
Carbohydrates >50% of energy content	Diet of unknown macronutrient composition
Proteins >19% of energy content	Use of HGI supplement such as sugars to non-HGI diet
In English	Review, conference abstract or supplementary stub article

Four databases, PubMed, Web of Science, Medline and Scopus, were searched for papers containing a HGI diet fed to either mice or rats (Table 2). Boolean operators, alternate phrases and spellings, truncated search terms and quotation marks were used to ensure the search completed was as comprehensive and specific as possible. This search was conducted on 3 January 2017.

Table 2. Search criteria and initial results. These specific terms were used to search four databases, with the number of results yielded stated. HGI, high glycaemic index; GI, glycaemic index.

Search Term	("high glycemic*" OR "high glycaemic*" OR "high-glycemic*" OR "high-glycaemic*" OR "HGI" OR "H-GI" OR "H GI" OR "high GI" OR "high-GI" OR "simple carbohydrate*" OR "simple-carbohydrate*" OR "high carbohydrate*" OR "high-carbohydrate*" OR "high glucose*" OR "high-glucose*" OR "higher GI*" OR "higher-GI*" OR "higher glycemic*" OR "higher-glycemic*" OR "higher glycaemic*" OR "higher-glycaemic*") AND (mice OR rat) AND ("GI" OR "glycemic*" OR "glycaemic*")				
Database	PubMed	Web of Science	Scopus	Medline	Total
Results	175	422	228	254	1079

Once duplicate articles had been removed, each article's abstract was subjected to the inclusion and exclusion criteria as described above. The remaining articles were then similarly screened based on their methodology. The reference lists of all articles that fit the inclusion and exclusion criteria were obtained and these articles similarly screened. Where information was missing, it was assumed that the article did not meet the specified standards on that point. For example, if the article did not state the macronutrient composition, it was assumed that it would not meet the requirements and was excluded.

2.2. Data Extraction

The full texts of the final articles were examined and requisite information was extracted, including the specific animals and diets used, and any measurement that could be used to determine the effects of the experiment on body weight, body adiposity, energy intake or glucose homeostasis. Glucose homeostasis measures included were fasting blood glucose levels, fasting plasma or serum insulin levels and the area under the curves (AUCs) for glucose and insulin during glucose tolerance tests (GTTs), leading to a total of seven traits. Data were collated primarily from text and/or tables, but where unavailable was extracted from graphs using GetData Graph Digitizer software v2.26.0.20 (GetData, Kogarah, Australia). To maximize the differences between HGI and LGI fed animals, the latest time point for each trait for each study, or subgroup of animals for studies with multiple cohorts, was used. Any data lacking sample size or a measure of sample distribution, such as standard deviation or standard error of the mean, were not included.

For each of the examined traits for each group of animals on a HGI or LGI diet, the mean trait value, a measure of error such as standard error of the mean, standard deviation or 95% confidence interval (CI) and the sample size were extracted. All measures of variation were back calculated to standard deviation for calculation of the effect size. Where a range for the sample size was given, the minimum value was used to ensure that the results were not given undeserved weighting.

Data for seven moderator variables were also extracted for each group of animals: species, sex, length of diet, dietary fat content, starting age, length of fasting time, and use of anaesthesia.

Three were categorical factors (species, sex and use of anaesthesia) and four were continuous variables (length of diet, dietary fat content, starting age and length of fasting time). All categorical factors were recorded as binary variables; mouse or rat, male or female, yes or no respectively. Continuous variables were recorded as an integer, with length of diet and starting age recorded in days, dietary fat content as a percentage of the dietary energy content, and length of fasting time in hours. Where a range was provided for any of the continuous factors, the midpoint was used. An age of "adult" was assumed to be 18 weeks or 126 days, and a fasting time of "overnight" was assumed to be 15 h unless other clarifying information was given. Two papers did not provide starting age and as such were not included in that particular meta-regression. The raw data are available in Table S1. All continuous variables were scaled to standard deviations and centred via z-transformation.

2.3. Effect Size Calculation and Meta-Analysis

For each experimental comparison within a study, we calculated Hedge's *d* (hereafter *d*) and its sampling variance following equations 1–4 in Nakagawa et al. 2015 [18]. We calculated *d* such that positive values suggest that the trait of interest was greater in the HGI group than the LGI group of animals. All effect size calculations and meta-analyses were implemented using the statistical programming language *R* v3.3.1 (The R Foundation, Vienna, Austria) [19]. The seven traits were analysed separately, but in an identical way. To estimate the overall effect size for each trait we first implemented a random-effects meta-analysis (REMA), using the "rma.mv" function in the *R* package *metafor* (Maastricht University, Maastricht, Netherlands) [20]. Each REMA fitted *d* as the response along with its sampling variance, as well as a matrix denoting the estimated covariance among effect sizes that were based on contrasts with the same control group of animals, following the method of Besson et al. 2016 [21]. This type of covariance is sometimes called non-independence due to shared control or "stochastic dependency" [22,23]. Statistical significance of the overall effect of the diet was identified when the 95% CI did not span 0. *d* values of 0.3, 0.5 and 0.8 were considered small, moderate and large differences, respectively [18]. Excluding *p*-values, all statistical results are reported to two decimal places. We report statistical heterogeneity, that is, variation among studies in the reported effect of diet on each trait, as τ and assess its statistical significance using Cochran's *Q* test [24]. τ is interpretable as the standard deviation of the effect sizes, that is, differences in study outcomes that is not attributable to sampling.

In a second step, we explored possible drivers of heterogeneity by fitting random-effects meta-regression (REMR) to explore factors that affected the magnitude of the reported effects. REMR were fitted in the same way as the REMA described above, but with moderator variables included. The effect of each moderator variable on each trait was evaluated in a separate model. The restriction of limited sample sizes precluded the consideration of multiple predictors simultaneously, thus each moderator was evaluated individually. The moderators chosen were those considered most crucial to the methodology, and as such were also those most likely reported. Due to limited sample sizes, only those moderators that were disclosed in most papers could be analysed.

We evaluated the impact of publication bias, which is a systematic under-representation of effect sizes owing to the process of peer-review and publishing, using a trim and fill analysis with the "trimfill" function in *metafor*. Trim and fill analyses were applied to the residuals of the REMA for each trait following the method of Nakagawa et al. 2012 [25]. All *R* code is presented in Text S1.

2.4. Methodological Quality Assessment (MQA)

To assess the quality and validity of the articles, we performed a MQA twice by independent researchers (G.J.C. and K.S.B.-A.). The criteria were obtained from Ainge et al. 2011 [13] and as such were a modified Downs and Black "Quality Index" (QI) (Table 3) [26]. There were two categories, each with three sections. These criteria were further modified for this review with an additional 4 criteria added to the Design and Outcomes section within the Reporting category, and one removed from the

Confounding section in the Internal Validity category. There were further slight modifications to the specific questions in order to tailor it to this review.

Table 3. Methodological Quality Assessment questions. MQA questions modified from Ainge et al. 2011 [13] and Downs and Black 1998 [26]. GTT, glucose tolerance test; GI, glycaemic index.

Modified Downs and Black Quality Index
Reporting
General
1. Were the hypotheses, aims or objectives of the study clearly described in the introduction? [a]   2. Were the main outcomes to be measured clearly described in the introduction or methods section?
Animal Characteristics

3. Was animal species/strain and sex specified? [a]
     4. Was the animal age at commencement of the study specified? [a]
     5. Have the animal weights at commencement of the study been specified or given graphically? [a]
     6. Have the animal starting numbers been specified? [a]
     7. Have the housing details been specified, including temperature, light cycle and group housing?
 Design and Outcomes
     8. Were the interventions of interest clearly described?
     9. Were the main findings of the study clearly described?
     10. Were estimates of the random variability in the data for the main outcomes clearly described, such as through standard deviation or standard error of the mean? [a]
     11. Have all important adverse events that may be a consequence of the intervention been reported?
     12. Have the actual probability values been reported for the main outcomes, except where probability value is less than 0.05? [a]
     13. Were all blood tests, GTTs performed without anaesthetic? [b]
     14. Were all GTTs performed after a maximum 5 h fast? [b]
     15. Were the diet matched both in terms of macronutrient composition and fibre content? [b]
     16. Was a difference in GI of the two diets shown, either quantitatively or qualitatively? [b]

Internal Validity
Bias

17. Was an attempt made to blind those measuring the main outcomes of the intervention?
     18. Were the statistical tests used to assess the main outcomes appropriate?
     19. Were the main outcomes measures used accurate (valid and reliable)?

Confounding
20. Was it stated in the text that the animals were randomised to intervention groups? [a]   21. Was there adequate adjustment for confounding in the analyses from which the main findings were drawn?
Power
22. Was the paper of sufficient power to detect a clinical important effect where the probability value for a difference being due to chance is less than 5%?

[a] Modified questions; [b] Additional questions.

For the quality assessment, if a specific point was not addressed in the article, such as blinding or the sample size for the power calculation, the article achieved a 0 for that question. The qualification of the final MQA question, concerning power, was determined using the software program G*Power v3.1.9.2 (Heinrich-Heine-Universität, Düsseldorf, Germany) [27]. A one tailed unpaired *t* test post-hoc analysis was performed with the effective size of a difference of 1.0 mM for plasma glucose, α, the probability of incorrectly rejecting the null hypothesis, set to 0.05, and the accepted power threshold of above 0.8. That is, an accepted threshold of β, the probability of incorrectly retaining the null hypothesis, of below 0.2.

3. Results

3.1. Article Selection

Twenty-two articles were identified from the initial search, with an additional eight from the manual search of reference lists. The systematic search procedure with the number of articles included and excluded at each point is depicted in Figure 1. The end result was 30 articles, with 15 [28–42] involving mice, 13 [43–55] involving rats and two [56,57] with experiments performed on both species. Of the articles that contained pertinent results, all contained information regarding the change or comparison of body weight or fat, and most examined some measure of glucose homeostasis. Each study also included a LGI diet as a comparison for the HGI diet; only two also included a chow diet as a control [28,46] and neither reported the in vivo postprandial glycaemic response to chow, which would have led to severely underpowered analyses; thus, only LGI control diets were analysed. No study used the standard rodent diets from the American Institute of Nutrition (AIN), AIN-93, as control diets. The average study quality was moderate with a range of 8–18 and a median of 14, and a range of 10–16 with a median of 12 for mice and rats, respectively.

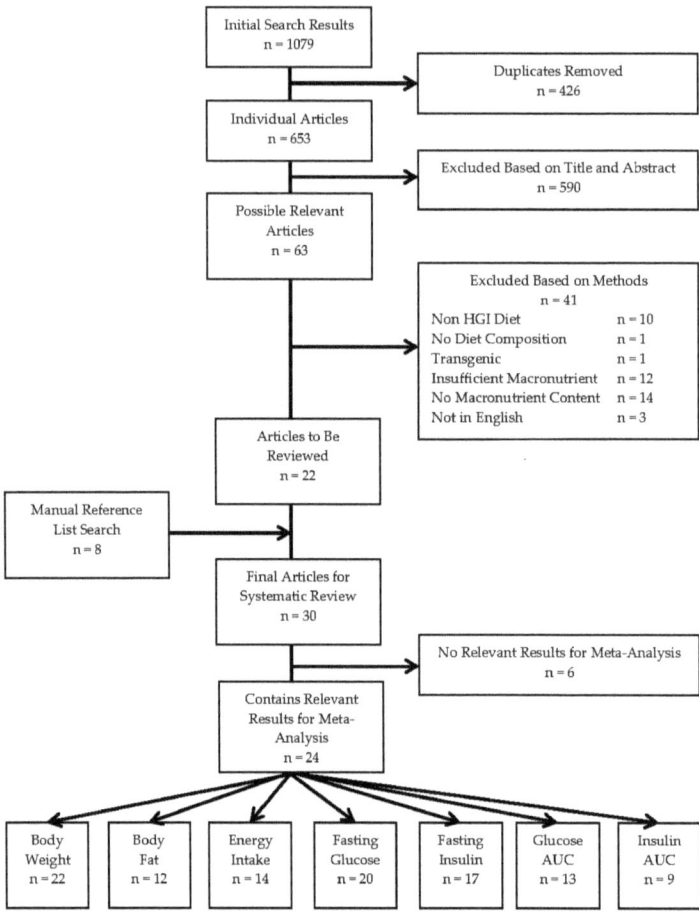

Figure 1. Flow of the study selection process for this systematic review. A flow diagram of article numbers included and excluded at each step of the systematic review process. Adapted from Ainge et al. 2011 [13]. *n*, number of articles; HGI, high glycaemic index; AUC, area under the curve.

3.2. Meta-Analysis

The meta-analyses of the extracted results, in terms of body weight, body adiposity and energy intake, and the results in relation to key measures of glucose homeostasis, are depicted in Figures 2 and 3, respectively. These results are shown dissected in terms of species, sex, length of diet, dietary fat content, starting age, fasting time and use of anaesthesia to illustrate some of the factors influencing results. Five of the seven traits showed an increased effect in male animals fed a HGI diet. Fasting blood glucose was only significantly increased by HGI feeding in male mice, and no significant effects were detected in energy intake. A funnel plot of the effect sizes for each trait is shown in Figure S1.

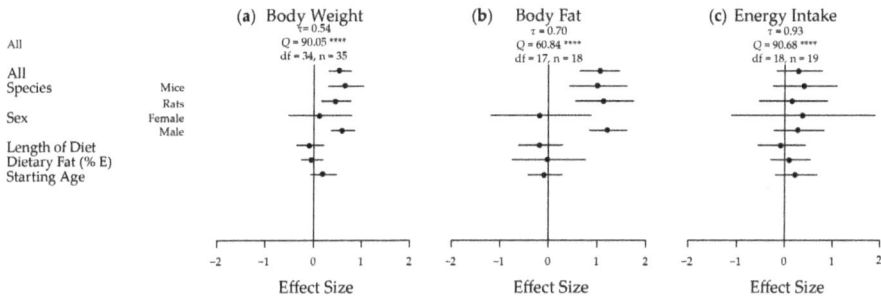

Figure 2. Meta-analysis of the extracted results for body weight, body fat and energy intake in high versus low glycaemic index fed animals. A plot of the calculated effect size and sampling variance for each of the three traits either as a whole or broken down by confounding factor: (**a**) Body Weight; All, Length of Diet and Dietary Fat n = 35, Mice n = 14, Rats n = 21, Female n = 4, Male n = 31, Starting Age n = 32; (**b**) Body Fat; All, Length of Diet and Dietary Fat n = 18, Mice n = 9, Rats n = 9, Female n = 2, Male n = 16, Starting Age n = 17; and (**c**) Energy Intake; All, Length of Diet and Dietary Fat n = 19, Mice n = 10, Rats n = 9, Female n = 2, Male n = 17, Starting Age n = 18. n, number of effect sizes; df, degrees of freedom; E, energy. **** $p < 0.0001$

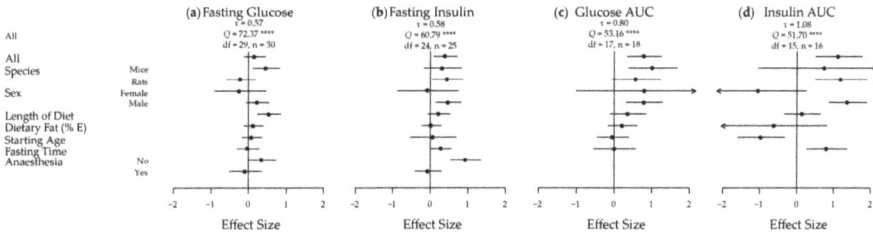

Figure 3. Meta-analysis of the extracted results for measures of glucose homeostasis in high versus low glycaemic index fed animals. A plot of the calculated effect size and sampling variance for each of the four traits either as a whole or broken down by confounding factor: (**a**) Fasting Glucose; All, Length of Diet, Dietary Fat and Fasting Time n = 30, Mice n = 16, Rats n = 14, Female n = 4, Male n = 26, Starting Age n = 27, No Anaesthesia n = 17, Anaesthesia n = 13; (**b**) Fasting Insulin; All, Length of Diet, Dietary Fat and Fasting Time n = 25, Mice n = 9, Rats n = 16, Female n = 3, Male n = 22, Starting Age, n = 23, No Anaesthesia n = 12, Anaesthesia n = 13; (**c**) Glucose AUC; All, Length of Diet, Dietary Fat and Fasting Time n = 18, Mice n = 9, Rats n = 9, Female n = 1, Male n = 17, Starting Age n = 17; and (**d**) Insulin AUC; All, Length of Diet, Dietary Fat and Fasting Time n = 16, Mice n = 2, Rats n = 14, Female n = 2, Male n = 14, Starting Age n = 14. AUC, area under the curve; n, number of effect sizes; df, degrees of freedom; E, energy. **** $p < 0.0001$

3.2.1. Body Weight

Overall, there was a moderate and statistically significant increase in body weight in HGI fed animals relative to LGI fed animals (d = 0.55; 95% CI: 0.31, 0.79) (Figure 2a). Significant heterogeneity was present across the studies (τ = 0.54, Q = 90.05, degrees of freedom (df) = 34, $p < 0.0001$) necessitating further analysis. When subset, this effect was significant and large in males (d = 0.61; 0.36, 0.86) but not in females (d = 0.13; −0.52, 0.47). The overall effect was also slightly greater in mice (d = 0.67; 0.30, 1.04) than in rats (d = 0.47; 0.16, 0.78). Starting age, length of diet and percentage of dietary fat as energy were not estimated to have any effect on the magnitude of the effect of diet on body weight. The effect on body weight was most prominent in ad libitum studies, and as most of the studies were indeed ad libitum, the results reflected this. For example, rats fed ad libitum HGI diets exhibited a much greater body weight than those fed a LGI diet in many of the studies [43,45,47,50,51,57], but, in contrast, rats fed energy-restricted HGI diets lost a greater proportion of body weight than those fed an energy-restricted LGI diet [50].

3.2.2. Body Fat Mass

Body fat mass also had a large and significant increase in HGI compared to LGI fed animals overall (d = 1.08; 0.67, 1.49) (Figure 2b), again there was significant heterogeneity amongst effects (τ = 0.70, Q = 60.84, df = 17, p < 0.0001). This effect was also large and significant in mice (d = 1.02; 0.43, 1.62), rats (d = 1.15; 0.56, 1.75) and males (d = 1.23, 0.84, 1.62). Again, there were no significant differences in females and no moderation of the effect was observed with continuous variables. In specific studies, a HGI diet induced significantly greater body fat mass [28–33,35–37,42] by as much as 93% [56] compared with mice on a LGI diet. Specifically, visceral and subcutaneous fat depots were heavier [30,31], and epididymal adipocytes were larger [56], in HGI fed mice.

3.2.3. Energy Intake

Energy intake was not significantly different in any subset (overall d = 0.31; −0.17, 0.78) (Figure 2c), and again no moderation was seen with the continuous variables. Significant heterogeneity existed amongst the effects (τ = 0.93, Q = 90.68, df = 18, p < 0.0001). Interestingly, a LGI diet increased energy output in faeces as measured by bomb calorimetry [29,32,42], compared to a HGI diet in the three papers that measured energy excretion to help inform energy balance.

3.2.4. Fasting Blood Glucose Levels

While fasting blood glucose levels were not significantly different overall (d = 0.17; −0.11, 0.45) (Figure 3a), significant heterogeneity was present (τ = 0.57, Q = 72.37, df = 29, p < 0.0001). Mice had a moderate significant high fasting blood glucose (d = 0.47; 0.12, 0.82) in response to HGI feeding, and a statistically significant moderately positive effect of length of time on the diet was also seen (d = 0.55; 0.25, 0.85). This suggests that the longer the animals were on the diet, the greater the difference in mean fasting blood glucose levels between the HGI and LGI fed animals.

3.2.5. Fasting Circulating Insulin Levels

Fasting insulin was moderately and significantly affected overall (d = 0.40; 0.09, 0.71) (Figure 3b), with significant heterogeneity (τ = 0.58, Q = 60.79, df = 24, p < 0.0001). Similar magnitude effects were seen in rats (d = 0.46; 0.04, 0.87) and males (d = 0.48; 0.14, 0.81). A large effect was seen in the subset of animals that were not anesthetized (d = 0.94; 0.54, 1.35). The difference in fasting insulin levels between diets also became greater with fasting time (d = 0.30; 0.03, 0.56).

3.2.6. Glucose and Insulin AUC

Glucose AUC in response to a GTT was significantly increased in HGI fed animals with a large difference seen overall (d = 0.80; 0.35, 1.25) (Figure 3c). Again, significant heterogeneity was present (τ = 0.80, Q = 53.17, df = 17, p < 0.0001). Large effects were observed in mice (d = 1.02; 0.38, 1.67) and in males (d = 0.80; 0.31, 1.29). Insulin AUC in response to a GTT was significantly and largely increased by a HGI diet overall (d = 1.14; 0.50, 1.77) (Figure 3d), in rats (d = 1.20; 0.50, 1.90) and in males (d = 1.39; 0.87, 1.90). Similar to fasting insulin, a large statistically significant effect was observed for fasting time on insulin AUC (d = 0.82; 0.27, 1.37), with significant heterogeneity (τ = 1.08, Q = 51.70, df = 15, $p < 0.0001$). A large negative significant effect was also seen for starting age on insulin AUC (d = −0.96; −1.60, −0.32). This implies that the older the animals are when they are placed on the experimental GI diets, the smaller the difference is between insulin AUC during a GTT between HGI and LGI diets, suggesting early exposure has a significantly more dramatic effect on insulin AUC.

3.2.7. Publication Bias

Trim and fill analysis detected no missing studies for four of the examined traits: body weight, energy intake, fasting glucose and glucose AUC. For both body fat and fasting insulin, it was calculated that there were four studies missing to the left side of the funnel (Figure S1b,e). After adjustment for these missing studies, the overall effects for body fat and fasting insulin where decreased by 0.25 to 0.84, and by 0.27 to 0.13, respectively. In the case of fasting insulin, this adjusted mean effect constitutes a small effect, as opposed to a moderate effect. Insulin AUC was estimated to have one missing study from the left of the funnel (Figure S1g), which reduced the effect size by 0.07 and led to an adjusted overall effect of 1.06, not altering the qualitative interpretation of the magnitude of the effect, similar to body fat.

3.3. MQA

The animal and diet characteristics for each of the studies are shown in Tables 4 and 5 for mice and rats, respectively, listed by year. The results for the MQA are shown in Tables 6 and 7 for mice and rats, respectively. The average score, out of 22, for each species was 13.3 and 12.2, respectively, to one decimal place. There was a small amount of variation in the quality of the articles, with the vast majority having a score between 11 and 16, inclusive [28–30,32,34–37,41–45,47,49–52,54–57], and a lowest score of 8 [38] and highest of 18 [31].

4. Discussion

This systematic review and meta-analysis provides evidence that male mice and rats fed HGI diets increase body weight, body adiposity, and parameters of glucose homeostasis compared to animals fed LGI diets. These findings highlight the importance of determining the effect of different carbohydrate subtypes on metabolic health, as while slowing carbohydrate bioavailability might be key, the LGI diets were also typically high in fibre such as RS which may exert mechanisms on metabolism independent of effects on postprandial glycaemia. Unfortunately, we were not able to analyse the effect of dietary fibre on metabolic outcomes across the studies in our meta-analysis as fibre content was typically not quantified in the original reporting articles, with only two paper specifying fibre content [34,57]. This lack of information was taken into consideration when determining the quality of the papers through question 15 of the MQA. This study, as a systematic review and meta-analysis, has examined the current state of rodent GI studies in an unbiased way, and has shown that, overall, better controlled studies are required in the future to properly elucidate the effect of GI alone on metabolism.

The current literature has been shown to be insufficient in regards to examining the specific effect of GI on metabolism. Only seven traits could be examined, and only seven moderators included in analyses due to the limit reported data. To examine any other effects or correlations, significant improvement in current reporting standards is required, as any other analyses conducted on the current literature would be severely underpowered. This reviews' calculated correlations are important new findings that can be used to plan future studies.

Table 4. Mouse animal and diet characteristics for intervention studies that compared LGI and HGI diets. The key information from each article using mice is shown. - denotes missing data. Approximate (~) starting weights are taken from graphs [a]. Carb, carbohydrates; HGI, high glycaemic index; LGI, low glycaemic index; E, energy; n, sample size; AP, amylopectin; AL, amylose; M, male; DS, digestible starch; RS, resistant starch; F, female.

Article	Carb in HGI Diet	Carb in LGI Diet	n	Starting Age	Sex	Strain	Starting Weight (g)	Length of Diet	Carb (% E)	Protein (% E)
Walker, 2002 [28]	Mazaca wax 100% AP	Hi-Maize 60% AL 40% AP	12	6 weeks	M	CBA/T6	24.6 ± 0.2	10 weeks	67	22
Pawlak, 2004 [56]	100% AP	60% AL 40% AP Hi-Maize	12	11 weeks	M	C57BL/6J	26.98 ± 0.37	9 weeks	69	20
Scribner, 2007 [29]	100% AP	60% AL 40% AP	9	5 weeks	M	129S2/SvPas	~17.5	25 weeks	68	19
So, 2007 [30]	DS Amioca 0% RS	RS Hi-Maize 60% RS	24 HGI or 16 LGI	3 weeks	M	C57BL/6	~17	8 weeks	69	20
Scribner, 2008 [31]	100% AP	60% AL 40% AP	16	5 weeks	M	129SvPas	~20	10 or 38 weeks	68	19
Zhou, 2008 [57]	Amioca 100% corn AP	Hi-Maize 260	-	Adult	M	C57BL/6J	-	19 days	64 [a]	19.3 [a]
Isken, 2009 [32]	100% AP	30% AP	10 or 8	16 or 44 weeks	M	C57BL/6	~23	20 or 26 weeks	65	23
Van Schothorst, 2009 [33]	100% AP	~60% AL 40% AP	50	9 weeks	M	C57BL/6JOlaHsol	25.68 ± 0.23 HGI or 25.63 ± 0.2 LGI	13-14 weeks	50	20
Anderson, 2010 [34]	Pregelatinized starch	Native starch	7	6-8 weeks	F	C57BL/6J	18.5 ± 0.2	22 weeks	68 HGI or 65 LGI	18 HGI or 22 LGI
Colbert Coate, 2010 [36]	Waxy maize 100% AP	high-AL-res 40% AP	6-8	6-8 weeks	M	C57BL/6	~22	16 weeks	65	20
Isken, 2010 [35]	100% AP	30% AP	10	16 weeks	M	C57BL/6J	~25	6 or 20 weeks	65	23
Van Schothorst, 2011 [37]	100% AP	~60% AL and ~40% AP	50	9 weeks	M	C57BL/6	25.7 ± 0.2 HGI or 25.6 ± 0.2 LGI	13-14 weeks	50	20
Uchiki, 2012 [38]	100% AP	30% AP 70% AL	-	16 months	-	C57BL/6	-	7.5 months	65	21
Weikel, 2012 [39]	100% AP (Amioca)	70% AL 30% AP (Hylon VII)	10	5 or 16 months	M	C57BL/6	-	26 or 46 weeks	65	21
Birarda, 2013 [40]	100% AP (Amioca)	70% AL 30% AP (Hylon VII)	-	5 months	M	C57BL/6	-	12 months	65	21
Rowan, 2014 [41]	100% AP (Amioca)	70% AL 30% AP (Hylon VII)	9-12	11 weeks	M	C57BL/6	~26	33 weeks	65	21
Kleckner, 2015 [42]	Amioca waxy 100% AP	H-Maize 260 60% AL	-	6-8 weeks	M	C57BL/6	~26	16 weeks	66.5	20.1

[a] Based on Ain-93G [58].

Table 5. Rat animal and diet characteristics for intervention studies that compared LGI and HGI diets. The key information from each article using rats is shown. - denotes missing data. Approximate (~) starting weights are taken from graphs. Articles using two different strains or sexes are across two rows, with the defining characteristics in bold. Only new information is entered in the second row. Carb, carbohydrates; HGI, high glycaemic index; LGI, low glycaemic index; E, energy; n, sample size; SHR, spontaneously hypertensive rats; AL, amylose; AP, amylopectin; SD, Sprague-Dawley; M, male; F, female.

Article	Carb in HGI Diet	Carb in LGI Diet	n	Starting Age	Sex	Strain	Starting Weight (g)	Length of Diet	Carb (% E)	Protein (% E)
Byrnes, 1995 [43]	Waxy cornstarch (0% AL)	Hi-Maize 60% AL	-	8 weeks	M	SD	250–300	8 weeks	60	22
				3–4 weeks	-	Wistar	50–90	4, 8 or 12 weeks		
Higgins, 1996 [44]	100% glucose or 100% AP (waxy corn)	Hi-Maize 60% AL	18	6 weeks	M	Wistar	200	8, 16 or 52 weeks	67	22
Lerer-Metzger, 1996 [45]	French toast (wheat starch)	Mung bean starch	14	6 weeks	M	SD	155	5 weeks	59	22 HGI or 24 LGI
Suga, 2000 [46]	60% glucose	60% fructose	-	-	F	SD	240–280	2 weeks	60	29
Widdup, 2000 [47]	Glucose	Hi-Maize 60% AL	-	6 weeks	M	Wistar	180 ± 2	6 weeks	65	22
Pawlak, 2001 [48]	Waxy corn 100% AP	Hi-Maize 60% AL	-	6–7 weeks	-	Wistar	215 ± 6	7 weeks	69	20
Pawlak, 2004 [56]	100% AP	60% AL Hi-Maize	12	6 weeks	M	SD	50–55	18 weeks	69	20
			7	7 weeks			-	3 weeks per diet		
Kopilas, 2007 [49]	50% Sucrose	50% LGI Starch	-	6 weeks	M	SHR	-	6 weeks	67.7 [b]	20.8 [b]
Zhou, 2008 [57]	Amioca 100% corn AP	Hi-Maize 260	50	-	M	SD	-	10 days	64 [a]	19.3 [a]
			10		F			32 days		
Aziz, 2009 [50]	100% AP	70% AL cornstarch or Cornstarch	10–12	Adult	M	SD	250	4 weeks	64 [a]	19.3 [a]
Belobrajdic, 2012 [51]	Low AL maize	High AL maize starch	8	9 weeks	M	SD	329 ± 5	4 weeks	64 [a]	19.3 [a]
Ble-Castillo, 2012 [52]	67% digestible cornstarch	67% native banana starch	30 total	7 weeks	M	Wistar	180–200	8 weeks	75.6	21.8
Stavrovskaya, 2013 [53]	High (65%) sucrose	Low (0%) sucrose	8	8 weeks	M	FBNF1	188 ± 2	8 weeks	68	21
Gugusheff, 2015 [54]	Dextrinised starch	Gel crisp starch	14	-	F	Wistar	200	70 weeks	63	21
Thompson, 2016 [55]	Amioca corn (high AP)	Resistant starch (high AL)	50	19 days	F	SD	~40	8 weeks	50	20

[a] Based on Ain-93G [58]; [b] Based on Ain-76A [58].

Table 6. Mouse Methodological Quality Assessment. Each article using mice was given a score of 0 or 1 for each question. MQA, methodological quality assessment.

Article	Reporting Questions															Internal Validity Questions							Total
Author, Year	1	2	3	4	5	6	7	8	9	10	11	12	13	14	15	16	17	18	19	20	21	22	
Walker, 2002 [28]	1	1	1	1	1	1	0	0	1	1	0	1	1	0	0	0	0	1	1	0	0	0	12
Pawlak, 2004 [56]	1	1	1	1	1	1	0	1	0	1	1	1	1	0	0	1	0	1	1	1	0	0	16
Scribner, 2007 [29]	0	1	1	1	1	1	0	1	1	1	0	1	1	1	0	0	0	1	1	1	1	0	15
So, 2007 [30]	1	1	1	1	1	1	1	1	1	1	0	1	1	0	0	0	1	1	1	0	0	1	16
Scribner, 2008 [31]	1	1	1	1	1	1	0	1	1	1	0	1	1	1	0	1	0	1	1	1	1	1	18
Zhou, 2008 [57]	0	1	1	0	1	0	1	1	0	1	0	1	1	1	0	0	0	1	1	0	0	0	11
Isken, 2009 [32]	1	1	1	1	1	1	1	1	1	1	0	1	0	0	0	1	0	1	1	0	0	0	14
Van Schothorst, 2009 [33]	1	1	1	1	1	1	1	1	1	1	0	1	1	0	0	1	0	1	1	0	0	1	17
Anderson, 2010 [34]	1	1	1	1	1	1	0	1	1	1	0	0	0	0	0	1	0	1	0	1	0	0	12
Colbert Coate, 2010 [36]	1	1	1	1	1	1	0	1	1	1	0	1	1	0	0	0	0	1	1	0	0	0	13
Isken, 2010 [35]	0	1	1	1	1	1	1	1	1	1	0	1	1	1	0	0	0	1	1	0	0	0	14
Van Schothorst, 2011 [37]	0	1	1	1	1	1	1	1	1	1	0	1	1	1	0	1	0	1	1	0	0	1	16
Uchiki, 2012 [38]	0	1	0	1	0	0	0	1	1	1	0	0	1	0	0	0	0	0	0	1	0	0	8
Weikel, 2012 [39]	0	1	1	1	0	1	0	1	1	1	0	0	1	0	0	0	0	1	0	0	0	0	9
Birarda, 2013 [40]	1	1	1	1	0	0	0	1	1	1	0	0	1	0	0	0	0	1	0	0	0	0	9
Rowan, 2014 [41]	1	1	1	1	0	0	1	0	0	1	0	1	1	0	0	0	0	1	1	0	0	0	11
Kleckner, 2015 [42]	1	1	1	1	1	0	1	1	1	1	0	1	1	0	0	1	0	1	1	0	1	0	15
Total	11	17	16	16	13	13	7	15	14	17	1	15	15	5	0	8	1	16	13	5	4	4	
17																							Mean
																							13.3

Table 7. Rat Methodological Quality Assessment; each article using rats was given a score of 0 or 1 for each question. MQA, methodological quality assessment.

Article	Reporting Questions																	Internal Validity Questions					Total
Author, Year	1	2	3	4	5	6	7	8	9	10	11	12	13	14	15	16	17	18	19	20	21	22	
Byrnes, 1995 [43]	1	1	1	1	1	0	0	1	1	1	0	1	1	0	0	1	0	1	0	0	0	0	12
Higgins, 1996 [44]	1	1	1	1	1	1	1	1	1	1	0	1	1	0	0	0	0	1	1	1	0	1	16
Lerer-Metzger, 1996 [45]	1	1	1	1	1	1	0	0	1	1	0	1	0	0	0	1	0	1	0	1	0	1	13
Suga, 2000 [46]	1	1	1	0	1	0	0	0	0	1	0	0	0	1	1	0	0	1	1	0	1	0	10
Widdup, 2000 [47]	1	1	1	1	1	0	1	1	1	1	0	1	1	0	0	0	0	1	1	0	0	0	13
Pawlak, 2001 [48]	1	1	0	1	1	0	0	1	0	1	0	1	1	0	0	0	0	1	0	1	0	0	10
Pawlak, 2004 [56]	1	1	1	1	0	0	0	1	1	0	1	1	1	0	0	0	0	0	1	1	1	0	12
Kopilas, 2007 [49]	1	1	1	1	0	0	1	1	1	1	0	1	0	0	0	0	0	1	0	0	1	0	11
Zhou, 2008 [57]	0	1	1	0	0	1	1	1	0	1	0	1	1	1	0	0	0	1	1	0	0	0	11
Aziz, 2009 [50]	1	1	1	0	1	1	1	1	1	1	0	1	1	0	0	1	0	1	0	0	0	0	13
Belobrajdic, 2012 [51]	1	1	1	1	1	1	1	1	1	1	0	1	1	0	0	0	0	1	1	1	0	0	14
Ble-Castillo, 2012 [52]	1	1	1	1	1	0	1	1	0	1	0	1	1	0	0	0	0	1	0	0	0	0	11
Stavrovskaya, 2013 [53]	0	0	1	1	1	1	0	1	1	0	1	0	0	0	0	0	0	1	1	0	0	0	10
Gugusheff, 2015 [54]	1	1	1	0	1	1	1	1	1	1	0	1	1	0	0	1	0	1	1	0	0	1	15
Thompson, 2016 [55]	1	1	1	1	1	1	1	0	0	1	0	1	0	0	0	1	0	0	0	1	0	1	12
Total																							Mean
15	13	14	14	11	12	8	9	12	10	14	1	14	9	2	1	5	0	12	8	7	3	4	12.2

Animal studies are key to dissecting the mechanisms underlying the health benefits of a LGI diet, but it seems isolating the effect of postprandial glycaemia while controlling for other physiological properties of LGI foods is not widespread. The glycaemic potential of most foods is indicated by starch digestibility and glucose absorption from the gut. A number of things could affect the GI including: type of starch, fibre viscosity, the food matrix, cooking, processing and macronutrient composition. The majority of articles in this review based their GI diets on the percentages of amylose or amylopectin in the starch. It has been reported that the commonly used low GI Hi-Maize amylose starch contains up to 60% RS [59]. RS reduces the available energy density of the diet and may have important effects on gut hormone secretion as it modifies the motility of the gastrointestinal tract and interacts with the gut microbiome. RS represents energy that is consumed but not available to the host, and would explain reports of increased energy content in faecal output in LGI fed animals [29,32,42]. Several articles also used RS and a control of a HGI starch. However, during the search for articles, we did not include a search term specific to RS as this review aimed to isolate articles that tested diets with differing GI, rather than RS content. This was regardless of the fact that many articles seemed to view these two concepts as synonymous. However, based on the premise for a systematic review, any articles that met the criteria were included in this review regardless of the RS content of their diets. It is important to note that most of the LGI diets contained RS at levels that could not be achieved in human diets, which may be a differentiator of human and animal studies.

There is no standard for GI testing in animals. As shown in the MQA, only eight mouse and five rat papers showed some difference in the GI of the diets, with most others merely assuming a difference in GI based on differing levels of RS as discussed above, or the GI of the carbohydrate sources as measured in humans. Some studies contained no basis for the high versus low GI diets, other than claiming them to be so. Of the studies that did discuss a proven difference in GI, most conducted a meal tolerance test that has similarities to the standard GI practice in humans, however they were key differences in methodology that showed no study accurately measured GI in vivo. In humans, ten subjects are tested in a cross-over design where they consume the tested meal over 12 min, have their blood glucose tested over the following 2 h, and the AUC compared to an identical test with glucose matched for the amount of carbohydrate in the meal by weight [60]. Three studies examined GI in vitro [34,45,54], which has been shown to be similar to GI measured in vivo in humans [6]. Of these three, one [34] used a reference food of white bread to calculate an approximate GI, another [45] also conducted a meal tolerance test, and the last [54] performed an additional in vivo GI test in rats, but did not state the methodology. Of the eleven papers that performed meal tolerance tests, two [31,50] performed the tests in a cross-over design as would be done in human. Three papers [33,37,50], with two of these from the same study, used a reference food, although one study used glucose in the diet frame as opposed to straight glucose and while the other did test straight glucose, it was compared to the carbohydrate source as opposed to the diet itself. Three provided the diets in terms of carbohydrate content [43,50,55], as is done in humans, as opposed to food weight. Only three papers used at least 10 animals for the test [33,37,45], and the time to eat ranged considerably with two as gavages [55], three eating for 15 min [33,37,45], and the remaining tests having the animals eat for 5 min [32,35,42,43]. Most studies did test the blood glucose over two hours [32,33,35,37,43,50,54,55]. While the meal tolerance tests were able to show a qualitative difference in glycaemic response, without following the standard methodology, particularly in terms of using a glucose reference, the GI could not be quantified for any diet. Regrettably no study examined the GI of standard diets, however, recently in our laboratory we successfully implemented GI testing in mice and found that both chow and high corn starch diets, similar to AIN-93, are actually relatively HGI [61].

Unfortunately, many articles lacked specific details relating to diet composition. We included six articles where the macronutrient content was stated, but not explicated to whether this percentage was of weight or energy [29–31,35,38,40]. Criteria for this review required carbohydrate contribution to energy greater than 50%, and, as protein, fat and carbohydrates have different energy densities, if the given content was of weight, some articles may not have actually met these criteria; accordingly, these articles were excluded. A further article [43] did not state the protein content at all, and it was only through being cited in another paper [44] that this information could be found and the original article [43] included. Future studies should report animal diet ingredients for confirmation that the diets meet all nutritional needs of the animals and enable reproducibility of experimental results.

Only one [46] of the HGI diets was paired with a control that was matched for both digestible macronutrient composition and fibre, including resistant starch, content, but this study had several limitations. Of all studies included in the meta-analysis, it was one of two studies that did not state the starting age of animals, one of only five using females, and was the shortest study overall, with rats exposed to the diet for only two weeks [46]. Of note, the rats from this paper were subjected to electrode induced lesions on the ventromedial nucleus of their hypothalamus, which would severely impact their metabolism [46]. Additionally, as this is a highly invasive surgery, even the sham rats would be significantly and adversely affected.

In the meta-regression, there were considerable differences between the sexes. Of the 30 articles included, five papers detailed experiments on female animals, one of which used mice [34] and four used rats [46,54,55,57]. This makes it difficult to be sure that there is really a sex difference in response to HGI feeding as the data suggest. Traditionally, female animals, as opposed to male, have been thought to present additional variation due to the continually changing hormonal environment through the oestrous cycle. One of the papers that used female rats also used male rats in a parallel experiment, however the sample size for female rats was five times smaller leading to insufficient power [57]. Ideally, equivalent studies should be performed, and results compared, in both sexes of mice and rats as it is becoming increasingly apparent that there are distinct metabolic differences between the sexes [62,63] necessitating the investigation of both sexes in order to gain full understanding of any metabolic factors. The dearth of female studies is most likely a crucial contributor to the lack of significant results for this subgroup for any of the traits.

Several articles did not use conventional animal strains. A "conventional" animal strain refers to those used in the majority of papers, such as C57BL/6 mice, and Albino Wistar or Sprague-Dawley rats, being the most well studied and hence understood. C57BL/6 mice are commonly used in metabolic studies, as they are susceptible to diet-induced obesity and hyperglycaemia. The animals studied within this systematic review that were not deemed "typical" include CBA/T6 [28] and 129S2/SvPas [33] mice, and spontaneously hypertensive [49], and partial pancreatomised [56] rats. The issue with the last is obvious, as any hormones the pancreas produces, not limited to insulin, would be decreased and therefore would alter the entire systemic environment. The experiments from that article [56] involving non-pancreatomised rats were still included in this meta-analysis, as were the experiments involving streptozotocin [45,57], gold thioglucose [28] or leptin [47] injections or surgery to induce lesions on the ventromedial nucleus of the hypothalamus [46] as a sensitivity test showed removing these data did not affect the analytical outcomes. A few studies [43,44,47,48] also involved inserting cannulas into the animals' jugular vein and/or carotid artery but as this is a relatively short surgery and the animals were given sufficient recovery time, the impact on metabolic results should be minimal. All of these animal strains and treatments have the potential to confound any and all results; regardless, results reported in these animals were generally consistent with those reported across most studies in this review. For this reason, as well as the limited number of total studies, it was decided not to investigate the effects broken down by the many different species during the meta-analysis.

The animal age at commencement of study and the length of study varied considerably. Most of the mice studies [32–42,56,57] utilized adult animals, ranging in age from 8 weeks to 16 months at the start of the study, while several rat studies used animals that were 6 weeks old or younger [43–45,47–49,55,56]. Two rat studies did not provide the starting age of animals, thus their developmental state is unknown [46,54]. One of these studies [54] is a study of the maternal effect of diet, so it could be assumed that the rats were mature, but the other [46] gave no age indication, and only lasted for two weeks, thus these rats may not be fully developed even at the completion of the study. Most of the mice studies were relatively long studies, consisting of interventions lasting at least 16 weeks [29,31,32,34–36,38–42] compared to the majority of rat studies lasting only 4–8 weeks, with only four lasting longer than eight weeks [43,44,54,56]. Due to the young age and short length of study, it is possible that the animals in some of these experiments [30,43,45–49,52,53,55,56] are still in an adolescence period and hence the results may not reflect the generalized effects of HGI or LGI diets. Indeed, the results of the meta-analysis show that early and longer exposure significantly increase the effect of HGI diets on insulin AUC and fasting glucose, respectively. This should be taken into account when planning a GI animal experiment.

The MQA question on power relied on sample size, hence the eight studies that did not provide sample size [38,40,42,43,46–49,57] received a potentially lower MQA score for this lack of information. Of all 30 articles, only eight had sufficient power [30,31,33,37,44,45,54,55] with a sample size of 14 or greater per group. Most of the studies had a sample size of 8–12, with the smallest sample size of six [36] and highest of 50 [33,37,55,57]. The insufficient replication in the papers with smaller sample sizes is indicative of less reliable results which require further investigation before a valid conclusion can be made and accepted. However, as the calculation for the effect size, and the associated estimate of sampling variance which was included in the analyses, was dependant on the sample size, our analyses accounted for differences among studies arising due to precision. Thus, all studies were included in the meta-analysis to avoid this bias.

Two questions we added to the MQA were regarding use of anaesthetic and fasting time. Most studies refrained from using anaesthesia, with only three studies using anaesthetic during non-terminal blood sampling [34,45,51]. No study used anaesthesia for GTTs. Anaesthesia affects brain metabolism and can affect serum concentrations of circulating proteins and metabolites [64], thus use of anaesthesia for metabolic studies is not desirable as it may impair the accurate measurement of postprandial changes in glucose and metabolic hormones such as insulin. The time frame used for acceptable fasting was 5 h. Excessive fasting, such as overnight for mice or greater than 18 h for rats, can artificially induce larger differences in the data [64]. Increased fasting time was shown to increase the effect of HGI on fasting insulin or insulin AUC to a small or large extent respectively. Fasting insulin was also affected by anaesthesia, with an absence of anaesthesia correlating with a large difference between HGI and LGI fed animals, and those subjected to anaesthesia showing no significant difference between the diets. Thus, use of anaesthesia may mask effects and fasting time may exaggerate effects on glucose metabolism and are important considerations in metabolic studies.

Most of the articles included measured some parameter of glucose homeostasis. The majority measured only blood glucose and insulin levels, with a few performing GTTs of some form. Only one paper conducted a euglycaemic clamp [48]; however, the findings are questionable given that the insulin levels reported were too high for animal survival, with a peak at 4200 pmol/L [48] compared with a typical peak being an average of 1200 pmol/L [65–67]. Additionally, the glucose metabolism calculations in this paper were not physiological [48]. All results in this paper therefore, must be seen as questionable.

The two articles by Van Schothorst et al. [33,37] contained one study, split over two papers. While the results were not duplicated as such, the continuation of a single study across the two articles is worth noting since it leads to a replication of similar findings included in this review from the two papers. The only relevant trait discussed in the latter paper [37], body weight, was already reported for the larger group of mice in the first article [33] and so was not included in the meta-analysis to avoid duplication.

5. Conclusions

This is the first systematic review and meta-analysis of the effects of GI on rodent metabolism. A HGI diet fed to male mice and rats increases body weight, adiposity, fasting insulin levels, and the AUCs for glucose and insulin levels during a GTT, to a greater extent than a LGI diet. There are too few studies in female animals to be confident of effects; future experiments should at least include females, if not specifically investigate the effects of GI diets in females given that the maternal nutritional environment is critical for the development of chronic diseases later in life. From these articles, it is difficult to conclude whether the beneficial health effects of a LGI diet are due to alterations in postprandial glycaemia per se, as dietary composition was considerably variable, particularly with respect to digestible macronutrient composition, fibre and resistant starch.

Supplementary Materials: The following are available online at www.mdpi.com/2072-6643/9/7/646/s1, Table S1: Raw data for meta-analysis, Text S1: R code for meta-analysis, Figure S1: Funnel plot of effect sizes.

Acknowledgments: The method and approach for this review were based on those of Ainge et al. 2011 [13] with assistance from the author Kieron Rooney. We thank Kieron Rooney, Jennie Brand-Miller and Damien Belobrajdic for providing critical reviews of this manuscript. This research received no specific grant from any funding agency, commercial or not-for-profit sectors.

Author Contributions: G.J.C. and K.S.B.-A. conceived, designed and performed the experiments; G.J.C. and A.M.S. analysed the data; and G.J.C. and K.S.B.-A. wrote the paper. All authors have read and approved the final manuscript.

Conflicts of Interest: The authors declare no conflict of interest.

Abbreviations

AIN	American Institute of Nutrition
AUC	area under curve
CI	confidence interval
df	degrees of freedom
DS	digestible starch
GI	glycaemic index
GTT	glucose tolerance test
HGI	high glycaemic index
LGI	low glycaemic index
MQA	methodological quality assessment
QI	quality index
RDS	rapidly digestible starch
REMA	random-effects meta-analysis
REMR	random-effects meta-regression
RQ	respiratory quotient
RS	resistant starch
SDS	slowly digestible starch

References

1. World Health Organization. *Overweight and Obesity*; WHO Press Global Health Observatory: Geneva, Switzerland, 2013.
2. Mann, J.; Cummings, J.H.; Englyst, H.N.; Key, T.; Liu, S.; Riccardi, G.; Summerbell, C.; Uauy, R.; van Dam, R.M.; Venn, B.; et al. FAO/WHO scientific update on carbohydrates in human nutrition: Conclusions. *Eur. J. Clin. Nutr.* **2007**, *61* (Suppl. S1), S132–S137. [CrossRef] [PubMed]
3. Trumbo, P.; Schlicker, S.; Yates, A.A.; Poos, M. Dietary reference intakes for energy, carbohydrate, fiber, fat, fatty acids, cholesterol, protein and amino acids. *J. Am. Diet. Assoc.* **2002**, *102*, 1621–1630. [CrossRef]
4. World Health Organization. *Guideline: Sugars Intake for Adults and Children*; WHO Press: Geneva, Switzerland, 2015.
5. Seal, C.J.; Nugent, A.P.; Tee, E.S.; Thielecke, F. Whole-grain dietary recommendations: The need for a unified global approach. *Br. J. Nutr.* **2016**, *115*, 2031–2038. [CrossRef] [PubMed]
6. Englyst, K.N.; Englyst, H.N.; Hudson, G.J.; Cole, T.J.; Cummings, J.H. Rapidly available glucose in foods: An in vitro measurement that reflects the glycemic response. *Am. J. Clin. Nutr.* **1999**, *69*, 448–454. [PubMed]
7. McCleary, B.V. Dietary fibre analysis. *Proc. Nutr. Soc.* **2003**, *62*, 3–9. [CrossRef] [PubMed]

8. Jenkins, D.J.A.; Kendall, C.W.C.; Augustin, L.S.A.; Franceschi, S.; Hamidi, M.; Marchie, A.; Jenkins, A.L.; Axelsen, M. Glycemic index: Overview of implications in health and disease. *Am. J. Clin. Nutr.* **2002**, *76*, 266S–273S. [PubMed]

9. Jenkins, D.J.A.; Wolever, T.M.S.; Taylor, R.H. Glycemic index of foods: A physiological basis for carbohydrate exchange. *Am. J. Clin. Nutr.* **1981**, *34*, 362–366. [PubMed]

10. Mackowiak, K.; Torlinska-Walkowiak, N.; Torlinska, B. Dietary fibre as an important constituent of the diet. *Postepy Higieny i Medycyny Doswiadczalnej* **2016**, *70*, 104–109. [CrossRef]

11. Brand-Miller, J.; Hayne, S.; Petocz, P.; Colagiuri, S. Low-glycemic index diets in the management of diabetes: A meta-analysis of randomized controlled trials. *Diabetes Care* **2003**, *26*, 2261–2267. [CrossRef] [PubMed]

12. Wolever, T.M.S.; Jenkins, D.J.A.; Vuksan, V.; Jenkins, A.L.; Buckley, G.C.; Wong, G.S.; Josse, R.G. Beneficial effect of a low glycaemic index diet in type 2 diabetes. *Diabet. Med.* **1992**, *9*, 451–458. [CrossRef] [PubMed]

13. Ainge, H.; Thompson, C.; Ozanne, S.E.; Rooney, K.B. A systematic review on animal models of maternal high fat feeding and offspring glycaemic control. *Int. J. Obes.* **2011**, *35*, 325–335. [CrossRef] [PubMed]

14. Higgins, J.; Green, S. *Cochrane Handbook for Systematic Reviews of Interventions*; Version 5.0.2; The Cochrane Collaboration: London, UK, 2009.

15. Goettsch, M. Comparative protein requirement of the rat and mouse for growth, reporduction and lactation using casein diets. *J. Nutr.* **1960**, *70*, 307–312.

16. Sheehan, P.M.; Clevidence, B.A.; Reynolds, L.K.; Thye, F.W.; Ritchey, S.J. Carcass nitrogen as a predictor of protein requirement for mature female rats. *J. Nutr.* **1981**, *111*, 1224–1230. [PubMed]

17. White, B.D.; Porter, M.H.; Martin, R.J. Effects of age on the feeding response to moderately low dietary protein in rats. *Physiol. Behav.* **2000**, *68*, 673–681. [CrossRef]

18. Nakagawa, S.; Poulin, R.; Mengersen, K.; Reinhold, K.; Engqvist, L.; Lagisz, M.; Senior, A.M.; O'Hara, R.B. Meta-analysis of variation: Ecological and evolutionary applications and beyond. *Methods Ecol. Evol.* **2015**, *6*, 143–152. [CrossRef]

19. R Core Team. *R: A Language and Environment for Statistical Computing*; R Foundation for Statistical Computing: Vienna, Austria, 2016.

20. Viechtbauer, W. Conducting meta-analyses in R with the metafor package. *J. Stat. Softw.* **2010**, *36*, 1–48. [CrossRef]

21. Besson, A.A.; Lagisz, M.; Senior, A.M.; Hector, K.L.; Nakagawa, S. Effect of maternal diet on offspring coping styles in rodents: A systematic review and meta-analysis. *Biol. Rev. Camb. Philos. Soc.* **2016**, *91*, 1065–1080. [CrossRef]

22. Nakagawa, S.; Noble, D.W.; Senior, A.M.; Lagisz, M. Meta-evaluation of meta-analysis: Ten appraisal questions for biologists. *BMC Biol.* **2017**, *15*, 18. [CrossRef] [PubMed]

23. Noble, D.W.; Lagisz, M.; O'Dea R, E.; Nakagawa, S. Nonindependence and sensitivity analyses in ecological and evolutionary meta-analyses. *Mol. Ecol.* **2017**, *26*, 2410–2425. [CrossRef] [PubMed]

24. Seehra, J.; Pandis, N.; Koletsi, D.; Fleming, P.S. Use of quality assessment tools in systematic reviews was varied and inconsistent. *J. Clin. Epidemiol.* **2015**, *69*, 179–184.e5. [CrossRef] [PubMed]

25. Nakagawa, S.; Santos, E.S.A. Methodological issues and advances in biological meta-analysis. *Evol. Ecol.* **2012**, *26*, 1253–1274. [CrossRef]

26. Downs, S.; Black, N. The feasibility of creating a checklist for the assessment of the methodological quality both of randomised and non-randomised studies of health care interventions. *J. Epidemiol. Community Health* **1998**, *52*, 377–384. [CrossRef] [PubMed]

27. Faul, F.; Erdfelder, E.; Buchner, A.; Lang, A.G. Statistical power analyses using G*power 3.1: Tests for correlation and regression analyses. *Behav. Res. Methods* **2009**, *41*, 1149–1160. [CrossRef] [PubMed]

28. Walker, C.G.; Bryson, J.M.; Phuyal, J.L.; Caterson, I.D. Dietary modulation of circulating leptin levels: Site-specific changes in fat deposition and ob mRNA expression. *Horm. Metab. Res.* **2002**, *34*, 176–181. [CrossRef] [PubMed]

29. Scribner, K.B.; Pawlak, D.B.; Ludwig, D.S. Hepatic steatosis and increased adiposity in mice consuming rapidly vs. slowly absorbed carbohydrate. *Obesity* **2007**, *15*, 2190–2199. [CrossRef]

30. So, P.W.; Yu, W.S.; Kuo, Y.T.; Wasserfall, C.; Goldstone, A.P.; Bell, J.D.; Frost, G. Impact of resistant starch on body fat patterning and central appetite regulation. *PLoS ONE* **2007**, *2*, e1309. [CrossRef] [PubMed]

31. Scribner, K.B.; Pawlak, D.B.; Aubin, C.M.; Majzoub, J.A.; Ludwig, D.S. Long-term effects of dietary glycemic index on adiposity, energy metabolism, and physical activity in mice. *Am. J. Physiol. Endocrinol. Metab.* **2008**, *295*, E1126–E1131. [CrossRef] [PubMed]

32. Isken, F.; Weickert, M.O.; Tschöp, M.H.; Nogueiras, R.; Möhlig, M.; Abdelrahman, A.; Klaus, S.; Thorens, B.; Pfeiffer, A.F. Metabolic effects of diets differing in glycaemic index depend on age and endogenous glucose-dependent insulinotrophic polypeptide in mice. *Diabetologia* **2009**, *52*, 2159–2168. [CrossRef]
33. Van Schothorst, E.M.; Bunschoten, A.; Schrauwen, P.; Mensink, R.P.; Keijer, J. Effects of a high-fat, low- versus high-glycemic index diet: Retardation of insulin resistance involves adipose tissue modulation. *FASEB J.* **2009**, *23*, 1092–1101. [CrossRef]
34. Andersson, U.; Rosén, L.; Wierup, N.; Östman, E.; Björck, I.; Holm, C. A low glycaemic diet improves oral glucose tolerance but has no effect on β-cell function in C57BL/6J mice. *Diabetes Obes. Metab.* **2010**, *12*, 976–982. [CrossRef] [PubMed]
35. Isken, F.; Klaus, S.; Petzke, K.J.; Loddenkemper, C.; Pfeiffer, A.F.H.; Weickert, M.O. Impairment of fat oxidation under high- vs. Low-glycemic index diet occurs before the development of an obese phenotype. *Am. J. Physiol. Endocrinol. Metab.* **2010**, *298*, E287–E295. [CrossRef] [PubMed]
36. Colbert Coate, K.; Huggins, K.W. Consumption of a high glycemic index diet increases abdominal adiposity but does not influence adipose tissue pro-oxidant and antioxidant gene expression in C57BL/6 mice. *Nutr. Res.* **2010**, *30*, 141–150. [CrossRef] [PubMed]
37. Van Schothorst, E.M.; Bunschoten, A.; Verlinde, E.; Schrauwen, P.; Keijer, J. Glycemic index differences of high-fat diets modulate primarily lipid metabolism in murine adipose tissue. *Physiol. Genom.* **2011**, *43*, 942–949. [CrossRef] [PubMed]
38. Uchiki, T.; Weikel, K.A.; Jiao, W.; Shang, F.; Caceres, A.; Pawlak, D.; Handa, J.T.; Brownlee, M.; Nagaraj, R.; Taylor, A. Glycation-altered proteolysis as a pathobiologic mechanism that links dietary glycemic index, aging, and age-related disease (in nondiabetics). *Aging Cell* **2012**, *11*, 1–13. [CrossRef] [PubMed]
39. Weikel, K.A.; FitzGerald, P.; Shang, F.; Andrea Caceres, M.; Bian, Q.; Handa, J.T.; Stitt, A.W.; Taylor, A. Natural history of age-related retinal lesions that precede amd in mice fed high or low glycemic index diets. *Investig. Ophthalmol. Vis. Sci.* **2012**, *53*, 622–632. [CrossRef] [PubMed]
40. Birarda, G.; Holman, E.A.; Fu, S.; Weikel, K.; Hu, P.; Blankenberg, F.G.; Holman, H.Y.; Taylor, A. Synchrotron infrared imaging of advanced glycation endproducts (AGEs) in cardiac tissue from mice fed high glycemic diets. *Biomed. Spectrosc. Imaging* **2013**, *2*, 301–315. [PubMed]
41. Rowan, S.; Weikel, K.; Chang, M.L.; Nagel, B.A.; Thinschmidt, J.S.; Carey, A.; Grant, M.B.; Fliesler, S.J.; Smith, D.; Taylor, A. Cfh genotype interacts with dietary glycemic index to modulate age-related macular degeneration-like features in mice. *Investig. Ophthalmol. Vis. Sci.* **2014**, *55*, 492–501. [CrossRef] [PubMed]
42. Kleckner, A.S.; Wong, S.; Corkey, B.E. The intra- or extracellular redox state was not affected by a high vs. Low glycemic response diet in mice. *PLoS ONE* **2015**, *10*, e0128380. [CrossRef] [PubMed]
43. Byrnes, S.E.; Miller, J.C.B.; Denyer, G.S. Amylopectin starch promotes the development of insulin resistance in rats. *J. Nutr.* **1995**, *125*, 1430–1437. [PubMed]
44. Higgins, J.A.; Brand Miller, J.C.; Denyer, G.S. Development of insulin resistance in the rat is dependent on the rate of glucose absorption from the diet. *J. Nutr.* **1996**, *126*, 596–602. [PubMed]
45. Lerer-Metzger, M.; Rizkalla, S.W.; Luo, J.; Champ, M.; Kabir, M.; Bruzzo, F.; Bornet, F.; Slama, G. Effects of long-term low-glycaemic index starchy food on plasma glucose and lipid concentrations and adipose tissue cellularity in normal and diabetic rats. *Br. J. Nutr.* **1996**, *75*, 723–732. [CrossRef] [PubMed]
46. Suga, A.; Hirano, T.; Kageyama, H.; Osaka, T.; Namba, Y.; Tsuji, M.; Miura, M.; Adachi, M.; Inoue, S. Effects of fructose and glucose on plasma leptin, insulin, and insulin resistance in lean and VMH-lesioned obese rats. *Am. J. Physiol. Endocrinol. Metab.* **2000**, *278*, E677–E683. [PubMed]
47. Widdup, G.; Bryson, J.M.; Pawlak, D.; Phuyal, J.L.; Denyer, G.S.; Caterson, I.D. In vivo and in vitro suppression by leptin of glucose-stimulated insulin hypersecretion in high glucose-fed rats. *Eur. J. Endocrinol.* **2000**, *143*, 431–437. [CrossRef] [PubMed]
48. Pawlak, D.B.; Bryson, J.M.; Denyer, G.S.; Brand-Miller, J.C. High glycemic index starch promotes hypersecretion of insulin and higher body fat in rats without affecting insulin sensitivity. *J. Nutr.* **2001**, *131*, 99–104. [PubMed]
49. Kopilas, M.A.; Dang, L.N.T.; Anderson, H.D.I. Effect of dietary chromium on resistance artery function and nitric oxide signaling in the sucrose-fed spontaneously hypertensive rat. *J. Vasc. Res.* **2007**, *44*, 110–118. [CrossRef] [PubMed]
50. Aziz, A.A.; Kenney, L.S.; Goulet, B.; Abdel-Aal, E.S. Dietary starch type affects body weight and glycemic control in freely fed but not energy-restricted obese rats. *J. Nutr.* **2009**, *139*, 1881–1889. [CrossRef] [PubMed]

51. Belobrajdic, D.P.; King, R.A.; Christophersen, C.T.; Bird, A.R. Dietary resistant starch dose-dependently reduces adiposity in obesity-prone and obesity-resistant male rats. *Nutr. Metab.* **2012**, *9*, 93. [CrossRef] [PubMed]

52. Ble-Castillo, J.L.; Aparicio-Trapala, M.A.; Juárez-Rojop, I.E.; Torres-Lopez, J.E.; Mendez, J.D.; Aguilar-Mariscal, H.; Olvera-Hernández, V.; Palma-Cordova, L.C.; Diaz-Zagoya, J.C. Differential effects of high-carbohydrate and high-fat diet composition on metabolic control and insulin resistance in normal rats. *Int. J. Environ. Res. Public Health* **2012**, *9*, 1663–1676. [CrossRef] [PubMed]

53. Stavrovskaya, I.G.; Bird, S.S.; Marur, V.R.; Sniatynski, M.J.; Baranov, S.V.; Greenberg, H.K.; Porter, C.L.; Kristal, B.S. Dietary macronutrients modulate the fatty acyl composition of rat liver mitochondrial cardiolipins. *J. Lipid Res.* **2013**, *54*, 2623–2635. [CrossRef] [PubMed]

54. Gugusheff, J.; Sim, P.; Kheng, A.; Gentili, S.; Al-Nussairawi, M.; Brand-Miller, J.; Muhlhausler, B. The effect of maternal and post-weaning low and high glycaemic index diets on glucose tolerance, fat deposition and hepatic function in rat offspring. *J. Dev. Orig. Health Dis.* **2015**, *7*, 320–329. [CrossRef] [PubMed]

55. Thompson, H.J.; Neuhouser, M.L.; Lampe, J.W.; McGinley, J.N.; Neil, E.S.; Schwartz, Y.; McTiernan, A. Effect of low or high glycemic load diets on experimentally induced mammary carcinogenesis in rats. *Mol. Nutr. Food Res.* **2016**, *60*, 1416–1426. [CrossRef] [PubMed]

56. Pawlak, D.B.; Kushner, J.A.; Ludwig, D.S. Effects of dietary glycaemic index on adiposity, glucose homoeostasis, and plasma lipids in animals. *Lancet* **2004**, *364*, 778–785. [CrossRef]

57. Zhou, J.; Martin, R.J.; Tulley, R.T.; Raggio, A.M.; McCutcheon, K.L.; Shen, L.; Danna, S.C.; Tripathy, S.; Hegsted, M.; Keenan, M.J. Dietary resistant starch upregulates total GLP-1 and PYY in a sustained day-long manner through fermentation in rodents. *Am. J. Physiol. Endocrinol. Metab.* **2008**, *295*, E1160–E1166. [CrossRef] [PubMed]

58. Reeves, P.G.; Nielsen, F.H.; Fahey, G.C., Jr. AIN-93 purified diets for laboratory rodents: Final report of the American institute of nutrition ad hoc writing committee on the reformulation of the AIN-76A rodent diet. *J. Nutr.* **1993**, *123*, 1939–1951. [PubMed]

59. Le Leu, R.K.; Hu, Y.; Brown, I.L.; Young, G.P. Effect of high amylose maize starches on colonic fermentation and apoptotic response to DNA-damage in the colon of rats. *Nutr. Metab.* **2009**, *6*, 11. [CrossRef] [PubMed]

60. Wolever, T.M.; Jenkins, D.J.; Jenkins, A.L.; Josse, R.G. The glycemic index: Methodology and clinical implications. *Am. J. Clin. Nutr.* **1991**, *54*, 846–854. [CrossRef] [PubMed]

61. Campbell, G.J.; Bell-Anderson, K.S. GI testing of standard and high sugar rodent diets in vivo in C57BL/6 mice. Unpublished work, **2017**.

62. Hedrington, M.S.; Davis, S.N. Sexual dimorphism in glucose and lipid metabolism during fasting, hypoglycemia, and exercise. *Front. Endocrinol.* **2015**, *6*, 61. [CrossRef] [PubMed]

63. Ter Horst, K.W.; Gilijamse, P.W.; de Weijer, B.A.; Kilicarslan, M.; Ackermans, M.T.; Nederveen, A.J.; Nieuwdorp, M.; Romijn, J.A.; Serlie, M.J. Sexual dimorphism in hepatic, adipose tissue, and peripheral tissue insulin sensitivity in obese humans. *Front. Endocrinol.* **2015**, *6*, 182. [CrossRef] [PubMed]

64. Passonneau, J.V.; Brunner, E.A.; Molstad, C.; Passonneau, R. The effects of altered endocrine states and of ether anaesthesia on mouse brain. *J. Neurochem.* **1971**, *18*, 2317–2328. [CrossRef] [PubMed]

65. Gauna, C.; Uitterlinden, P.; Kramer, P.; Kiewiet, R.M.; Janssen, J.A.; Delhanty, P.J.; van Aken, M.O.; Ghigo, E.; Hofland, L.J.; Themmen, A.P.; et al. Intravenous glucose administration in fasting rats has differential effects on acylated and unacylated ghrelin in the portal and systemic circulation: A comparison between portal and peripheral concentrations in anesthetized rats. *Endocrinology* **2007**, *148*, 5278–5287. [CrossRef] [PubMed]

66. Kassis, N.; Bernard, C.; Pusterla, A.; Casteilla, L.; Penicaud, L.; Richard, D.; Ricquier, D.; Ktorza, A. Correlation between pancreatic islet uncoupling protein-2 (UCP2) mRNA concentration and insulin status in rats. *Int. J. Exp. Diabetes Res.* **2000**, *1*, 185–193. [CrossRef] [PubMed]

67. McArthur, M.D.; You, D.; Klapstein, K.; Finegood, D.T. Glucose effectiveness is the major determinant of intravenous glucose tolerance in the rat. *Am. J. Physiol.* **1999**, *276*, E739–E746. [PubMed]

nutrients

Article

LXRα Regulates Hepatic ChREBPα Activity and Lipogenesis upon Glucose, but Not Fructose Feeding in Mice

Qiong Fan [1,*], Rikke C. Nørgaard [1], Christian Bindesbøll [2], Christin Lucas [1], Knut Tomas Dalen [1], Eshrat Babaie [3], Harri M. Itkonen [4], Jason Matthews [1,5], Hilde I. Nebb [1] and Line M. Grønning-Wang [1,*]

[1] Department of Nutrition, Institute of Basic Medical Sciences, University of Oslo, 0317 Oslo, Norway; r.c.norgaard@medisin.uio.no (R.C.N.); christin.lucas@medisin.uio.no (C.L.); k.t.dalen@medisin.uio.no (K.T.D.); jason.matthews@medisin.uio.no (J.M.); h.i.nebb@medisin.uio.no (H.I.N.)

[2] Department of Molecular Medicine, Institute of Basic Medical Sciences, University of Oslo, 0317 Oslo, Norway; christian.bindesboll@medisin.uio.no

[3] Centre for Molecular Medicine Norway, University of Oslo, 0318 Oslo, Norway; eshrat.babaie@ncmm.uio.no

[4] Prostate Cancer Research Group, Centre for Molecular Medicine (Norway), University of Oslo and Oslo University Hospitals, 0318 Oslo, Norway; Harri_Itkonen@hms.harvard.edu

[5] Department of Pharmacology and Toxicology, University of Toronto, Toronto, ON M5S1A8, Canada

* Correspondence: qiong.fan@medisin.uio.no (Q.F.); l.m.gronning-wang@medisin.uio.no (L.M.G.-W.); Tel.: +47-9047-3674 (Q.F.)

Received: 31 May 2017; Accepted: 26 June 2017; Published: 29 June 2017

Abstract: Liver X receptors (LXRα/β) and carbohydrate response element-binding proteins (ChREBPα/β) are key players in the transcriptional control of hepatic *de novo* lipogenesis. LXRα/β double knockout (LXRα$^{-/-}$/β$^{-/-}$) mice have reduced feeding-induced nuclear *O*-linked *N*-acetylglucosamine (*O*-GlcNAc) signaling, ChREBPα activity, and lipogenic gene expression in livers, suggesting important roles for LXRs in linking hepatic glucose utilization to lipid synthesis. However, the role of LXRs in fructose-induced ChREBP activation and lipogenesis is currently unknown. In this study, we studied the effects of high fructose or high glucose feeding on hepatic carbohydrate metabolism and lipogenic gene expression in livers from fasted (24 h) and fasted-refed (12 h) wild type and LXRα knockout (LXRα$^{-/-}$) mice. Hepatic lipogenic gene expression was reduced in glucose fed, but not fructose fed LXRα$^{-/-}$ mice. This was associated with lower expression of liver pyruvate-kinase (*L-pk*) and *Chrebpβ*, indicating reduced ChREBPα activity in glucose fed, but not fructose fed mice. Interestingly, ChREBP binding to the *L-pk* promoter was increased in fructose fed LXRα$^{-/-}$ mice, concomitant with increased glucose-6-phosphatase (*G6pc*) expression and *O*-GlcNAc modified LXRβ, suggesting a role for LXRβ in regulating ChREBPα activity upon fructose feeding. In conclusion, we propose that LXRα is an important regulator of hepatic lipogenesis and ChREBPα activity upon glucose, but not fructose feeding in mice.

Keywords: LXR; ChREBP; *de novo* lipogenesis (DNL); *O*-GlcNAc

1. Introduction

Diets rich in the simple sugars glucose and fructose stimulate hepatic *de novo* lipogenesis (DNL) and increase circulating triglycerides in humans and rodents [1–4]. Many of the enzymes involved in DNL and triglyceride synthesis are primarily regulated at the transcriptional level in a coordinate manner through multiple transcription factors in response to glucose and insulin [5]. Three transcription factors have been identified as particularly important for regulation of lipogenesis: the liver X receptors (LXRα; Nuclear Receptor Subfamily 1 Group H Member 3 (NR1H3) and LXRβ;

Nuclear Receptor Subfamily 1 Group H Member 2 (NR1H2)), sterol regulatory element-binding protein 1c (SREBP-1c), and carbohydrate response element-binding protein-α (ChREBPα) [5].

The LXRs are classically known as oxysterol-activated nuclear transcription factors and members of the nuclear receptor family. LXRs heterodimerize with retinoic X receptor (RXR; Nuclear Receptor Subfamily 2 Group B (NR2B)) family members to regulate the expression of genes involved in cholesterol homeostasis, lipogenesis, glucose metabolism, and inflammation [6]. LXRα is the predominantly expressed isoform in lipogenic tissues such as liver and adipose, whereas LXRβ is ubiquitously expressed [7]. In response to dietary cholesterol, glucose, and insulin, hepatic LXRs, in particular LXRα, activate transcription of the two other lipogenic transcription factors SREBP-1c and ChREBPα, which alone or in concert with LXRs induce expression of glycolytic and lipogenic enzymes in hepatic DNL, such as glucokinase (*Gk*), liver pyruvate kinase (*L-pk*), ATP citrate lyase (*Acl*), acetyl-CoA carboxylase (*Acc*), fatty acid synthase (*Fasn*), and stearoyl-CoA desaturase-1 (*Scd1*) [6,8]. In addition, LXRs improve glucose tolerance by negatively regulating hepatic glucose-6-phosphate (*G6pc*) expression and glucose production [9–11].

We and others have shown that LXRs and ChREBPα are post-translationally modified by *O*-linked *N*-acetylglucosamine (*O*-GlcNAc) in response to high glucose by *O*-GlcNAc transferase (OGT), which increases their lipogenic potential [12,13]. OGT uses UDP-*N*-acetylglucosamine (UDP-GlcNAc), the high energy product of the hexosamine biosynthetic pathway (HBP), as a substrate for reversible *O*-linked GlcNAcylation of nuclear, cytoplasmic, and mitochondrial target proteins affecting transcription, metabolism, apoptosis, organelle biogenesis, and transport [14]. The HBP is a branch of the glycolytic pathway that couples nutrient sensing to cellular metabolism and signaling via activation of OGT [15]. *O*-GlcNAcylation regulates transcription through OGT-associated chromatin-modifying complexes [16]. In this way, glucose not only serves as an energy source and substrate for lipogenesis, but also acts as a signaling molecule in the regulation of glycolytic and lipogenic gene expression [17]. In line with this notion, *O*-GlcNAc signaling is associated with increased ChREBPα activity, enhanced glycogenic and lipogenic gene expression, and hepatic steatosis [13]. We recently reported that insulin-independent glucose-*O*-GlcNAc signaling potentiates LXR-mediated transactivation of the SREBP-1c and ChREBPα promoters, linking glucose metabolism to LXR activation and lipogenesis [12,18]. Furthermore, we reported that LXRs are important for nuclear *O*-GlcNAc signaling, ChREBPα *O*-GlcNAcylation and *L-pk* promoter binding activity, and glycogenic and lipogenic gene expression, including expression of the newly discovered *Chrebpβ* isoform in mouse livers. Collectively, these data suggest that LXRs connect hepatic glucose utilization to lipogenesis via regulation of nuclear OGT and ChREBPα activity [18,19]. However, the specific roles of LXRα and LXRβ in this process are currently unknown.

ChREBPβ is derived from an alternative promoter within exon 1b of the ChREBP gene, resulting in a shorter constitutively nuclear protein lacking most of the low glucose inhibitory domain (LID) in *N*-terminus [20]. *Chrebpβ* expression is low during prolonged fasting and strongly induced following high carbohydrate refeeding in mice [21]. ChREBPα mediates this response in a tissue-specific manner via transactivation of carbohydrate response elements (ChoREs) upstream of and in exon 1b [21]. Interestingly, ChREBPβ conferred a higher transcriptional activity than ChREBPα under both low and high glucose conditions and appears to be the major regulator of lipogenesis in response to dietary carbohydrates [20,22]. Recently, a role for ChREBP and in particular ChREBPβ, in fructose-induced *de novo* lipogenesis was suggested [21,23]. ChREBP null mice are intolerant to high fructose diet, in part by blunted gene expression of fructose-metabolizing enzyme genes, suggesting also a crucial role for ChREBP in fructose metabolism [24]. Interestingly, a recent study showed that ChREBP induces hepatic *G6pc* expression and glucose production by short-term fructose feeding in mice [25], suggesting a role for ChREBP in contributing to selective hepatic insulin resistance.

The objectives of the present study were to investigate the LXRα dependent effect of dietary fructose and glucose on hepatic ChREBPα activity, glycogenic and lipogenic gene expression, intermediate carbohydrate metabolism, and *O*-GlcNAc levels by using wild type and LXRα deficient mice. The results from this study indicate that LXRα is important in hepatic DNL and ChREBPα activity upon glucose, but not fructose feeding in mice.

2. Materials and Methods

2.1. Materials

Formaldehyde (F1635) and UDP-GlcNAc (U4375) were purchased from Sigma Aldrich (St. Louis, MO, USA). UltraPure™ Phenol: Chloroform:Isoamyl Alcohol (15593-031) was from Invitrogen Aldrich (Thermo Fisher Scientific, Waltham, MA, USA). All other chemicals were of the highest quality available from commercial vendors.

2.2. Animals and Treatment

Wildtype (LXRα$^{+/+}$) and LXRα deficient (LXRα$^{-/-}$) mice were housed in a temperature-controlled (22 °C) facility with a strict 12 h light/dark cycle. Mice had free access to standard chow diet (SDS diets, RM3, #801190, consisting of 12% calories from fat, 27% from protein, and 61% from carbohydrate) and water at all times prior to experiments. The generation of the LXRα$^{-/-}$ mice has been described previously [26]. The LXRα$^{+/+}$ and LXRα$^{-/-}$ mice used were of mixed genetic background (129/Sv/C57BL/6) backcrossed into the C57BL/6N strain for six generations. Twelve-week-old male mice (*n* = 5) were fasted for 24 h or fasted for 24 h and subsequently refed for 12 h on an isocaloric diet (3.99 kcal/g) containing 60.8% calories from fructose (5BN7) or glucose (5BN8) (TestDiet), 22.6% fat, and 16.7% protein. The mice were sacrificed in a mixed order between fasted and refed groups by cervical dislocation at 7–9 a.m., and tissues were weighed and snap frozen in liquid nitrogen and stored at −80 °C until further analysis. All use of animals was registered and approved by the local veterinary and the Norwegian Animal Research authority (FOTS #5457 and #6378).

2.3. Blood Chemistries

Plasma was separated from blood by centrifugation. Plasma insulin was measured using the Ultrasensitive Insulin Kit from Mercodia (Mercodia AB, Uppsala, Sweden) according to the manufacturer's instructions. Plasma triglycerides (TGs) were determined with a Triglycerides Enzymatic PAP 150 kit (TGPAP 150; BioMérieux, Marcy-l'Étoile, France).

2.4. Metabolomics

Liver tissues (*n* = 5 mice for each group) were sent to Metabolon, Inc. (Research Triangle Park, Durham, NC, USA) for metabolomics analysis as described [27]. The metabolomics data is included in Supplementary Table S1.

2.5. RNA Extraction, cDNA Synthesis and Real-Time Quantitative PCR (RT-qPCR)

RNA was isolated by phenol chloroform extraction followed by high salt precipitation (0.8 M sodium acetate, 1.5 M NaCl) to avoid contaminating polysaccharides to co-precipitate with RNA. Extracted RNA was further purified using RNeasy spin columns (#74104; QIAGEN, Hilden, Germany). Isolated RNA (500 ng) was reverse transcribed into cDNA using SuperScript III Reverse Transcriptase (Invitrogen, Thermo Fisher Scientific, Waltham, MA, USA) and random hexamer primers. qPCR was performed with 1 μL of the cDNA synthesis reaction using Kapa SYBR FAST qPCR Master Mix (KapaBiosystems, Roche, Basel, Switzerland) on a Bio-Rad CFX96 Touch™ Real-Time PCR Detection System. Gene expression was normalized against the expression of TATA-binding protein (*Tbp*). Assay primers were designed with Primer-BLAST software (NCBI, Bethesda, MD, USA) [28]. Sequences are listed in Supplementary Table S1.

2.6. Liver Extracts and Immunoblot Analysis

Nuclear and cytoplasmic proteins were prepared using NE-PER^{TM} Nuclear and Cytoplasmic Extraction kit (Pierce Biotechnology, Thermo Fisher Scientific, Waltham, MA, USA), with the following inhibitors added to the buffers: 1 mM NaF, 1 mM Na$_3$VO$_4$, 1 mM β-glycerophosphate, 1 μM *O*-GlcNAcase inhibitor GlcNAc-thiazoline, and Complete^{TM} protease inhibitors (Roche Applied Science, Penzberg, Germany). Proteins were separated by Sodium Dodecyl Sulfate Polyacrylamide Gel Electrophoresis (SDS-PAGE) (Bio-Rad, Hercules, CA, USA) and blotted onto Polyvinylidene fluoride (PVDF) membrane (Merck Millipore, Billerica, MA, USA). Primary antibodies used were rabbit anti-mouse LXR (1:500) [19], ChREBP (1:1000; #NB400-135; Novus Biologicals, Littleton, CO, USA), FASN (1:500; #sc-55580; Santa Cruz Biotechnology, Dallas, TX, USA), SCD1 (1:2000; #sc-14719; Santa Cruz Biotechnology, Dallas, TX, USA), α-Tubulin (1:20,000; #T5168; Sigma-Aldrich, St. Louis, MO, USA), Lamin A (1:1000; #L1293; Sigma-Aldrich, St. Louis, MO, USA), OGT (1:1000; #AL25) [29], RL2 (1:1000, #MA1-072, Invitrogen, Thermo Fisher Scientific, Waltham, MA, USA), SREBP-1 (1:1000) [30], and L-PK (1:2000; #MABS148; Merck Millipore, Billerica, MA, USA). Secondary horseradish peroxidase-conjugated anti-mouse (#115-035-174) and anti-rabbit (#211-032-171; Jackson ImmunoResearch Laboratories, West Grove, PA, USA); anti-goat (#605-4302; Rockland, Limerick, PA, USA) antibodies were used at 1:10,000 dilutions. Anti-mouse IgM (#A8786; Sigma-Aldrich, St. Louis, MO, USA) was used at 1:5000 dilutions. Blots were quantified from five mice for each experimental group, using the ImageJ software (NIH, Bethesda, Maryland). All lanes were normalized to loading controls as indicated in figure text (α-tubulin or Lamin A).

2.7. Chromatin Immunoprecipitation (ChIP)

ChIP experiments were performed as described previously with modest changes [19]. Briefly, liver tissue was homogenized and crosslinked with 1% formaldehyde/Phosphate-buffered saline (PBS) for 10 min at room temperature. Crosslinking was stopped by 3 min incubation with 125 mM glycine. Samples were washed twice in cold PBS and resuspended in lysis buffer (0.1% SDS, 1% Triton X-100, 0.15 M NaCl, 1 mM Ethylenediaminetetraacetic acid (EDTA), and 20 mM Tris (pH 8.0). Lysed tissue was sonicated to an average size of 200–500 bp fragments using a bioruptor (Diagenode, Seraing, Belgium). Chromatin was immunoprecipitated with 2 μg antibody against ChREBP (NB400-135; Novus Biologicals, Littleton, CO, USA) or rabbit IgG (011-000-002; Jackson ImmunoResearch Laboratory, West Grove, PA, USA) over night at 4 °C. Dynabeads Protein A (Invitrogen) were washed four times in lysis buffer, and then added to the chromatin and rotated at 4 °C for 2 h. Dynabeads were then washed three times with wash buffer 1 (0.1% SDS, 1% Triton X-100, 0.15 M NaCl, 1 mM EDTA, 20 mM Tris (pH 8)); followed by washing once in wash buffer 2 (0.1% SDS, 1% Triton X-100, 0.5 M NaCl, 1 mM EDTA, 20 mM Tris (pH 8)), and then once in wash buffer 3 (0.25 M LiCl, 1% NaDOC, 1% NP-40, 1 mM EDTA, 20 mM Tris (pH 8)), and then once in wash buffer 1. All washing steps were done for five minutes at room temperature. DNA-protein complexes were eluted with 1% SDS and reverse cross-linked overnight at 65 °C. DNA was purified by using the QIAquick PCR Purification Kit (#28104; QIAGEN, Hilden, Germany). DNA enrichment was quantified by quantitative RT-PCR. The binding of ChREBP to the carbohydrate response element (ChoRE) in the promoter-proximal enhancer of *L-pk* has been described previously [13] and the ChIP primers used were as follows: *L-pk* ChoRE (5′-GTCCCACACTTTGGAAGCAT, 5′-CCCAACACTGATTCTACCC). The negative control primers located 2.2 kp downstream from the ChoRE were as follows: 5′-TGCAACTGGGGAACTAGCCA, 5′-AGCTTTGTGTGATGGCTGAAG.

2.8. Wheat Germ Agglutinin (WGA) Pulldown

Nuclear extracts (100 μg) were incubated with protein A/G-agarose beads (sc-2003; Santa Cruz Biotechnology) for 1 h at 4 °C. Cleared extracts were transferred to new tubes and incubated with 30 μL of WGA-agarose (Vector Lab, Burlingame, CA, USA) overnight at 4 °C. After four washes (PBS,

0.2% NP-40), proteins were eluted from the beads in 2× Laemmli buffer and separated by SDS-PAGE. The captured proteins were analyzed by immunoblotting.

2.9. UDP-GlcNAc Measurements

UDP-GlcNAc in liver tissue was extracted and analyzed using described methods [31] with minor changes. Liver tissue (20 µg) was homogenized in liquid nitrogen-chilled CryoGrinder (OPS Diagnostics, Lebanon, NJ, USA), and resuspended in 0.15 mL cold PBS. Ice-cold ethanol (0.45 mL) was added to the samples, followed by sonication in Diagenode Bioruptor to lyse the cells. The lysate was centrifuged at $16,000 \times g$ for 10 min at 4 °C, and a third of the supernatant was used for determination of protein concentration using the bicinchoninic acid (BCA) assay kit (Pierce, Thermo Fisher Scientific, Waltham, MA, USA), and the rest was vacuum dried in a Savant DNA100 SpeedVac to measure levels of nucleotide sugars with ion-pair reversed phase High-performance liquid chromatography (HPLC). The preparatory columns for HPLC were done as previously reported [32], while the run-time in HPLC was extended up to 2 h. A standard UDP-GlcNAc (100 µM) was spiked into samples to verify the accurate peak.

2.10. Statistical Analysis

Statistical analyses were performed using GraphPad Prims 6 (GraphPad Software Inc., San Diego, CA, USA). All data were presented as means and standard error of the mean (SEM), and error bars for all results were derived from biological replicates rather than technical replicates. Statistical differences between groups were determined by two-way analysis of variance (ANOVA) followed by Tukey's multiple comparison tests. For all statistical tests $p < 0.05$ was considered statistically significant. For metabolomics data, statistical differences between groups were determined by repeated measures two-way ANOVA.

3. Results

3.1. Regulation of Genes Involved in Carbohydrate Metabolism by Dietary Fructose and Glucose-Role of LXRα

To examine LXRα-dependent regulation of genes encoding for glucose and fructose metabolizing enzymes in response to fructose and glucose feeding, wild type and LXRα$^{-/-}$ mice were divided into three groups: fasted (24 h), fasted-refed with 60% fructose, or fasted-refed with 60% glucose for 12 h. No statistically significant differences in food intake or body weight were found among the fructose and glucose fed wild type and LXRα$^{-/-}$ mice (Table 1). A schematic representation of the enzymes involved in carbohydrate metabolism (glycolysis, fructolysis, and gluconeogenesis) is presented in Figure 1A. The majority of the analyzed genes were similarly regulated in wild type and LXRα$^{-/-}$ mice. Both diets upregulated glycolytic *Gk* mRNA to a similar degree, and fructose feeding upregulated the ChREBP target gene *L-pk* mRNA more strongly than glucose feeding, which is in agreement with previous observations [33,34] (Figure 1B). Interestingly, glucose-, but not fructose-mediated induction of *L-pk* was significantly attenuated in LXRα$^{-/-}$ mice. Figure 1C shows genes in the fructolysis pathway, fructokinase (*Fk*) and aldolase B (*AldoB*). Only fructose feeding upregulated Fk and AldoB, as previously reported [33]. We observed increased expression of gluconeogenic genes *G6pc* and glucose transporter 2 (*Glut2*) with fructose feeding in the LXRα$^{-/-}$ mice (Figure 1B,D). These observations were concomitant with an approximately 2-fold increase in plasma insulin levels in fructose fed LXRα$^{-/-}$ mice compared to wild type mice on the same diet (Table 1). These data suggest increased glucose production and glucose output in fructose fed mice lacking LXRα.

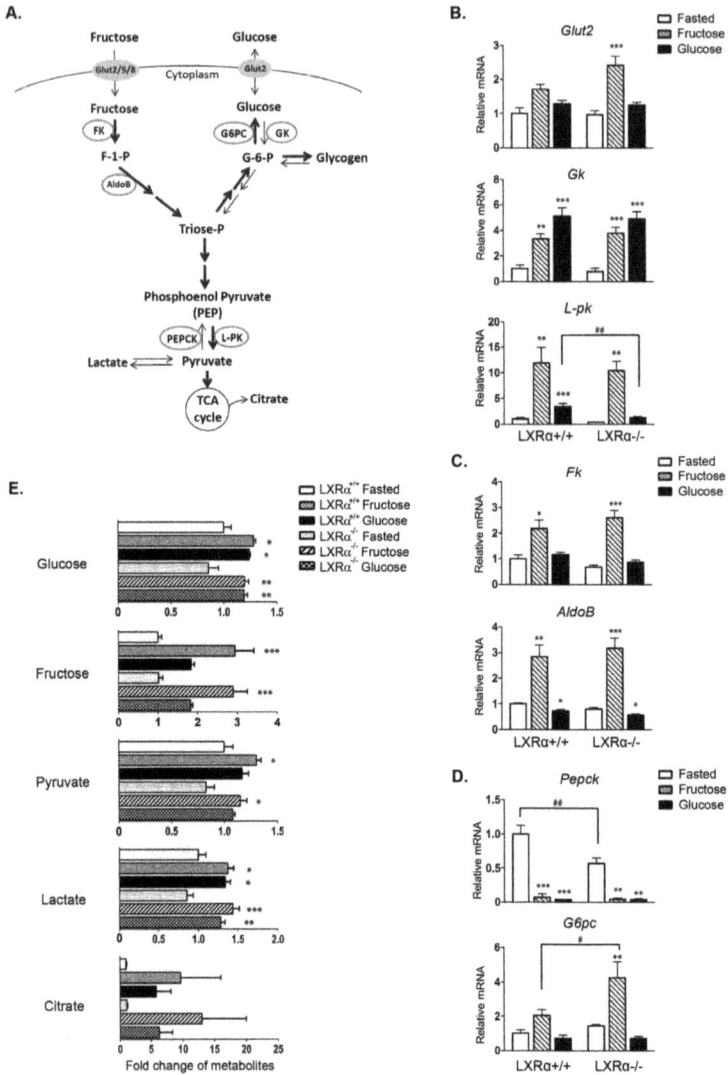

Figure 1. Hepatic liver pyruvate-kinase (*L-pk*) expression by dietary glucose is reduced in LXRα knockout (LXRα$^{-/-}$) mice. The male mice of inbred strain C57 Black 6 (C57BL/6) were fasted 24 h (white bars) or fasted-refed for 12 h on a 60% fructose diet (gray bars) or 60% glucose diet (black bars) as described in Materials and Methods. (**A**) Simplified schematic overview of the genes involved in glycolysis, fructolysis, and gluconeogenesis. F-1-P: fructose-1-phosphate; G-1-P: glucose-1-phosphate; Triose-P: triose phosphate. TCA: tricarboxylic acid; (**B**) Hepatic gene expression of the glycolytic genes glucose transporter 2 (*Glut2*), glucokinase (*Gk*) and *L-pk*; (**C**) Hepatic gene expression of the fructolytic genes fructokinase (*Fk*) and aldolase B (*AldoB*); (**D**) Hepatic gene expression of the gluconeogenic genes phosphoenolpyruvate carboxykinase (*Pepck*) and glucose-6-phosphatase (*G6pc*). Expression of above genes (B-D) were analyzed by real-time quantitative PCR (RT-qPCR) and normalized to TATA-binding protein (*Tbp*); (**E**) Relative metabolite expression levels of hepatic glucose, fructose, pyruvate, lactate, and citrate in response to fructose and glucose feeding in wild type and LXRα$^{-/-}$ mice. Data represent the mean ± standard error of the mean (SEM) (*n* = 5). Significant differences were found using two-way analysis of variance (ANOVA) followed by Tukey's multiple comparison test (fasted vs. fructose fed and fasted vs. glucose fed were analyzed separately). * $p < 0.05$, ** $p < 0.01$, *** $p < 0.001$ compared to fasted. # $p < 0.05$, ## $p < 0.01$, compared to LXRα$^{+/+}$ 106 mice.

Table 1. Body weight, food intake, and plasma insulin of fasted and refed mice.

Genotype	Treatment	Body Weight (g)	Food Intake (g)	Plasma Insulin (µg/dL)
LXRα$^{+/+}$	Fasted	30.8 ± 2.0		0.7 ± 0.1
	Refed-Fructose	28.4 ± 1.7	2.4 ± 0.4	2.0 ± 0.7
	Refed-Glucose	29.6 ± 2.3	3.3 ± 0.2	7.0 ± 1.0 ***
LXRα$^{-/-}$	Fasted	32.7 ± 1.5		0.7 ± 0.1
	Refed-Fructose	31.1 ± 2.9	2.8 ± 0.2	5.0 ± 1.5 *
	Refed-Glucose	32.1 ± 3.5	2.7 ± 0.4	7.1 ± 0.8 ***

Wild type (LXRα$^{+/+}$) and LXRα knockout (LXRα$^{-/-}$) mice were treated as explained in Materials and Methods. Values are mean ± standard error of the mean (SEM) (n = 5). Significant differences were calculated by two-way analysis of variance (ANOVA) followed by Tukey's multiple comparisons test. * $p < 0.05$, *** $p < 0.001$ compared to fasted within same genotype.

Metabolite analysis showed significantly increased hepatic levels of glucose and also increased hepatic fructose levels in high glucose fed mice (Figure 1E), suggesting increased activity of aldose reductase upon high glucose consumption, as previously reported [35]. Hepatic glucose levels were also higher in fructose fed mice, in agreement with increased glycogen synthesis and turnover [36,37]. We observed elevated hepatic levels of glycogenic metabolites upon both fructose and glucose feeding, likely derived from the continuous production and turnover of glycogen (Supplementary Table S2). Hepatic pyruvate, lactate, and citrate levels were increased in wild type and LXRα$^{-/-}$ mice upon fructose and glucose feeding (Figure 1E), which correspond with increased carbohydrate oxidation after digestion of these carbohydrate rich diets.

3.2. Expression of Hepatic de novo Lipogenic Genes Mediated by Dietary Glucose Is Reduced in LXRα$^{-/-}$ Compared to Wild Type Mice

We next addressed the gene expression of central lipogenic enzymes in fasted and refed mice. A schematic representation of DNL and the enzymes involved is presented in Figure 2A. All lipogenic genes assessed, except for *Scd1*, were more strongly upregulated by fructose compared to glucose feeding (Figure 2B), in agreement with previous reports [21,38,39]. Glucose-induced lipogenic gene expression was reduced in LXRα$^{-/-}$ mice (Figure 2B). In line with previous observations in our laboratory [19], *Scd1* expression was not upregulated after high sugar feeding. SCD1 mRNA and protein expression were almost completely abolished in LXRα$^{-/-}$ fasted mice (Figure 2B,C), consistent with a previous study reporting that LXR directly regulate SCD1 expression [40]. Although lipogenic gene expression was significantly induced by both diets in wild type mice, plasma triglyceride levels were not significantly elevated after 12 h refeeding (Figure 2D).

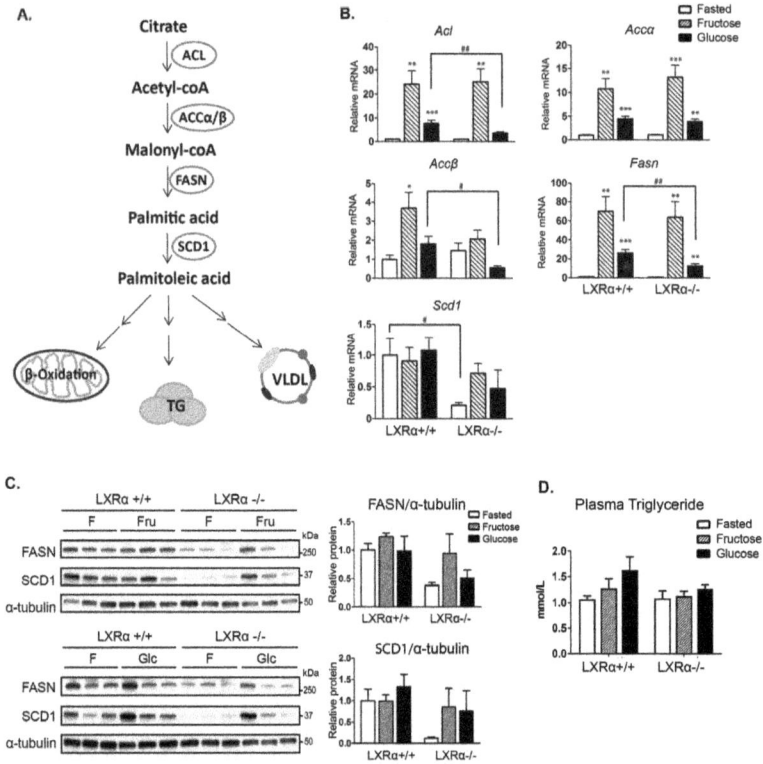

Figure 2. Hepatic *de novo* lipogenic gene expression is dependent on LXRα in response to dietary glucose. (**A**) Simplified schematic overview of genes involved in hepatic *de novo* lipogenesis. VLDL: very low-density lipoprotein; (**B**) Hepatic gene expression of lipogenic genes ATP citrate lyase (*Acl*), acetyl-CoA carboxylase (*Acc*), fatty acid synthase (*Fasn*), and stearoyl-CoA desaturase-1 (*Scd1*) were analyzed by quantitative RT-PCR and normalized to *Tbp*; (**C**) Cytosolic lysates were immunoblotted with antibodies against FASN and SCD1 with α-Tubulin as loading control. Each lane represents independent mice from each experimental group. One representative western blot is shown (*n* = 3). Quantification of cytosolic FASN and SCD1 proteins was analyzed by ImageJ (*n* = 5). F: fasted mice; Fru: fructose-fed mice; Glc: glucose-fed mice; (**D**) Plasma triglycerides (TG). Data represent the mean ± SEM (*n* = 5). Significant differences were found using two-way ANOVA followed by Tukey's multiple comparison test (fasted vs. fructose fed and fasted vs. glucose fed RNA data were analyzed separately). * $p < 0.05$, ** $p < 0.01$, *** $p < 0.001$ compared to fasted. [#] $p < 0.05$, [##] $p < 0.01$, compared to LXRα$^{+/+}$ mice.

3.3. Regulation of Srebp-1 and Chrebpβ Expression Mediated by Dietary Glucose Is Dependent of LXRα

LXRs, and particularly LXRα, are known as upstream regulators of SREBP-1c and CHREBPα expression in response to glucose and insulin, and all three transcription factors are involved in the regulation of hepatic lipogenic enzyme genes described above [6,19]. Recent investigations have shown that ChREBPα is activated by dietary fructose, which in turn induces the expression of *Chrebpβ* [21,23]. We did not observe significant changes in *Chrebpα* mRNA levels by both diets in wild type mice, while high fructose feeding more potently induced *Chrebpβ* mRNA expression as compared to glucose (Figure 3A). Compared to wild type mice, *Chrebpβ* mRNA expression was significantly reduced in glucose fed, but not in fructose fed LXRα$^{-/-}$ mice, concomitant with reduced nuclear LXRβ protein expression in glucose-fed LXRα$^{-/-}$ mice (Figure 3B, quantification is shown in Supplementary Figure S1B). This supports our previous observations that both LXR isoforms are necessary to maintain hepatic ChREBPα activity in response to glucose [19], and suggests that LXRβ is able to compensate for the lack of LXRα in regulation of ChREBPα in fructose fed mice. SREBP-1 mRNA and protein (proform) expression were significantly reduced in glucose fed LXRα$^{-/-}$ mice (Figure 3A,B, quantification in Supplementary Figure S1A), supporting the role of LXRα as a central regulator of hepatic *Srebp-1*expression. Notably, we could only detect the ChREBPα protein but not ChREBPβ in our in vivo liver samples, suggestive of a high turnover (rapid degradation) of the constitutively active nuclear ChREBPβ protein.

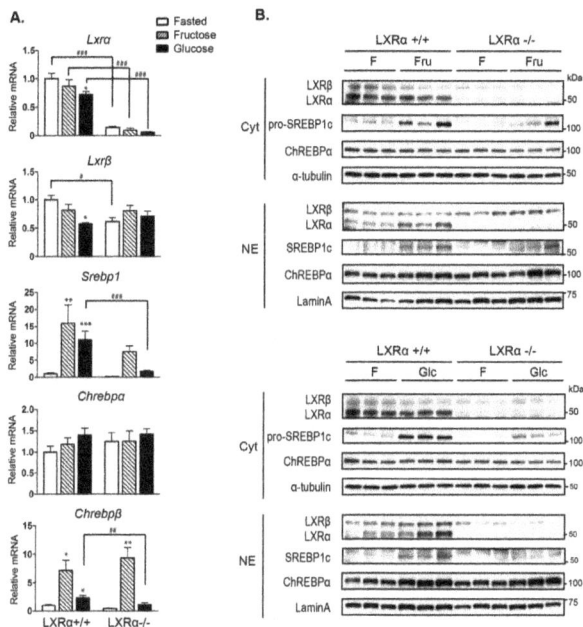

Figure 3. Induction of hepatic *Srebp-1* and carbohydrate response element-binding protein (*Chrebp*)β expression by dietary glucose is reduced in LXRα$^{-/-}$ mice. (**A**) Hepatic gene expression of *Lxrα/β*, *Srebp-1*, and *Chrebpα/β* was analyzed by quantitative RT-PCR and normalized to *Tbp*; (**B**) Cytosolic and nuclear lysates were immunoblotted with antibodies against LXR, SREBP-1, and ChREBP with α-Tubulin and Lamin A as loading controls. Each lane represents independent mice from each group. One representative western blot is shown ($n = 3$). Data represent the mean $\pm$ SEM ($n = 5$). Significant differences were found using two-way ANOVA followed by Tukey's multiple comparison test (fasted vs. fructose fed and fasted vs. glucose fed RNA data were analyzed separately). * $p < 0.05$, ** $p < 0.01$, *** $p < 0.001$ compared to fasted. # $p < 0.05$, ## $p < 0.01$, ### $p < 0.001$ compared to LXRα$^{+/+}$ mice.

3.4. Dietary Fructose Induces Nuclear O-GlcNAc Signaling

O-GlcNAcylation of ChREBPα has been shown to potentiate its activity, stability, and binding to the *L-pk* promoter in response to refeeding in normal mice and more so in hyperglycemic diabetic mice [13]. The hexosamine biosynthetic pathway (HBP) involving activation of OGT is depicted in Figure 4A. Metabolomics analysis revealed lower levels of N-acetylglucosamine-1-P (GlcNAc-1-P) in refed mice with both diets compared to fasting (Supplementary Table S2), suggesting high glutamine-fructose-6-phosphate amidotransferase 2 (GFAT2) activity during fasting, possibly via glucagon-PKA-mediated activation [41]. Although HPLC analysis showed no significant increase in total liver UDP-GlcNAc levels in refed compared to fasted mice (Figure 4B), the fraction of UDP-GlcNAc that is utilized for *O*-GlcNAc modification may be increased towards the nuclear pool after refeeding. Ogt mRNA levels were strongly downregulated upon refeeding (Figure 4C), which is suggestive of a negative feedback regulation due to high OGT activity, as reported previously [42]. We observed increased nuclear protein *O*-GlcNAcylation levels upon fructose refeeding in wild type mice (Figure 4D, quantification in Supplementary Figure S2B). Notably, the *O*-GlcNAcylation of LXRβ was significantly increased in fructose-fed LXRα$^{-/-}$ mice (representative western blots in Figure 4E, quantification (*n* = 4) in Supplementary Figure S2C).

Because ChREBP binding to its cognate DNA-binding site in the *L-pk* promoter is affected by high glucose feeding, *O*-GlcNAc and LXRα/β signaling in rat and mice livers [13,19,43], we next investigated ChREBP recruitment to the *L-pk* promoter in chromatin immunoprecipitation (ChIP) assays. We did not observe a significant increase in ChREBP recruitment after refeeding in wild type mice, but observed a significant increase after fructose feeding in LXRα$^{-/-}$ mice (Figure 4F). This observation, together with increased *O*-GlcNAcylation of LXRβ, suggests that LXRβ may compensate for the lack of LXRα in regulating ChREBPα activity upon fructose feeding.

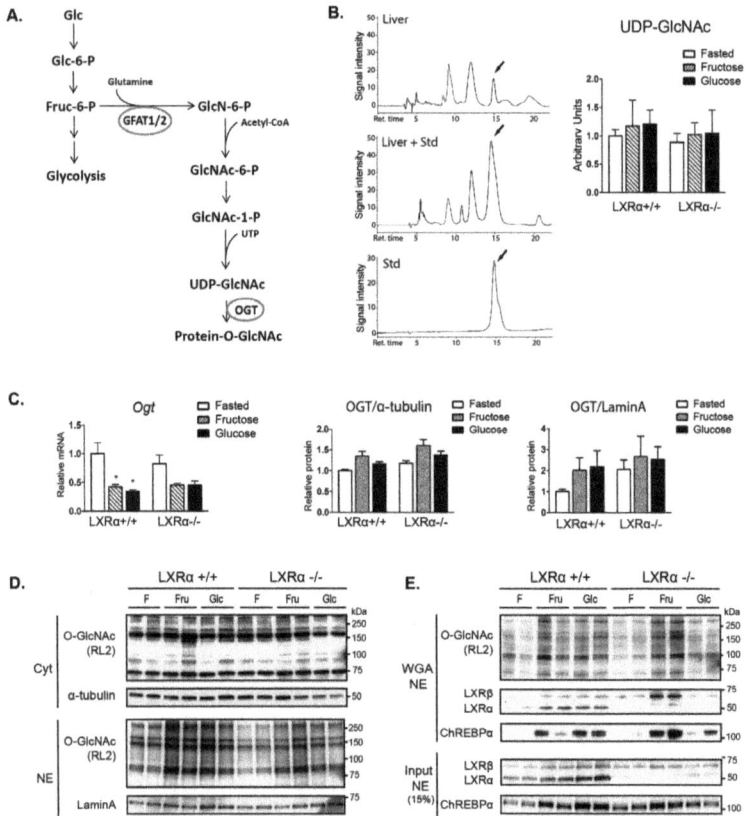

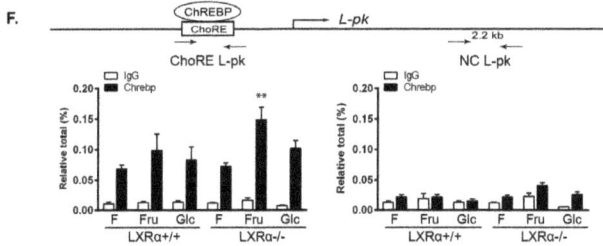

Figure 4. Dietary fructose induces nuclear *O*-GlcNAc signaling. (**A**) Simplified schematic overview of the hexosamine signaling pathway including the rate limiting enzyme GFAT and *O*-GlcNAc transferase (OGT); (**B**) High-performance liquid chromatography (HPLC) analysis of hepatic UDP-GlcNAc levels. Left panel: Example of a typical running profile with or without injection of UDP-GlcNAc standard (Std.) is shown. Right panel: Quantification of the HPLC data normalized to protein concentration; (**C**) Hepatic expression of the *Ogt* gene and cytosolic and nuclear OGT protein were analyzed by quantitative RT-PCR and western blotting/Image J and normalized to *Tbp*, α-tubulin, and Lamin A, respectively. Data represent the mean ± SEM ($n = 5$); (**D**) Cytosolic and nuclear lysates were immunoblotted with anti-*O*-GlcNAc antibody (RL2) with α-Tubulin and Lamin A as loading controls. Each lane represents independent mice from each group. One representative western blot is shown ($n = 2$); (**E**) Nuclear lysates were subjected to wheat germ agglutinin (WGA) beads to precipitate *O*-GlcNAcylated proteins. WGA enriched samples (upper panel) and input lysates (bottom panel) were immunoblotted with antibodies detecting *O*-GlcNAcylated proteins (RL2), LXR and ChREBP. Each lane represents independent mice from experimental groups. Representative western blots are shown ($n = 2$); (**F**) ChREBP binding to the carbohydrate response element (ChoRE) containing region of the *L-pk* promoter and negative control sequence (NC) 2216–2288 bp into the *L-pk* gene after ChoRE sequence was detected by chromatin immunoprecipitation (ChIP) using antibodies against ChREBP or IgG as a control. Data represent the mean ± SEM ($n = 5$). Significant differences were found using two-way ANOVA followed by Tukey's multiple comparison test. ** $p < 0.01$ compared to fasted.

4. Discussion and Conclusions

The results of this study show that LXRα is important for ChREBPα activity upon glucose, but not fructose feeding in the livers of mice, as visualized by lower expression of ChREBPα specific target genes *L-pk* and *Chrebpβ* in LXRα$^{-/-}$ compared to wild type mice. LXRα is also important for the transcriptional control of classical hepatic DNL genes upon glucose feeding, including *Fasn* and *Accβ*. This suggests that LXRα is important in integrating nutritional cues upon glucose feeding upstream of ChREBPα.

In the human diet, fructose and glucose are rarely consumed in isolation. Studies comparing more commonly consumed sucrose and high fructose corn syrup have yielded different results compared to studies with pure fructose or glucose [37]. However, long-term feeding studies (12 h or more) in rat and mice have shown similar fold induction of hepatic lipogenic genes by high fructose compared to sucrose and always above levels in mice fed high glucose [21,33,34,38]. This is likely due to most of the dietary fructose being taken up by the liver whereas only 30% of dietary glucose is metabolized by the liver; the remaining glucose is metabolized in muscle, adipose tissues, brain, kidney, and red blood cells [44]. However, approximately 30–50% of fructose taken up by the liver is converted to glucose and 15% to lactate in part as a fuel source for extrahepatic tissues [37,45]. This suggests additive effects of hepatic glucose and fructose metabolism and signaling in the control of DNL in response to dietary fructose, at least in the late refed phase. This is supported by a study by Matsuzaka et al. [34] that showed the expression of SREBP-1 and FASN above glucose-induced levels after 9 h fructose feeding concomitant with increased plasma glucose and insulin levels. At 6 h refeeding, however, glucose, but not fructose, strongly induced SREBP-1 and FASN expression. Stamatikos et al. reported increased fructose-induced *Accα* and *Fasn* gene expression following inhibition of Fructose-1,6-biphosphatase (FBPase) in human hepatocyte carcinoma (HepG2) cells, suggesting that dietary fructose induces DNL gene expression independently of its ability to generate glucose [21]. However, these results were not verified in primary hepatocytes, suggesting that this effect may be specific for hepatoma cells.

In the present study, we observed similar hepatic glucose levels in fructose and glucose fed mice, but significantly higher pyruvate levels after fructose feeding. This observation is concomitant with more strongly induced expression of the ChREBP target gene *L-pk* upon fructose feeding than glucose feeding, in agreement with previous observations [34]. This supports the notion that fructose activates ChREBP and *L-pk* expression to a higher degree than glucose, thus generating more substrate to DNL [39,46,47]. Notably, *L-pk* expression was almost back to fasting levels in LXRα depleted glucose fed mice, but not in fructose fed mice, suggesting differential roles of LXRα and ChREBP in response to early phase of glucose and fructose metabolism.

We have recently shown that LXRα$^{-/-}$/β$^{-/-}$ mice have reduced nuclear *O*-GlcNAc levels, ChREBPα activity, and lipogenic gene expression in livers as compared wild type mice upon refeeding with a chow diet [19]. Herein, we provide evidence that LXRα is important for ChREBPα activity upon glucose feeding. The effects on hepatic DNL is less striking when depleting LXRα alone, and LXRα was dispensable in fructose fed, but not glucose fed mice, which were also low in LXRβ protein levels. Levels of nuclear LXRβ protein were maintained in fructose fed but surprisingly strongly reduced in glucose fed LXRα$^{-/-}$ mice, suggesting compensation by LXRβ in regulating SREBP-1c and lipogenic gene expression in fructose fed mice. This may be explained, at least in part, by redundancy between the LXRs [48]. Notably, increased WGA-recovery of LXRβ and ChREBPα, which is indicative of increased *O*-GlcNAc modifications of these proteins, was observed in fructose fed LXRα$^{-/-}$ mice along with increased ChREBP binding to the *L-pk* promoter in these mice. It is thus possible that LXRα, LXRβ, and ChREBPα collectively integrate different sugar metabolites with lipogenesis in a wild type context. Acetyl-CoA generated by the ACL enzyme, which was strongly upregulated by fructose feeding (Figure 1B), also acts as a substrate for protein acetylation [49]. Metabolic sensing by LXRs and ChREBP involves modification by *O*-GlcNAcylation and acetylation [50,51] and there seems to be interplay between OGT and the acetylating p300 transcriptional coregulator [52], which is likely to impact LXR and ChREBP activity [53].

We did not observe reduced levels of nuclear *O*-GlcNAcylated proteins when knocking out LXRα, suggesting that both LXRα and β must be depleted for this phenotype. Because *O*-GlcNAc modification of ChREBPα is important for its activity and because ChREBPα was more modified in fructose fed LXRα$^{-/-}$ mice, it would be interesting to study if the dietary effects of fructose on DNL gene expression and ChREBPα activity would have been more pronounced in an animal model where both LXR isoforms were depleted.

In summary, we provide evidence that hepatic expression of *de novo* lipogenic genes require LXRα activity in response to dietary glucose and that LXRβ may compensate for the lack of LXRα in fructose fed mice. As aberrant *O*-GlcNAc signaling during nutrient stress and diabetes leads to excessive glucose production and lipid accumulation in the liver [54], studies including mutated *O*-GlcNAc residues in LXRs and ChREBPα will provide a better understanding of the relevance of coordinated *O*-GlcNAcylation of these transcriptional regulators under physiological and pathophysiological conditions.

Supplementary Materials: The following are available online at www.mdpi.com/2072-6643/9/7/678/s1. Figure S1: Quantification of cytosolic and nuclear proteins LXRα/β, SREBP-1 and ChREBPα. Western blot quantification of cytosolic (A) and nuclear (B) proteins was analyzed by ImageJ. Data represent the mean ± SEM (*n* = 3–4). Significant differences were found using two-way ANOVA followed by Tukey's multiple comparison test. ** *p* < 0.01, *** *p* < 0.001 compared to fasted. # *p* < 0.05, ## *p* < 0.01, ### *p* < 0.001 compared to LXRα$^{+/+}$ mice. Figure S2: Quantification of nuclear lysates input and WGA enriched proteins LXRα/β, ChREBPα and *O*-GlcNAcylation levels (RL2). Western blot quantification of cytosolic lysates (A), nuclear lysates input (B), and WGA enriched (C) proteins were analyzed by ImageJ. Data represent the mean ± SEM (*n* = 4). Significant differences were found using two-way ANOVA followed by Tukey's multiple comparison test. * *p* < 0.05 compared to fasted. # *p* < 0.05,

p < 0.01, ### p < 0.001 compared to LXRα+/+ mice. Table S1: Mouse SYBR Green Primers sequences. Table S2: Metabolomics analysis of liver tissues of wild type and LXRα−/− mice.

Acknowledgments: This work was supported by grants from the University of Oslo, the Research Council of Norway, the Johan Throne Holst Foundation, the Novo Nordic Foundation, and the Anders Jahre Foundation. The authors thank Timothy F. Osborne (University of California) for providing the SREBP-1 antibody and Gerald Hart (Johns Hopkins University) for providing the OGT AL25 antibody.

Author Contributions: L.M.G., H.I.N., K.T.D. and C.B. conceived and designed the experiments; Q.F., R.C.N., C.L., C.B., K.T.D. and E.B. performed experiments; Q.F., R.C.N., C.L., H.M.I. and J.M. analyzed data; Q.F., H.M.I., L.M.G. and J.M. interpreted results of experiments; Q.F. and L.M.G. prepared figures; Q.F. and L.M.G. drafted manuscript; Q.F. and L.M.G. edited and revised the manuscript; all co-authors approved the final version of manuscript.

References

1. Bantle, J.P.; Raatz, S.K.; Thomas, W.; Georgopoulos, A. Effects of dietary fructose on plasma lipids in healthy subjects. *Am. J. Clin. Nutr.* **2000**, *72*, 1128–1134. [PubMed]
2. Parks, E.J.; Hellerstein, M.K. Carbohydrate-induced hypertriacylglycerolemia: Historical perspective and review of biological mechanisms. *Am. J. Clin. Nutr.* **2000**, *71*, 412–433. [PubMed]
3. Faeh, D.; Minehira, K.; Schwarz, J.M.; Periasamy, R.; Park, S.; Tappy, L. Effect of fructose overfeeding and fish oil administration on hepatic de novo lipogenesis and insulin sensitivity in healthy men. *Diabetes* **2005**, *54*, 1907–1913. [CrossRef] [PubMed]
4. Caton, P.W.; Nayuni, N.K.; Khan, N.Q.; Wood, E.G.; Corder, R. Fructose induces gluconeogenesis and lipogenesis through a SIRT1-dependent mechanism. *J. Endocrinol.* **2011**, *208*, 273–283. [CrossRef] [PubMed]
5. Strable, M.S.; Ntambi, J.M. Genetic control of de novo lipogenesis: Role in diet-induced obesity. *Crit. Rev. Biochem. Mol. Biol.* **2010**, *45*, 199–214. [CrossRef] [PubMed]
6. Grønning-Wang, L.M.; Bindesbøll, C.; Nebb, H.I. *The Role of Liver X Receptor in Hepatic de novo Lipogenesis and Cross-Talk with Insulin and Glucose Signaling*; INTECH Open Access Publisher: Rijeka, Croatia, 2013.
7. Repa, J.J.; Mangelsdorf, D.J. The role of orphan nuclear receptors in the regulation of cholesterol homeostasis. *Annu. Rev. Cell Dev. Biol.* **2000**, *16*, 459–481. [CrossRef] [PubMed]
8. Filhoulaud, G.; Guilmeau, S.; Dentin, R.; Girard, J.; Postic, C. Novel insights into ChREBP regulation and function. *Trends Endocrinol. Metab.* **2013**, *24*, 257–268. [CrossRef] [PubMed]
9. Laffitte, B.A.; Chao, L.C.; Li, J.; Walczak, R.; Hummasti, S.; Joseph, S.B.; Castrillo, A.; Wilpitz, D.C.; Mangelsdorf, D.J.; Collins, J.L.; et al. Activation of liver X receptor improves glucose tolerance through coordinate regulation of glucose metabolism in liver and adipose tissue. *Proc. Natl. Acad. Sci. USA* **2003**, *100*, 5419–5424. [CrossRef] [PubMed]
10. Cao, G.; Liang, Y.; Broderick, C.L.; Oldham, B.A.; Beyer, T.P.; Schmidt, R.J.; Zhang, Y.; Stayrook, K.R.; Suen, C.; Otto, K.A.; et al. Antidiabetic action of a liver x receptor agonist mediated by inhibition of hepatic gluconeogenesis. *J. Boil. Chem.* **2003**, *278*, 1131–1136. [CrossRef] [PubMed]
11. Grempler, R.; Gunther, S.; Steffensen, K.R.; Nilsson, M.; Barthel, A.; Schmoll, D.; Walther, R. Evidence for an indirect transcriptional regulation of glucose-6-phosphatase gene expression by liver X receptors. *Biochem. Biophys. Res. Commun.* **2005**, *338*, 981–986. [CrossRef] [PubMed]
12. Anthonisen, E.H.; Berven, L.; Holm, S.; Nygard, M.; Nebb, H.I.; Gronning-Wang, L.M. Nuclear Receptor Liver X Receptor Is O-GlcNAc-modified in Response to Glucose. *J. Biol. Chem.* **2010**, *285*, 1607–1615. [CrossRef] [PubMed]
13. Guinez, C.; Filhoulaud, G.; Rayah-Benhamed, F.; Marmier, S.; Dubuquoy, C.; Dentin, R.; Moldes, M.; Burnol, A.F.; Yang, X.; Lefebvre, T.; et al. O-GlcNAcylation increases ChREBP protein content and transcriptional activity in the liver. *Diabetes* **2011**, *60*, 1399–1413. [CrossRef] [PubMed]
14. Bond, M.R.; Hanover, J.A. A little sugar goes a long way: The cell biology of O-GlcNAc. *J. Cell Biol.* **2015**, *208*, 869–880. [CrossRef] [PubMed]
15. Hart, G.W.; Slawson, C.; Ramirez-Correa, G.; Lagerlof, O. Cross talk between O-GlcNAcylation and phosphorylation: Roles in signaling, transcription, and chronic disease. *Annu. Rev. Biochem.* **2011**, *80*, 825–858. [CrossRef] [PubMed]

16. Gambetta, M.C.; Muller, J. A critical perspective of the diverse roles of *O*-GlcNAc transferase in chromatin. *Chromosoma* **2015**, *124*, 429–442. [CrossRef] [PubMed]

17. Rumberger, J.M.; Wu, T.; Hering, M.A.; Marshall, S. Role of hexosamine biosynthesis in glucose-mediated up-regulation of lipogenic enzyme mRNA levels: Effects of glucose, glutamine, and glucosamine on glycerophosphate dehydrogenase, fatty acid synthase, and acetyl-CoA carboxylase mRNA levels. *J. Biol. Chem.* **2003**, *278*, 28547–28552. [CrossRef] [PubMed]

18. Bindesbøll, C.; Grønning-Wang, L.M. Liver X receptors connect nuclear *O*-GlcNAc signaling to hepatic glucose utilization and lipogenesis. *Recept. Clin. Investig.* **2015**. [CrossRef]

19. Bindesboll, C.; Fan, Q.; Norgaard, R.C.; MacPherson, L.; Ruan, H.B.; Wu, J.; Pedersen, T.A.; Steffensen, K.R.; Yang, X.; Matthews, J.; et al. Liver X receptor regulates hepatic nuclear *O*-GlcNAc signaling and carbohydrate responsive element-binding protein activity. *J. Lipid Res.* **2015**, *56*, 771–785. [CrossRef] [PubMed]

20. Herman, M.A.; Peroni, O.D.; Villoria, J.; Schon, M.R.; Abumrad, N.A.; Bluher, M.; Klein, S.; Kahn, B.B. A novel ChREBP isoform in adipose tissue regulates systemic glucose metabolism. *Nature* **2012**, *484*, 333–338. [CrossRef] [PubMed]

21. Stamatikos, A.D.; da Silva, R.P.; Lewis, J.T.; Douglas, D.N.; Kneteman, N.M.; Jacobs, R.L.; Paton, C.M. Tissue Specific Effects of Dietary Carbohydrates and Obesity on ChREBPα and ChREBPβ Expression. *Lipids* **2016**, *51*, 95–104. [CrossRef] [PubMed]

22. Zhang, P.; Kumar, A.; Katz, L.S.; Li, L.; Paulynice, M.; Herman, M.A.; Scott, D.K. Induction of the ChREBPβ Isoform Is Essential for Glucose-Stimulated β-Cell Proliferation. *Diabetes* **2015**, *64*, 4158–4170. [CrossRef] [PubMed]

23. Erion, D.M.; Popov, V.; Hsiao, J.J.; Vatner, D.; Mitchell, K.; Yonemitsu, S.; Nagai, Y.; Kahn, M.; Gillum, M. P.; Dong, J.; et al. The role of the carbohydrate response element-binding protein in male fructose-fed rats. *Endocrinology* **2013**, *154*, 36–44. [CrossRef] [PubMed]

24. Iizuka, K.; Bruick, R.K.; Liang, G.; Horton, J.D.; Uyeda, K. Deficiency of carbohydrate response element-binding protein (ChREBP) reduces lipogenesis as well as glycolysis. *Proc. Natl. Acad. Sci. USA* **2004**, *101*, 7281–7286. [CrossRef] [PubMed]

25. Kim, M.S.; Krawczyk, S.A.; Doridot, L.; Fowler, A.J.; Wang, J.X.; Trauger, S.A.; Noh, H.L.; Kang, H.J.; Meissen, J.K.; Blatnik, M.; et al. ChREBP regulates fructose-induced glucose production independently of insulin signaling. *J. Clin. Investig.* **2016**, *126*, 4372–4386. [CrossRef] [PubMed]

26. Hessvik, N.P.; Boekschoten, M.V.; Baltzersen, M.A.; Kersten, S.; Xu, X.; Andersen, H.; Rustan, A.C.; Thoresen, G.H. LXRβ is the dominant LXR subtype in skeletal muscle regulating lipogenesis and cholesterol efflux. *Am. J. Physiol. Endocrinol. Metab.* **2010**, *298*, E602–E613. [CrossRef] [PubMed]

27. Rull, A.; Geeraert, B.; Aragones, G.; Beltran-Debon, R.; Rodriguez-Gallego, E.; Garcia-Heredia, A.; Pedro-Botet, J.; Joven, J.; Holvoet, P.; Camps, J. Rosiglitazone and fenofibrate exacerbate liver steatosis in a mouse model of obesity and hyperlipidemia. A transcriptomic and metabolomic study. *J. Proteome Res.* **2014**, *13*, 1731–1743. [CrossRef] [PubMed]

28. Ye, J.; Coulouris, G.; Zaretskaya, I.; Cutcutache, I.; Rozen, S.; Madden, T.L. Primer-BLAST: A tool to design target-specific primers for polymerase chain reaction. *BMC Bioinform.* **2012**, *13*, 134. [CrossRef] [PubMed]

29. Iyer, S.P.; Akimoto, Y.; Hart, G.W. Identification and cloning of a novel family of coiled-coil domain proteins that interact with *O*-GlcNAc transferase. *J. Biol. Chem.* **2003**, *278*, 5399–5409. [CrossRef] [PubMed]

30. Seo, Y.K.; Chong, H.K.; Infante, A.M.; Im, S.S.; Xie, X.; Osborne, T.F. Genome-wide analysis of SREBP-1 binding in mouse liver chromatin reveals a preference for promoter proximal binding to a new motif. *Proc. Natl. Acad. Sci. USA* **2009**, *106*, 13765–13769. [CrossRef] [PubMed]

31. Itkonen, H.M.; Engedal, N.; Babaie, E.; Luhr, M.; Guldvik, I.J.; Minner, S.; Hohloch, J.; Tsourlakis, M.C.; Schlomm, T.; Mills, I.G. UAP1 is overexpressed in prostate cancer and is protective against inhibitors of *N*-linked glycosylation. *Oncogene* **2015**, *34*, 3744–3750. [CrossRef] [PubMed]

32. Nakajima, K.; Kitazume, S.; Angata, T.; Fujinawa, R.; Ohtsubo, K.; Miyoshi, E.; Taniguchi, N. Simultaneous determination of nucleotide sugars with ion-pair reversed-phase HPLC. *Glycobiology* **2010**, *20*, 865–871. [CrossRef] [PubMed]

33. Koo, H.Y.; Wallig, M.A.; Chung, B.H.; Nara, T.Y.; Cho, B.H.; Nakamura, M.T. Dietary fructose induces a wide range of genes with distinct shift in carbohydrate and lipid metabolism in fed and fasted rat liver. *Biochim. Biophys. Acta* **2008**, *1782*, 341–348. [CrossRef] [PubMed]

34. Matsuzaka, T.; Shimano, H.; Yahagi, N.; Amemiya-Kudo, M.; Okazaki, H.; Tamura, Y.; Iizuka, Y.; Ohashi, K.; Tomita, S.; Sekiya, M.; et al. Insulin-independent induction of sterol regulatory element-binding protein-1c expression in the livers of streptozotocin-treated mice. *Diabetes* **2004**, *53*, 560–569. [CrossRef] [PubMed]

35. Lanaspa, M.A.; Ishimoto, T.; Li, N.; Cicerchi, C.; Orlicky, D.J.; Ruzycki, P.; Rivard, C.; Inaba, S.; Roncal-Jimenez, C.A.; Bales, E.S.; et al. Endogenous fructose production and metabolism in the liver contributes to the development of metabolic syndrome. *Nat. Commun.* **2013**, *4*, 2434. [CrossRef] [PubMed]

36. Tounian, P.; Schneiter, P.; Henry, S.; Jequier, E.; Tappy, L. Effects of infused fructose on endogenous glucose production, gluconeogenesis, and glycogen metabolism. *Am. J. Physiol.* **1994**, *267*, E710–E717. [CrossRef]

37. Rippe, J.M.; Angelopoulos, T.J. Sucrose, high-fructose corn syrup, and fructose, their metabolism and potential health effects: What do we really know? *Adv. Nutr. Int. Rev. J.* **2013**, *4*, 236–245. [CrossRef] [PubMed]

38. Janevski, M.; Ratnayake, S.; Siljanovski, S.; McGlynn, M.A.; Cameron-Smith, D.; Lewandowski, P. Fructose containing sugars modulate mRNA of lipogenic genes ACC and FAS and protein levels of transcription factors ChREBP and SREBP1c with no effect on body weight or liver fat. *Food Funct.* **2012**, *3*, 141–149. [CrossRef] [PubMed]

39. Koo, H.Y.; Miyashita, M.; Cho, B.H.; Nakamura, M.T. Replacing dietary glucose with fructose increases ChREBP activity and SREBP-1 protein in rat liver nucleus. *Biochem. Biophys. Res. Commun.* **2009**, *390*, 285–289. [CrossRef] [PubMed]

40. Chu, K.; Miyazaki, M.; Man, W.C.; Ntambi, J.M. Stearoyl-coenzyme A desaturase 1 deficiency protects against hypertriglyceridemia and increases plasma high-density lipoprotein cholesterol induced by liver X receptor activation. *Mol. Cell. Biol.* **2006**, *26*, 6786–6798. [CrossRef] [PubMed]

41. Hu, Y.; Riesland, L.; Paterson, A.J.; Kudlow, J.E. Phosphorylation of mouse glutamine-fructose-6-phosphate amidotransferase 2 (GFAT2) by cAMP-dependent protein kinase increases the enzyme activity. *J. Biol. Chem.* **2004**, *279*, 29988–29993. [CrossRef] [PubMed]

42. Taylor, R.P.; Geisler, T.S.; Chambers, J.H.; McClain, D.A. Up-regulation of *O*-GlcNAc transferase with glucose deprivation in HepG2 cells is mediated by decreased hexosamine pathway flux. *J. Boil. Chem.* **2009**, *284*, 3425–3432. [CrossRef] [PubMed]

43. Ishii, S.; Iizuka, K.; Miller, B.C.; Uyeda, K. Carbohydrate response element binding protein directly promotes lipogenic enzyme gene transcription. *Proc. Natl. Acad. Sci. USA* **2004**, *101*, 15597–15602. [CrossRef] [PubMed]

44. Moore, M.C.; Coate, K.C.; Winnick, J.J.; An, Z.; Cherrington, A.D. Regulation of hepatic glucose uptake and storage in vivo. *Adv. Nutr. (Bethesda, Md.)* **2012**, *3*, 286–294. [CrossRef] [PubMed]

45. Sun, S.Z.; Empie, M.W. Fructose metabolism in humans—What isotopic tracer studies tell us. *Nutr. Metab.* **2012**, *9*, 89. [CrossRef] [PubMed]

46. Parks, E.J.; Skokan, L.E.; Timlin, M.T.; Dingfelder, C.S. Dietary sugars stimulate fatty acid synthesis in adults. *J. Nutr.* **2008**, *138*, 1039–1046. [PubMed]

47. Moore, J.B.; Gunn, P.J.; Fielding, B.A. The role of dietary sugars and *de novo* lipogenesis in non-alcoholic fatty liver disease. *Nutrients* **2014**, *6*, 5679–5703. [CrossRef] [PubMed]

48. Quinet, E.M.; Savio, D.A.; Halpern, A.R.; Chen, L.; Schuster, G.U.; Gustafsson, J.A.; et al. Liver X receptor (LXR)-β regulation in LXRα-deficient mice: Implications for therapeutic targeting. *Mol. Pharmacol.* **2006**, *70*, 1340–1349. [CrossRef] [PubMed]

49. Choudhary, C.; Weinert, B.T.; Nishida, Y.; Verdin, E.; Mann, M. The growing landscape of lysine acetylation links metabolism and cell signalling. *Nat. Rev. Mol. Cell Biol.* **2014**, *15*, 536–550. [CrossRef] [PubMed]

50. Li, X.; Zhang, S.; Blander, G.; Tse, J.G.; Krieger, M.; Guarente, L. SIRT1 deacetylates and positively regulates the nuclear receptor LXR. *Mol. Cell* **2007**, *28*, 91–106. [CrossRef] [PubMed]

51. Bricambert, J.; Miranda, J.; Benhamed, F.; Girard, J.; Postic, C.; Dentin, R. Salt-inducible kinase 2 links transcriptional coactivator p300 phosphorylation to the prevention of ChREBP-dependent hepatic steatosis in mice. *J. Clin. Investig.* **2010**, *120*, 4316–4331. [CrossRef] [PubMed]

52. Allison, D.F.; Wamsley, J.J.; Kumar, M.; Li, D.; Gray, L.G.; Hart, G.W.; Jones, D.R.; Mayo, M.W. Modification of RelA by *O*-linked *N*-acetylglucosamine links glucose metabolism to NF-κB acetylation and transcription. *Proc. Natl. Acad. Sci. USA* **2012**, *109*, 16888–16893. [CrossRef] [PubMed]

53. Oosterveer, M.H.; Schoonjans, K. Hepatic glucose sensing and integrative pathways in the liver. *Cell. Mol. Life Sci. CMLS* **2014**, *71*, 1453–1467. [CrossRef] [PubMed]
54. Zhang, K.; Yin, R.; Yang, X. *O*-GlcNAc: A Bittersweet Switch in Liver. *Front. Endocrinol.* **2014**, *5*, 221. [CrossRef] [PubMed]

nutrients

MDPI

Article

The 1-Week and 8-Month Effects of a Ketogenic Diet or Ketone Salt Supplementation on Multi-Organ Markers of Oxidative Stress and Mitochondrial Function in Rats

Wesley C. Kephart [1], Petey W. Mumford [1], Xuansong Mao [1], Matthew A. Romero [1],
Hayden W. Hyatt [1], Yufeng Zhang [2], Christopher B. Mobley [1], John C. Quindry [3],
Kaelin C. Young [1,4], Darren T. Beck [1,4], Jeffrey S. Martin [1,4], Danielle J. McCullough [1,4],
Dominic P. D'Agostino [5], Ryan P. Lowery [6], Jacob M. Wilson [6], Andreas N. Kavazis [1,4,*]
and Michael D. Roberts [1,4,*]

[1]   School of Kinesiology, Auburn University, Auburn, AL 36849, USA; wck0007@auburn.edu (W.C.K.);
      pwm0009@auburn.edu (P.W.M.); xzm0012@auburn.edu (X.M.); mzr0049@auburn.edu (M.A.R.);
      hwh0001@auburn.edu (H.W.H.); moblecb@auburn.edu (C.B.M.); kyoung@auburn.vcom.edu (K.C.Y.);
      dbeck@auburn.vcom.edu (D.T.B.); jmartin@auburn.vcom.edu (J.S.M.);
      dmccullough@auburn.vcom.edu (D.J.M.)
[2]   Department of Biological Sciences, Auburn University, Auburn, AL 36849, USA; yzz0095@auburn.edu
[3]   Department of Human Health Performance, University of Montana, Missoula, MT 59812, USA;
      john.quindry@mso.umt.edu
[4]   Department of Cell Biology and Physiology, Edward via College of Osteopathic Medicine—
      Auburn Campus, Auburn, AL 36849, USA
[5]   Department of Molecular Pharmacology and Physiology, University of South Florida,
      Tampa, FL 33620, USA; dagostino.dominic1@gmail.com
[6]   Applied Sports Performance Institute, Tampa, FL 33607, USA; rlowery@theaspi.com (R.P.L.);
      jwilson@theaspi.com (J.M.W.)
*   Correspondence: ank0012@auburn.edu (A.N.K.); mdr0024@auburn.edu (M.D.R.);
    Fax: +1-334-844-1479 (A.N.K.); +1-334-844-1467 (M.D.R.)

Received: 28 August 2017; Accepted: 13 September 2017; Published: 15 September 2017

Abstract: We determined the short- and long-term effects of a ketogenic diet (KD) or ketone salt (KS) supplementation on multi-organ oxidative stress and mitochondrial markers. For short-term feedings, 4 month-old male rats were provided isocaloric amounts of KD ($n = 10$), standard chow (SC) ($n = 10$) or SC + KS (~1.2 g/day, $n = 10$). For long-term feedings, 4 month-old male rats were provided KD ($n = 8$), SC ($n = 7$) or SC + KS ($n = 7$) for 8 months and rotarod tested every 2 months. Blood, brain (whole cortex), liver and gastrocnemius muscle were harvested from all rats for biochemical analyses. Additionally, mitochondria from the brain, muscle and liver tissue of long-term-fed rats were analyzed for mitochondrial quantity (maximal citrate synthase activity), quality (state 3 and 4 respiration) and reactive oxygen species (ROS) assays. Liver antioxidant capacity trended higher in short-term KD- and SC + KS-fed versus SC-fed rats, and short-term KD-fed rats exhibited significantly greater serum ketones compared to SC + KS-fed rats indicating that the diet (not KS supplementation) induced ketonemia. In long term-fed rats: (a) serum ketones were significantly greater in KD- versus SC- and SC + KS-fed rats; (b) liver antioxidant capacity and glutathione peroxidase protein was significantly greater in KD- versus SC-fed rats, respectively, while liver protein carbonyls were lowest in KD-fed rats; and (c) gastrocnemius mitochondrial ROS production was significantly greater in KD-fed rats versus other groups, and this paralleled lower mitochondrial glutathione levels. Additionally, the gastrocnemius pyruvate-malate mitochondrial respiratory control ratio was significantly impaired in long-term KD-fed rats, and gastrocnemius mitochondrial quantity was lowest in these animals. Rotarod performance was greatest in KD-fed rats versus all other groups at 2, 4 and 8 months, although there was a significant age-related decline in performance existed in KD-fed rats which was not evident in the other two groups. In conclusion, short- and long-term KD improves select markers of liver oxidative stress compared to SC feeding, although long-term KD feeding may negatively affect skeletal muscle mitochondrial physiology.

Keywords: ketogenic dieting; ketone salts; skeletal muscle; brain; liver; oxidative stress; mitochondria

1. Introduction

Ketogenic diets (KD) are high fat, moderate protein and low carbohydrate diets that have been associated with a myriad of health benefits including weight loss/management, neurological improvements (e.g., treatment of epilepsy and certain brain cancers) and longevity [1,2]. Rodent models have also indicated that ketogenic dieting increases lifespan [3,4], although the specific mechanisms underpinning these observations have not been well elucidated. Excessive tissue accretion of oxidative stress due to repetitive insults from reactive oxygen species (ROS) are a hallmark signature of the aging process [5,6]. Mitochondria bear the brunt of oxidative damage in the lesions to mitochondrial DNA and are reported to be as much as 10-fold higher than in nuclear DNA [7]. Hence, dietary manipulations that mitigate mitochondrial oxidative stress may serve to mitigate the aging process and optimize cellular function.

Interestingly, oxidative stress defense and improved mitochondrial quantity/quality are potential mechanisms through which KD feeding may confer physiological benefits [8,9]. For instance, hippocampal mitochondrial biogenesis in rodents has been reported after four weeks of KD feeding [10]; this is a phenomenon which could occur via KD-induced AMPK pathway signaling [11,12]. KD feeding has also been shown to decrease ROS and hydrogen peroxide (H_2O_2) production in rodent tissues by increasing uncoupling protein-mediated proton conductance and/or increasing glutathione biosynthesis [10,13–15]. With regard to the later mechanism, Milder and Patel [8] have theorized that KD-induced increases in ROS defense is related to an acute stimulation of 4-hydroxy-2-nonenal (4-HNE) or H_2O_2 which, in turn, stimulates the nuclear factor (erythroid-derived 2)-like 2 (Nrf2) transcription factor to translocate into the nucleus. Nrf2 has specifically been shown to upregulate the mRNA expression glutamate-cysteine ligase (GCL) subunits which is the rate limiting enzyme in glutathione biosynthesis [16,17]. In support of this model, Milder and colleagues [18] reported that KD feeding stimulates H_2O_2 and 4-HNE production in the hippocampus of rodents within a 1–3 days period, and this leads to Nrf2 nuclear translocation within 7 days as well as subsequent increases in the expression of ROS-protective genes that promote oxidative stress resistance during 1–3 weeks of feeding. While it is currently unclear whether these redox adaptations are due to a direct effect of the diet (i.e., dietary ketosis) or an indirect effect of the diet (i.e., weight loss, lower insulin levels, etc.), it is notable that several beneficial physiological effects of KD feeding are independent of weight loss. In this regard, potential antioxidant effects of ketogenic dieting may be stimulated by the consequent metabolites of ketone bodies, beta-hydroxybutyrate (BHB) and acetoacetate (AcAc). For instance, ROS production is decreased and resistance to a H_2O_2-induced stress is conferred when ketones are introduced in cell culture models [19,20]. Thus, it is also possible that exogenous ketone supplementation may also have a direct effect in terms of attenuating ROS production and quenching free radical chain reactions.

Based upon the aforementioned supporting literature, it stands to reason that KD feeding or dietary ketone salt supplementation may mitigate ROS production or attenuate oxidative stress over long-term periods which, in turn, may improve mitochondrial quality. Therefore, the purpose of this investigation was to examine if short-term (i.e., 1 week) and/or long-term (i.e., 8 months) KD feeding or BHB salt supplementation affect skeletal muscle, brain and/or liver measures of oxidative stress and mitochondrial function. Our a priori hypothesis was that KD feeding and ketone salt supplementation would acutely and chronically enhance the expression of endogenous antioxidants and, in long-term rats, this would translate to a reduction in tissue oxidative stress markers and enhanced mitochondrial quantity and quality.

2. Materials and Methods

2.1. Rats in 1 Week Experiment

All experimental procedures were approved by Auburn University's Institutional Animal Care and Use Committee (IACUC, protocol # 2016-2814, approval date 6 January 2016). Male Fisher 344 rats at 4 months of age (~360 g) were purchased (Harlan Laboratories, Indianapolis, IN, USA) and allowed to acclimate in the animal housing facility at Auburn University for 1 week prior to experimentation. During acclimation, rats were provided standard rodent chow (SC; 24% (% kcal) protein, 58% CHO (2.8% fiber w/w), 18% fat; Teklad Global #2018 Diet, Harlan Laboratories) and water ad libitum in a maintained ambient temperature and constant 12 h light: 12 h dark cycle. For a 1 week period following acclimation, rats were provided isocaloric-isonitrogenous-isofibrous amounts of one of three diets:

(1) 10 rats (SC) were provided with 20 g/day of the aforementioned SC given during the acclimation phase.
(2) 10 rats (KD) were provided with 16 g/day of a commercially designed KD (Harlan Tekland diet #10787; Harlan Laboratories, Indianapolis, IN, USA) that was designed to induce nutritional ketosis and has been used previously by our laboratory [21]. Casein protein (Optimum Nutrition Inc., Downers Grove, IL, USA) and cellulose powder (Allergy Research Group, Alameda, CA, USA) were added to better compensate for between group differences in protein and fiber content (added protein and fiber were 23.5% of the modified diet). The diet specifications (following modifications) were as follows: 4.15 kcal/g, 23% protein, 10% carbohydrate (2.9% fiber w/w), and 67% fat. Medium chain triglycerides, flaxseed oil and canola oil were prominent fat sources in the parent KD.
(3) 10 rats (SC + KS) were provided with 20 g/day of the aforementioned SC, along with sodium BHB salt (DL-3 sodium hydroxybutyric acid, 5.8 kcal/g; NNB Nutrition, Lewisville, TX, USA) which were added to drinking water for ad libitum consumption with the intent to deliver ~ 2.2 g/day. This dosing schedule was designed with the intent of delivering a human-equivalent dose of 40 g/day dose as a "loading phase" per the body surface area rat-to-human conversions of Reagan-Shaw et al. [22] assuming the average rat weight of 400 g and the average human weight of 80 kg.

Notably, body masses were measured every other day, and food weights were measured daily.

2.2. Rats in 8 Months Experiment

As with the procedures outlined above, 8 months experimental procedures were approved by Auburn University's IACUC (protocol # 2016-2814, approval date 6 January 2016). Male Fisher 344 rats 4 months of age (~360 g) were purchased (Harlan Laboratories, Indianapolis, IN, USA) and allowed to acclimate in the animal housing facility for 1 week prior to experimentation in the same manner as described above. For an 8 months period following acclimation, rats were provided with isocaloric-isonitrogenous-isofibrous amounts of the three diets described above (SC $n = 8$, KD $n = 8$, SC + KS $n = 8$). In the SC + KS group, KS were added to water bottles attempting to deliver ~2.2 g/day for the first week as a "loading phase", then 0.3 g/day as a "maintenance phase" for the remaining duration. This dosing schedule was designed with the intent of delivering a human-equivalent dose of 40 g/day dose during the loading phase and a 10 g/day dose during the maintenance phase per the body surface area rat-to-human conversions of Reagan-Shaw et al. described above. One SC + KS rat inexplicably lost >20% body mass during the first two months of treatment and, thus, was euthanized for humane reasons and not included in the analyses. One SC rat experienced rapid weight loss

119

towards the end of the 8 months intervention. This rat presented large intra-abdominal tumors upon necropsy and was also not included in the analysis. Thus, final n-sizes were SC $n = 7$, KD $n = 8$ and SC + KS $n = 7$. Notably, bodyweights were recorded weekly, food intakes were recorded daily, and water was measured daily in order to ascertain the amount of KS ingested by the SC + KS group.

2.3. Rotarod Performance in 8 Month-Fed Rats

Rotarod performance is used in rodent studies to assess a combination of balance, grip strength, motor coordination and muscular endurance [23]. In 8 month-fed rats, rotarod performance was assessed at 2, 4, 6 and 8 months into the intervention using a single-lane device (Product#: ENV-571R; Med Associates Inc., Saint Albans City, VT, USA). Briefly, all assessments took place during the beginning of the rat light cycle (i.e., 0600–0800) whereby rats were placed on the device and the motorized rotor was initiated at a progressive speed from 4.0 to 40.0 r/min. An automated timer tracked time spent on the rod and, once the rats fatigued and dismounted from the rod, a laser beam break stopped the timer.

2.4. Necropsies and Tissue Preparation in Rats from Both Feeding Experiments

On the morning of necropsies (0500–0600), rats were food-deprived but provided water ad libitum. Rats were then transported from the campus vivarium to the School of Kinesiology and allowed to acclimate for 2–3 h. Thereafter, rats were euthanized under CO_2 gas in a 2 L induction chamber (VetEquip, Pleasanton, CA, USA). Following euthanasia, a final body mass was recorded, and blood was collected from the heart using a 22 gauge syringe. Collected blood was placed in a 6 mL serum separator tube and allowed to clot, centrifuged at 3500 g for 10 min, and resultant serum was aliquoted into 2.7 mL microcentrifuge tubes for storage at $-80\,°C$ until analysis. The gastrocnemius, brain, and liver were dissected out. In 1 week-fed rats, approximately 50 mg from each tissue was immediately placed in RNA/DNA Shield (Zymo Research, Irvine, CA, USA) and stored at 4 °C until RNA isolation using Trizol-based methods (described in Section 2.5), whereas the remainder of the tissue was flash frozen in liquid nitrogen and stored at $-80\,°C$ until total antioxidant analysis. In 8 month-fed rats, necropsies were carried out exactly as detailed above with the exception being that approximately 800 mg from the right gastrocnemius muscle, 1000 mg of liver and 500 mg of whole-brain (without cerebellum) was immediately used for mitochondria isolation as described below. The remainder of the tissue was flash frozen in liquid nitrogen and stored at $-80\,°C$ until Western blotting, GSH/GSSG and total antioxidant capacity analyses described below.

2.5. RNA Isolation, cDNA Synthesis and Real-Time Polymerase Chain Reaction (RT-PCR) for 1 Week-Fed Rat Tissues

Muscle/brain/liver stored in RNA/DNA Shield described above were placed in 10 volumes (500 µL) Ribozol (Ameresco, Solon, OH, USA) in a 2.7 mL microcentrifuge tube and were homogenized with a tight-fitting pestle. Phase separation (for RNA isolation) was achieved according to manufacturer's instructions. Following RNA precipitation and pelleting, pellets were resuspended in 30 µL of RNase-free water, and RNA concentrations were determined in duplicate at an absorbance of 260 nm by using a NanoDrop Lite (Thermo Scientific, Waltham, MA, USA). For cDNA synthesis, 1 µg of muscle/liver/brain RNA was reverse transcribed into cDNA for RT-PCR analyses with a commercial qScript cDNA SuperMix (Quanta Biosciences, Gaithersburg, MD, USA). RT-PCR was performed with gene-specific primers and SYBR green-based methods in a RT-PCR thermal cycler (Bio-Rad, Hercules, CA, USA). Primers were designed with open-sourced software (Primer3Plus, Cambridge, MA, USA), and melt curve analyses demonstrated that one PCR product was amplified per reaction. The forward and reverse primer sequences are as follows: Glutamate-Cysteine Ligase Modifier Subunit (Gclm): forward primer 5'-ACATTGAAGCCCAGGAGTGG-3', reverse primer 5'-CGATGACCGAGTACCTCAGC-3'; Glutamate-Cysteine Ligase Catalytic Subunit (Gclc): forward primer 5'-GAGATGCCGTCTTACAGGGG-3', reverse primer 5'-TTGCTACACCCATCC

ACCAC-3'. Catalase (Cat): forward primer 5'-TTAACGCGCAGATCATGCA-3', reverse primer 5'-CAAGTTTTTGATGCCCTGGT-3'. Glutathione peroxidase (Gpx): forward primer 5'-TCTGCACA CTCCCAGACAAG-3', reverse primer 5'-AGTCACCCATCACCGTCTTC-3'. Superoxide dismutase 2 (Sod2): forward primer 5'-TTAACGCGCAGATCATGCA-3', reverse primer 5'-CCTCGGTGAC GTTCAGATTGT-3'. Fold-change values from SC rats were performed using the $2^{-\Delta\Delta CT}$ method where $2^{\Delta CT}$ = (housekeeping gene (HKG) CT − gene of interest CT) and $2^{-\Delta\Delta CT}$ (or fold-change) = ($2^{\Delta CT}$ value for each rat/$2^{\Delta CT}$ group average of SC). Of note, 18S ribosomal rRNA (18S) was used as a HKG given that it remained stable across all treatments (primer sequence: forward primer 5'-AAACGGCTACCACATCCAAG-3', reverse primer 5'-CCTCCAATGGATCCTCGTTA-3').

2.6. Tissue Total Antioxidant and Serum BHB Assays for All Rats

Commercial colorimetric assay kits were used to determine muscle/liver/brain total antioxidant capacity (Antioxidant assay kit #709001; Cayman Chemical, Ann Arbor, MI, USA) and serum BHB levels (BHB colorimetric assay kit #700190; Cayman Chemical), respectively, according to manufacturer's instructions. For total antioxidant capacity determination, approximately 100 mg of frozen muscle/brain/liver was homogenized in kit assay buffer and centrifuged according to manufacturer's instructions, and supernatants were assayed. Given that brain tissue in 8 month-fed rats was devoted to mitochondrial assays and Western blotting, brain total antioxidant capacity determination was not performed because of lack of available tissue. Following assay execution, all plates were read in a UV-Vis microplate reader (BioTek Synergy H1 Multi-Mode Reader; BioTek, Winooski, VT, USA) at absorbance readings according to manufacturer's recommendations.

2.7. Western Blot Analysis in 8 Month-Fed Rat Tissues

Muscle/brain/liver was removed from −80 °C storage and crushed on a liquid nitrogen-cooled stage. Approximately 50 mg of tissue from each tissue sample was placed in 500 μL of 1× non-denaturing cell lysis buffer (Cell Signaling, Danvers, MA, USA) with added protease inhibitors (1 μg/mL leupeptin) and phosphatase inhibitors (2.5 mM sodium pyrophosphate, 1 mM β-glycerophoshate, 1 mM sodium orthovanadate) and was homogenized in microcentrifuge tubes using tight-fitting pestles. Samples were centrifuged at 500 *g* for 5 min at 4 °C. Supernatants were then subjected to a protein assay using a commercial bicinchoninic acid assay (Thermo Scientific) and were prepared for Western blotting using 4× Laemmli reducing buffer at 2 μg/μL. Subsequently, 20 μL of prepped samples were loaded onto precast 12% SDS-polyacrylamide gels (Bio-Rad) and were subjected to electrophoresis (200 V at 75 min) using premade 1× SDS-PAGE running buffer (CBS Scientific, San Diego, CA, USA). Proteins were transferred to polyvinylidene difluoride membranes (Bio-Rad), and membranes were stained with Ponceau S following transfers to ensure even loading and transfer between samples. Membranes were then blocked with 5% nonfat milk powder diluted in TBS with 0.1% Tween-20 (TBST) for 1 h at room temperature. Primary antibodies directed against the proteins of interest were incubated with membranes overnight at 4 °C in TBST with 5% BSA added. Primary antibodies were used to detect whole-tissue 4-HNE (Abcam, Cambridge, MA, USA), superoxide dismutase 2 (SOD2; GeneTex, Irvine, CA, USA), Catalase (Cat; GeneTex), glutathione peroxidase (Gpx; GeneTex) and protein carbonyls (Oxyblot kit; EMD Millipore; Bellirica, MA, USA). On the following day, membranes were incubated with anti-rabbit or anti-mouse IgG secondary antibodies diluted in TBST with 5% BSA added (1:2000; Cell Signaling) at room temperature for 1 h prior to membrane development. Membrane development was accomplished by using an enhanced chemiluminescent reagent (EMD Millipore), and band densitometry was achieved with the use of a digitized gel documentation system and associated densitometry software (UVP, Upland, CA, USA). All protein band densities were normalized to Ponceau stain densities. All Western blot analysis data are presented in terms of fold-change from SC rats. Notably, Ponceau staining was used for normalization due to the unknown effects that long-term KD or KS supplementation exerted on putative housekeeping proteins involved in metabolic processes (e.g., GAPDH). Notwithstanding, we have allocated this methodology in past studies examining the effects of KD feeding on skeletal muscle and adipose tissue physiology [9,21,24], and several commentaries exist suggesting that Ponceau normalization for Western blot data is appropriate for models which may affect putative housekeeping protein expression levels [25–27].

2.8. Tissue Mitochondrial Glutathione Assays in 8 Month-Fed Rat Tissues

Muscle/brain/liver mitochondrial oxidized (GSSG) and total glutathione levels were determined using a commercial colorimetric assay (kit #700190; Cayman Chemical) according to the manufacturer's instructions, and reduced glutathione (GSH) levels was extrapolated from these values.

2.9. Mitochondrial Isolation, Respiration Assays, and Mitochondrial ROS Determination in 8 Month-Fed Rat Tissues

The day of necropsies, differential centrifugation was used to isolate gastrocnemius, brain, and liver mitochondria from fresh tissue using techniques described previously [28,29]. Mitochondrial oxygen consumption was measured as described by Messer et al. [30] in a respiration chamber maintained at 37 °C (Hansatech Instruments, Pentney, King's Lynn, UK). Isolated mitochondria were incubated with 1 mL of respiration buffer containing (in mM) 100 KCl, 5 KH_2PO_4, 1 EGTA, 50 MOPS, 10 $MgCl_2$, and 0.2% BSA at 37 °C in a water-jacketed respiratory chamber with continuous stirring. Flux through complex I was measured using 2 mM pyruvate and 2 mM malate, whereas flux through complex II was measured using 5 mM succinate. Rotenone (5 μM) was added to prevent electron backflow to complex I in the succinate-driven experiments. The maximal respiration (state 3), defined as the rate of respiration in the presence of ADP, was initiated by adding 0.25 mM ADP to the respiration chamber containing mitochondria and respiratory substrates. State 4 respiration was recorded following the phosphorylation of ADP. The respiratory control ratio (RCR) was calculated by dividing state 3 by state 4 respiration.

Mitochondrial ROS production was determined using Amplex red (Molecular Probes, Eugene, OR, USA). The assay was performed at 37 °C in 96-well plates with succinate as the substrate. Specifically, this assay was developed on the concept that horseradish peroxidase catalyzes the H_2O_2-dependent oxidation of nonfluorescent Amplex red to fluorescent resorufin red, and it is used to measure H_2O_2 as an indicator of superoxide production. SOD was added at 40 U/mL to convert all superoxide to H_2O_2. Using a multiwell-plate reader fluorometer (BioTek Synergy H1 Multi-Mode Reader; BioTek), we monitored resorufin formation at an excitation wavelength of 545 nm and a production wavelength of 590 nm. The level of resorufin formation was recorded every 5 min for 15 min, and H_2O_2 production was calculated with a standard curve.

2.10. Citrate Synthase Activity Assays in 8 Month-Fed Rat Tissues

Muscle/brain/liver tissue homogenate citrate synthase activities were performed as previously described by our laboratory [9]. Briefly, 40 μg of tissue lysate (obtained from tissue lysis described in Section 2.7) were loaded onto 96-well plates. Subsequently, citrate synthase activity was measured as a function of the increase in absorbance from 5,5'-dithiobis-2-nitrobenzoic acid reduction following the methods described elsewhere [31].

2.11. Statistical Analysis

All statistical analyses were performed using IBM SPSS version 22.0 (IBM, Armonk, North Castle, NY, USA) and all data are presented as means ± standard error. Statistics on most dependent variables in both experiments were performed using one-way analysis of variance (ANOVA) tests with Tukey post hoc tests being performed when ANOVA $p < 0.05$. A two-way repeated measures ANOVA was performed for rotarod performance data in 8 month-fed rats. If group $p < 0.05$ then a Tukey post hoc test was performed to determine main group effects. If a group × time interaction $p < 0.05$ then: (a) one-way ANOVAs were performed with Tukey post hoc tests in order to determine between-group differences at each time point; and (b) repeated measures ANOVAs were performed within each group and Bonferroni post hoc tests were performed if $p < 0.05$ in order to determine within-group changes over time.

3. Results

3.1. Effects of Short-Term Feedings on Body Mass Change, Feed Efficiency and Serum BHB Levels

Days 3, 5 and 7 body masses were significantly different between treatments (all ANOVA $p \leq 0.001$). Specifically, change in body mass at day 3 was significantly lower in KD versus SC and SC + KS rats ($p < 0.001$ and $p < 0.001$, respectively), and this difference persisted at day 5 ($p = 0.010$ and $p = 0.002$, respectively) and day 7 ($p = 0.036$ and $p < 0.001$, respectively) (Figure 1a). Feed efficiency (g body mass gained/kcal consumed) over the 7 days feeding period was significantly different between groups (ANOVA $p = 0.001$). Specifically, KD rats presented significantly lower values compared to SC + KS rats ($p < 0.001$) and differences between KD and SC rats approached significance ($p = 0.070$) (Figure 1b). Serum BHB levels were significantly different between groups (ANOVA $p = 0.035$). Specifically, serum BHB levels were higher in KD versus SC + KS rats ($p = 0.034$), but not different between KD and SC rats ($p = 0.140$) (Figure 1c). Notably, SC + KS rats consumed an average of 0.85 ± 0.03 g/day of sodium beta-hydroxybutyrate over the 1-week intervention through their drinking water.

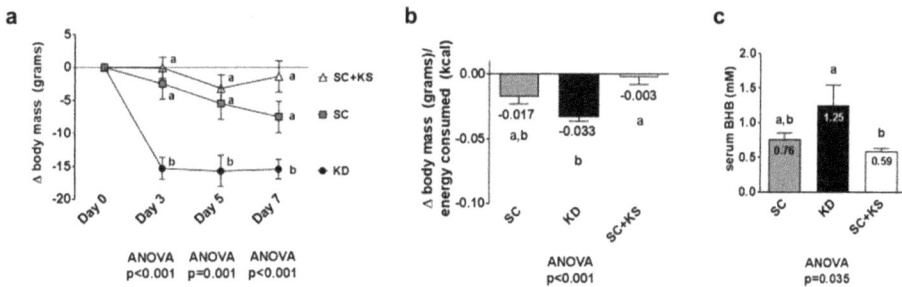

Figure 1. Change in body mass, overall feed efficiency and serum beta-hydroxybutyrate (BHB) levels in 1 week-fed rats. Legend: Change in body mass is presented in panel (**a**). Feed efficiency is presented in panel (**b**). Serum BHB levels are presented in panel (**c**). Data in panel (**a**) are presented as mean ± standard error values ($n = 10$ per group). In panels (**b,c**), all bars are presented as mean ± standard error values and group means are indicated within each bar. For all panels, ANOVA p-values are presented below data, and different lettering ("a" versus "b") indicates significant between-group differences whereas shared lettering ("a" as well as "b" or "a, b") indicates lack of significant between-group differences. Abbreviations: SC, standard chow-fed rats; KD, ketogenic diet-fed rats; SC + KS, standard chow-fed rats with supplemental sodium beta-hydroxybutyrate.

3.2. Effects of Short-Term Feedings on Muscle/Brain/Liver Oxidative Stress-Related mRNAs

Oxidative stress-related genes (i.e., Gclm, Gclc, Gsr, Gpx1, Cat or Sod2) were not differentially expressed in the gastrocnemius, brain or liver (all ANOVA $p > 0.05$; Figure 2a–c). Gastrocnemius antioxidant capacity was significantly different between groups (ANOVA $p = 0.025$; Figure 2d). Specifically, gastrocnemius antioxidant capacity was significantly lower in KD versus SC + KS rats ($p = 0.019$) but was not different between the KD and SC groups ($p = 0.275$) or SC and SC + KS groups ($p = 0.366$). Brain antioxidant capacity was not different between treatments (ANOVA $p = 0.920$; Figure 2d). Liver antioxidant capacity was significantly different between groups (ANOVA $p = 0.040$), although lower values in the SC group approached but were not significant compared to KD ($p = 0.056$) and SC + KS rats ($p = 0.083$) (Figure 2d).

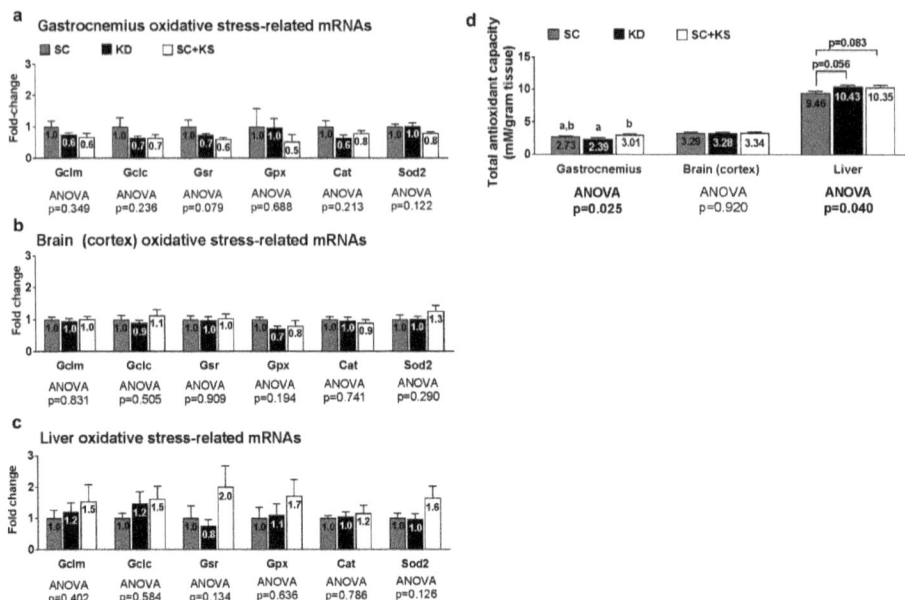

Figure 2. Oxidative stress and endogenous antioxidant-related markers in 1 week-fed rats. Legend: Oxidative stress and endogenous antioxidant related mRNAs are shown for gastrocnemius panel (**a**); brain panel (**b**); and liver panel (**c**). Tissue total antioxidant capacity levels are presented in panel (**d**). Fold-change for each mRNA are expressed relative to the SC group, all bars are presented as mean ± standard error values (*n* = 10 per group), group means are indicated within each bar, ANOVA *p*-values are presented below data, and different lettering in panel D ("a" versus "b") indicates significant between-group differences whereas shared lettering ("a" as well as "b" or "a, b") indicates lack of significant between-group differences. Abbreviations: Gclm, glutamate-cysteine ligase modulatory subunit; Gclc, glutamate-cysteine ligase catalytic subunit; Gsr, glutathione reductase; Gpx1, glutathione peroxidase; Cat, catalase; Sod2, supoeroxide dismutase 2; SC, standard chow-fed rats; KD, ketogenic diet-fed rats; SC + KS, standard chow-fed rats with supplemental sodium beta-hydroxybutyrate.

3.3. Effects of Long-Term Feedings on Body Masses, Feed Efficiency and Serum BHB

Regarding weekly body masses, there were significant differences between groups from weeks 10–30 (all ANOVA *p* < 0.05). Specifically, KD rats weighed less than SC and SC + KS rats from 10 weeks to 30 weeks (*p* < 0.05 at all time points; Figure 3a). Feed efficiency over the 8 months intervention (i.e., g body mass gained/kcal consumed) was significantly different between groups (ANOVA *p* < 0.001). Specifically, KD rats presented significantly feed efficiency values compared to SC + KS rats (*p* < 0.001) and SC rats (*p* < 0.001) (Figure 3b). Serum BHB levels were significantly different between groups (ANOVA *p* < 0.001). Specifically, serum BHB values were higher in KD versus SC rats (*p* < 0.001) and SC + KS rats (*p* < 0.001) (Figure 3c). Notably, SC + KS rats consumed an average of 0.21 ± 0.02 g/day of sodium beta-hydroxybutyrate salts through their drinking water over the 8 months intervention.

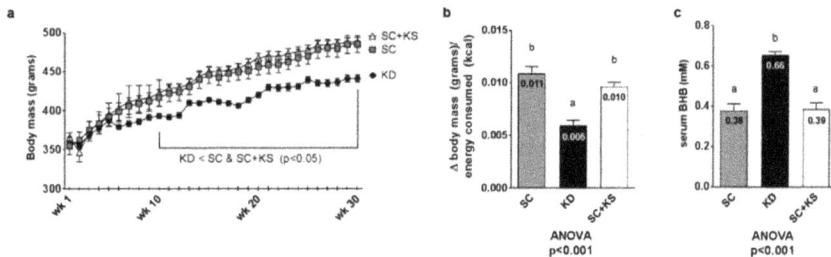

Figure 3. Change in body mass, overall feed efficiency and serum BHB in 8 month-fed rats. Legend: Body mass changes over the intervention are presented in panel (**a**), feed efficiency is presented in panel (**b**) and serum BHB levels are presented in panel (**c**). Data in panel (**a**) are presented as mean ± standard error values (*n* = 7–8 per group). In panels (**b**,**c**), all bars are presented as mean ± standard error values and group means are indicated within each bar. In panels B and C ANOVA *p*-values are presented below data, and different lettering ("a" versus "b") indicates significant between-group differences whereas shared lettering ("a" as well as "b" or "a, b") indicates lack of significant between-group differences. Abbreviations: SC, standard chow-fed rats; KD, ketogenic diet-fed rats; SC + KS, standard chow-fed rats with supplemental sodium beta-hydroxybutyrate.

3.4. Effects of Long-Term Feedings on Rotarod Performance

There was a significant main group effect ($p < 0.001$) and group × time interaction ($p = 0.005$) for rotarod performance (Figure 4). Regarding the main group effect, rotarod performance was significantly higher in KD rats versus SC ($p = 0.001$) and SC + KS rats ($p = 0.005$). Regarding the group × time interaction: (a) at 2 months, performance was significantly higher in the KD versus the SC ($p < 0.001$) and SC + KS groups ($p = 0.009$); (b) at 4 months, performance was significantly higher in the KD versus the SC ($p = 0.017$) group, and values between the KD versus SC + KS group approached significance ($p = 0.060$); (c) at 8 months, performance was significantly higher in the KD versus the SC ($p = 0.008$) and SC + KS groups ($p = 0.049$); and (d) within the KD group there was a significant time effect ($p = 0.001$) whereby values at 6 months ($p = 0.006$) and 8 months ($p = 0.021$) were significantly lower than 2 months values (notably, there was no time effect within the SC ($p = 0.531$) and SC + KS groups ($p = 0.382$).

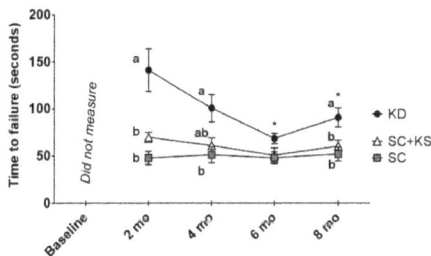

Figure 4. Change in rotarod performance in 8 month-fed rats. Legend: Rotarod performance every 2 months are presented herein. All data are presented as mean ± standard error values (*n* = 7–8 per group). Given the significant group × tine interaction, post hoc analyses at each time point and within each group were performed. Specifically, different lettering ("a" versus "b") indicates significant between-group differences whereas shared lettering ("a" as well as "b" or "a, b") indicates lack of significant between-group differences, and asterisks indicate significant decreases within the KD rats at 6 and 8 months compared to 2 months values ($p < 0.05$). Abbreviations: SC, standard chow-fed rats; KD, ketogenic diet-fed rats; SC + KS, standard chow-fed rats with supplemental sodium beta-hydroxybutyrate.

3.5. Effects of Long-Term Feedings on Gastrocnemius Oxidative Stress-Related Proteins and Markers

Gastrocnemius protein expression levels of Cat (ANOVA $p = 0.393$), Gpx (ANOVA $p = 0.702$) and Sod2 (ANOVA $p = 0.925$) were not differentially expressed between groups (Figure 5a). Likewise, gastrocnemius 4-HNE (ANOVA $p = 0.725$) and protein carbonyl levels (ANOVA $p = 0.504$) were not differentially expressed groups (Figure 5b).

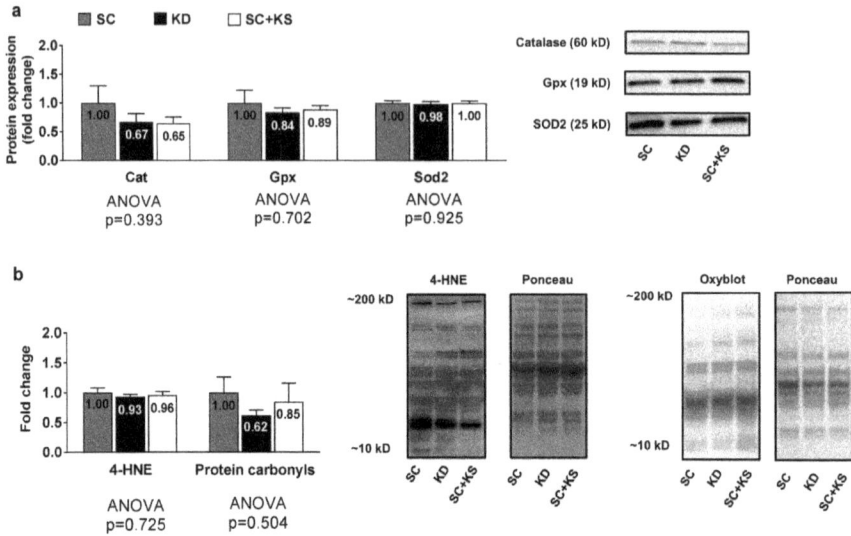

Figure 5. Gastrocnemius oxidative stress-related proteins and markers in 8 month-fed rats. Legend: Gastrocnemius catalase (Cat), glutathione peroxidase (Gpx), and mitochondrial superoxide dismutase (Sod2) protein levels are presented with representative images in panel (**a**). In panel (**b**), tissue 4-hydroxynonenal (4-HNE) and protein carbonyls (Oxyblot) are presented with representative images. Fold-change for each protein are expressed relative to the SC group, all bars are presented as mean ± standard error values (*n* = 7–8 per group), group means are indicated within each bar and ANOVA *p*-values are presented below data. Abbreviations: SC, standard chow-fed rats; KD, ketogenic diet-fed rats; SC + KS, standard chow-fed rats with supplemental sodium beta-hydroxybutyrate.

3.6. Effects of Long-Term Feedings on Brain Oxidative Stress-Related Proteins and Markers

Brain protein expression levels of Cat (ANOVA $p = 0.331$), Gpx (ANOVA $p = 0.084$) and Sod2 (ANOVA $p = 0.741$) were not differentially expressed between groups (Figure 6a). Likewise, brain 4-HNE (ANOVA $p = 0.676$) and protein carbonyl levels (ANOVA $p = 0.905$) were not differentially expressed between groups (Figure 6b).

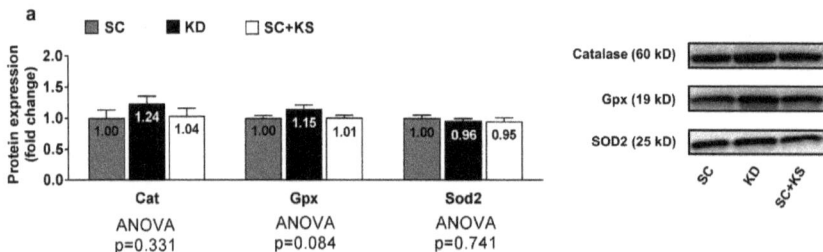

Figure 6. *Cont.*

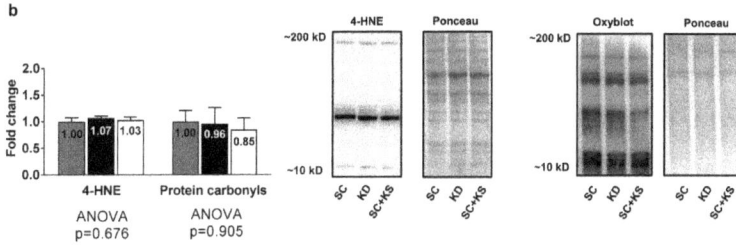

Figure 6. Brain oxidative stress-related proteins and markers in 8 month-fed rats. Legend: Brain catalase (Cat), glutathione peroxidase (Gpx), and mitochondrial superoxide dismutase (Sod2) protein levels are presented with representative images in panel (**a**). In panel (**b**), tissue 4-Hydroxynonenal (4-HNE), and protein carbonyls (Oxyblot) are presented with representative images. Fold-change for each protein are expressed relative to the SC group, all bars are presented as mean ± standard error values (*n* = 7–8 per group), group means are indicated within each bar and ANOVA *p*-values are presented below data. Abbreviations: SC, standard chow-fed rats; KD, ketogenic diet-fed rats; SC + KS, standard chow-fed rats with supplemental sodium beta-hydroxybutyrate.

3.7. Effects of Long-Term Feedings on Liver Oxidative Stress-Related Proteins and Markers

Liver protein expression levels of Gpx were greater in the KD versus the SC group (*p* = 0.004) and high levels approached significance in the KD versus the SC + KS group (*p* = 0.068) (Figure 7a). Liver protein expression levels of Cat (ANOVA *p* = 0.932) and Sod2 (ANOVA *p* = 0.253) were not differentially expressed between groups (Figure 7a). Liver 4-HNE levels were not differentially expressed between groups (ANOVA *p* = 0.684), although liver protein carbonyls were greater in the SC + KS versus SC (*p* = 0.034) and KD (*p* = 0.001) groups (Figure 7b).

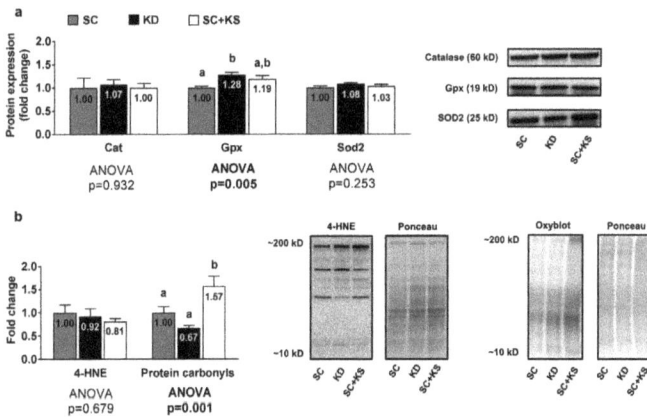

Figure 7. Liver oxidative stress-related proteins and markers in 8 month-fed rats. Legend: Liver catalase (Cat), glutathione peroxidase (Gpx), and mitochondrial superoxide dismutase (Sod2) protein levels are presented with representative images in panel (**a**). In panel (**b**), tissue 4-Hydroxynonenal (4-HNE), and protein carbonyls (Oxyblot) are presented with representative images. Fold-change for each protein are expressed relative to the SC group, all bars are presented as mean ± standard error values (*n* = 7–8 per group), group means are indicated within each bar, ANOVA *p*-values are presented below data and different lettering ("a" versus "b") indicates significant between-group differences whereas shared lettering ("a" as well as "b" or "a, b") indicates lack of significant between-group differences. Abbreviations: SC, standard chow-fed rats; KD, ketogenic diet-fed rats; SC + KS, standard chow-fed rats with supplemental sodium beta-hydroxybutyrate.

3.8. Effects of Long-Term Feedings on Tissue Mitochondrial Glutathione, Total Antioxidant Capacity, and ROS Levels

Gastrocnemius mitochondrial oxidized (GSSG) glutathione, reduced (GSH) glutathione and total glutathione were differentially expressed between groups (ANOVA p values = 0.001–0.017; Figure 8a). Specifically, the following was found to occur: (a) GSSG levels were greater in SC-fed versus KD-fed rats (p = 0.018) and trended higher in SC-fed versus SC + KS-fed rats (p = 0.063); (b) GSH levels were greater in the SC versus the KD (p = 0.001) and SC + KS (p = 0.004) groups; and (c) total glutathione levels were greater in the SC versus the KD (p = 0.001) and SC + KS (p = 0.005) groups. Brain mitochondrial GSSG (ANOVA p = 0.294), GSH (ANOVA p = 0.191) and total glutathione (ANOVA p = 0.200) were not differentially expressed between groups (Figure 8b). Liver mitochondrial GSSG (ANOVA p = 0.117), GSH (ANOVA p = 0.699) and total glutathione (ANOVA p = 0.321) were not differentially expressed between groups (Figure 8c). Gastrocnemius total antioxidant capacity levels were not different between groups (ANOVA p = 0.146; Figure 8d). Liver total antioxidant capacity levels were significantly higher in KD versus SC rats (p = 0.028), but not KD versus SC + KS rats (p = 0.216; Figure 8d). Gastrocnemius mitochondrial ROS production was significantly different between groups (ANOVA p = 0.007; Figure 8e). Specifically, ROS values were significantly higher in KD rats versus the SC (p = 0.014) and SC + KS groups (p = 0.015). Mitochondrial ROS production in the brain (ANOVA p = 0.162) and liver (ANOVA p = 0.222) were not different between groups (Figure 8e).

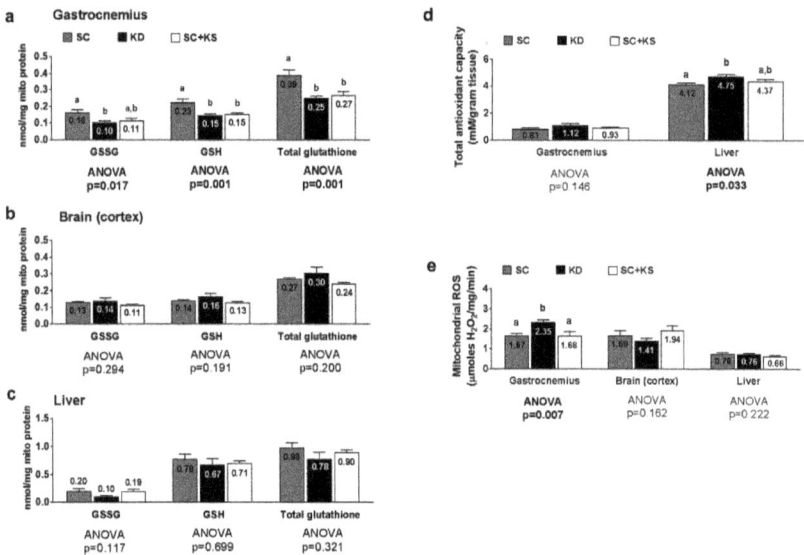

Figure 8. Tissue mitochondrial glutathione, total antioxidant capacity, and ROS levels in 8 month-fed rats. Legend: Gastrocnemius mitochondrial oxidized (GSSG), reduced (GSH) and total glutathione are presented in panel (**a**); Brain mitochondrial oxidized (GSSG), reduced (GSH) and total glutathione are presented in panel (**b**); Liver mitochondrial oxidized (GSSG), reduced (GSH) and total glutathione are presented in panel (**c**); Tissue total antioxidant capacity levels are presented in panel (**d**); Tissue mitochondrial reactive oxygen species production levels are presented in panel (**e**). All bars are presented as mean ± standard error values (*n* = 7–8 per group), group means are indicated within each bar and ANOVA *p*-values are presented below data, and different lettering ("a" versus "b") indicates significant between-group differences whereas shared lettering ("a" as well as "b" or "a, b") indicates lack of significant between-group differences. Abbreviations: SC, standard chow-fed rats; KD, ketogenic diet-fed rats; SC + KS, standard chow-fed rats with supplemental sodium beta-hydroxybutyrate.

3.9. Effects of Long-Term Feedings on Gastrocnemius Mitochondrial Function

Between-group differences in gastrocnemius pyruvate-malate state 3 approached significance (ANOVA $p = 0.072$) whereby lower values in KD versus SC rats approached significance ($p = 0.064$) (Figure 9a). Gastrocnemius pyruvate-malate state 4 was not different between groups ($p = 0.119$; Figure 9b). Gastrocnemius pyruvate-malate RCR values were significantly different between groups (ANOVA $p = 0.018$). Specifically, values were significantly lower in KD versus SC rats ($p = 0.030$) and SC + KS versus SC rats ($p = 0.035$) (Figure 9c). Gastrocnemius succinate state 3 (ANOVA $p = 0.706$; Figure 9d), succinate state 4 (ANOVA $p = 0.500$; Figure 9e) and succinate RCR values (ANOVA $p = 0.582$; Figure 9f) were not different between groups.

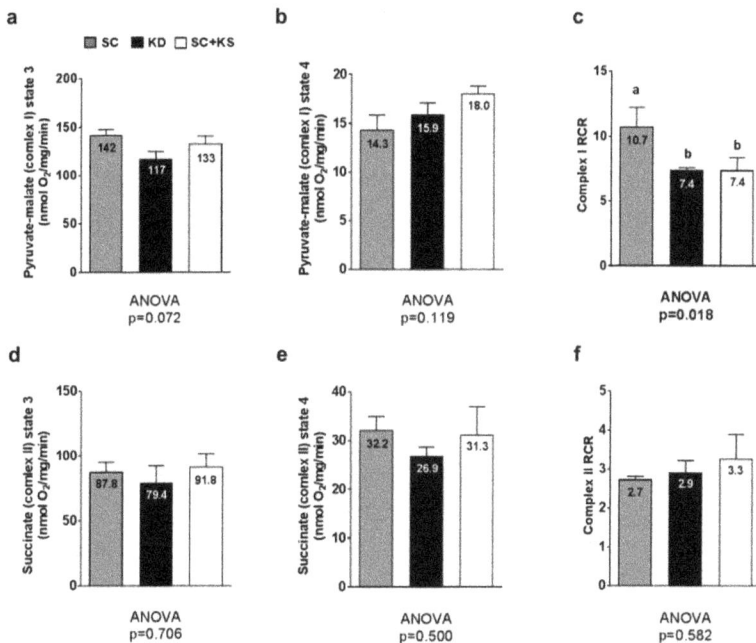

Figure 9. Gastrocnemius mitochondrial function in 8 month-fed rats. Legend: State 3 pyruvate-malate is presented in panel (**a**); State 4 pyruvate-malate is presented in panel (**b**); Pyruvate-malate RCR is presented in panel (**c**); State 3 succinate is presented in panel (**d**); State 4 succinate is presented in panel (**e**); Succinate RCR is presented in panel (**f**). All bars are presented as mean ± standard error values ($n = 7$–8 per group), group means are indicated within each bar and ANOVA p-values are presented below data, and different lettering ("a" versus "b") indicates significant between-group differences whereas shared lettering ("a" as well as "b" or "a, b") indicates lack of significant between-group differences. Abbreviations: SC, standard chow-fed rats; KD, ketogenic diet-fed rats; SC + KS, standard chow-fed rats with supplemental sodium beta-hydroxybutyrate; RCR, respiratory exchange ratio.

3.10. Effects of Long-Term Feedings on Brain and Liver Mitochondrial Function

Brain pyruvate-malate state 3 (ANOVA $p = 0.706$), pyruvate-malate state 4 (ANOVA $p = 0.500$) and pyruvate-malate RCR values (ANOVA $p = 0.582$; Figure 10a) were not different between groups. Notably, brain succinate state 3, state 4 and RCR values are not reported given that a reliable state 4 could not be obtained.

Liver pyruvate-malate state 3 (ANOVA $p = 0.466$), pyruvate-malate state 4 (ANOVA $p = 0.364$) and pyruvate-malate RCR values (ANOVA $p = 0.714$; Figure 10b) were not different between groups.

Liver succinate state 3 (ANOVA $p = 0.246$), succinate state 4 (ANOVA $p = 0.552$) and succinate RCR values (ANOVA $p = 0.112$; Figure 10c) were not different between groups.

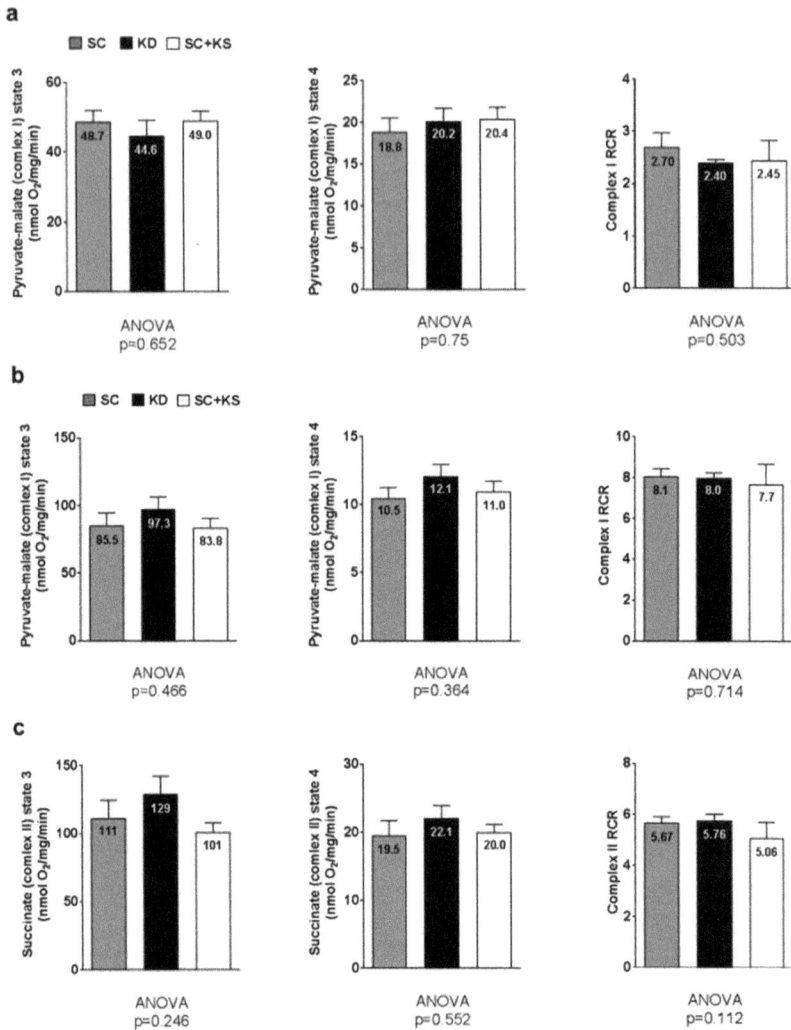

Figure 10. Brain and liver mitochondrial function in 8 month-fed rats. Legend: Brain state 3 and 4 pyruvate-malate and respiratory control ratio (RCR) values are presented in panel (**a**); Liver state 3 and 4 pyruvate-malate and RCR values are presented in panel (**b**); Liver state 3 and 4 succinate and RCR values are presented in panel (**c**). All bars are presented as mean ± standard error values (n = 7–8 per group), group means are indicated within each bar and ANOVA p-values are presented below data. Abbreviations: SC, standard chow-fed rats; KD, ketogenic diet-fed rats; SC + KS, standard chow-fed rats with supplemental sodium beta-hydroxybutyrate; RCR, respiratory exchange ratio.

3.11. Effects of Long-Term Feedings on Tissue Citrate Synthase Activity

Gastrocnemius citrate synthase activity levels were significantly different between groups (ANOVA $p < 0.001$). Specifically, values were lower in KD versus SC + KS ($p < 0.001$) and SC ($p = 0.009$)

rats (Figure 11). Citrate synthase activity levels in the brain (ANOVA $p = 0.386$) and liver (ANOVA $p = 0.548$) were similar between groups (Figure 11).

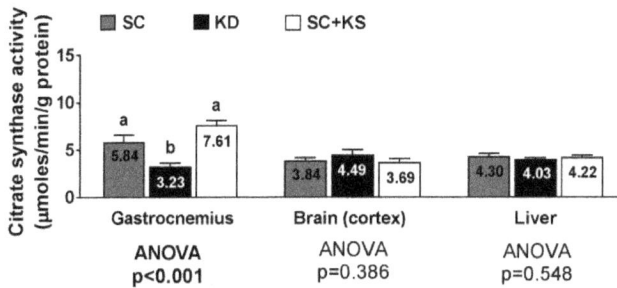

Figure 11. Tissue citrate synthase activity in 8 month-fed rats. Legend: Tissue citrate synthase activity is presented. All bars are presented as mean ± standard error values ($n = 7$–8 per group), group means are indicated within each bar and ANOVA p-values are presented below data, and different lettering ("a" versus "b") indicates significant between-group differences whereas shared lettering ("a" as well as "b" or "a, b") indicates lack of significant between-group differences. Abbreviations: SC, standard control; KD, ketogenic diet; SC + KS, standard control with supplemental sodium beta-hydroxybutyrate.

4. Discussion

Overall, KD-fed rats exhibited more weight loss (1 week feedings) or an attenuation of weight gain (8 months feedings) compared to the SC and SC + KS groups. In short-term fed rats, serum ketones and liver antioxidant capacity trended higher in KD- versus SC-fed rats, although all muscle, brain and liver oxidative stress-related mRNAs were similar between treatments. In long term-fed rats, serum ketones were greater in KD- versus SC- and SC + KS-fed rats ($p < 0.05$) suggesting that diet, but not the employed KS dosage, induces ketonemia. Moreover, liver total antioxidant capacity and glutathione peroxidase protein levels were 15% and 28% higher, respectively, in long-term KD-versus SC-fed rats ($p < 0.05$), while liver protein carbonyls were lowest in KD-fed rats. Contrary to our hypothesis, gastrocnemius mitochondrial ROS production was ~40% higher in KD-fed rats versus other groups ($p = 0.007$) and this may have been related to significantly lower mitochondrial glutathione levels. Further, pyruvate-malate mitochondrial respiratory control ratio was lower in KD- and SC + KS- versus SC-fed rats ($p < 0.05$), and gastrocnemius mitochondrial quantity was 81% and 136% greater in SC- and SC + KS-fed rats versus KD-fed rats ($p < 0.001$). Rotarod performance was greatest in KD-fed rats versus other groups at 2, 4 and 8 months ($p < 0.05$), although an age-related decline in performance existed in KD-fed rats whereas this was not evident in the other two groups.

4.1. Ketogenic Diet Feeding, But Not Ketone Salt Supplementation, Elevates Serum BHB Levels and Produces Short-Term Weight Loss as Well as a Long-Term Attenuation of Weight Gain

Herein, we report that KD feeding induces short-term weight loss and reduced feed efficiency relative to standard chow feeding. Notably, SC-fed rats also presented weight loss during the 1-week feeding experiment, although KD-induced weight loss was significantly greater. While this finding is difficult to reconcile, we believe that the initial handling of rats (e.g., body mass assessment every other day and food weighing every day) was likely a stressor to all groups of rats in the 1-week study. Hence, we are uncertain as to whether or not this stressor appreciably compromised the molecular markers examined in the 1-week animals. In spite of the aforementioned limitation, our group has previously observed similar outcomes in a separate cohort of rats fed the unmodified KD (i.e., no additional protein or fiber added) over a 6-week period when compared to SC-fed and Western diet-fed rats [24]. Likewise, other studies have reported similar findings using ketogenic diets in rodents [32–34]. While the particular biochemical cascade and subsequent hormonal milieu that occurred during this timeframe was beyond the scope of this study, KD-induced weight loss (or weight gain prevention) may be prompted by chronically low insulin levels and/or due to the bodily loss of biochemical energy via ketone body excretion in the urine [35]. Notably, ketone salt supplementation did not affect body weight or feed efficiency when comparing SC + KS versus SC-fed rats. Ketone ester feedings in mice over a 4-week period has been shown to robustly elevate serum BHB levels, stimulate brown fat activation, elevate resting energy expenditure and promote weight loss in mice [36]. Kesl et al. [37] also reported that 5 and 10 g/kg of KS, which is 8-to-16-fold greater than our employed KS dosage, prevents weight gain in rats although these dosages did not alter serum BHB levels. Providing a mechanistic explanation as to why serum BHB levels are not responsive to exogenous KS versus ketone ester supplementation in rats is difficult to reconcile, but we posit that the lack of significant physiological effects of KS supplementation in both feeding experiments are likely due to the relatively lower KS doses employed in the current study.

4.2. Short-Term Ketogenic Diet Feeding or Ketone Salt Supplementation Do Not Alter Oxidative Stress-Related Gene Expression in Muscle/Brain/Liver Tissue

The purpose of the 1 week experiment was to test the theory suggested by Milder and Patel [8] whereby ketogenic diet feeding is posited to increase the mRNA expression of glutathione-related mRNAs (i.e., Gclc and Gclm) after a one week period leading to an enhanced endogenous antioxidant defense system. Additionally, we were interested in examining the effects of KS supplementation given recent findings like that of Shimazu et al. [38] who have reported that BHB acts as an antioxidant in vitro. The data presented herein do not demonstrate altered mRNA expression of either Gclc or Gclm in skeletal muscle, whole brain or liver tissue after 1 week of KD or KS feeding in rats. In addition, we did not observe mRNA alterations in other endogenous antioxidants (i.e., Gsr, Gpx1 and Sod2) or total antioxidant capacity in the assayed tissues. In contrast to our findings, Jarrett and colleagues [14] observed increased GCL activity and elevated protein levels in response to a 3 weeks of KD feeding in rat hippocampal tissue. Discrepancies in our outcomes related to KD feeding could be due to: (a) our investigation of mRNA markers following a one week time course; and/or (b) with regard to our findings in the brain, discordant findings may be related to the region of the brain that was assayed. Notably, we performed all assays in the whole cortical brain while others examining KD-induced alterations in mRNA and protein/enzyme activity have examined the hippocampus. For instance, Ziegler and colleagues [15] also reported that KD feeding over an 8 week period in rats increased Gpx activity increased in the hippocampus, reduced Cat activity in the cerebellum and did not alter these markers in the cortex. Thus, these data collectively suggest that: (a) different brain regions likely express different biochemical responses to KD feeding; and (b) short- or long-term KD feeding does not appreciably alter oxidative stress-related gene expression in skeletal muscle or liver tissue. The lack of appreciable antioxidant effects regarding short-term KS supplementation is likely due to the employed dosage not eliciting increases in serum BHB levels as stated above.

4.3. Long-Term Ketogenic Diet Feeding Positively Impacts Select Markers of Oxidative Stress in the Liver But Does Not Alter Mitochondrial Quality in Liver or Brain Tissue

Our findings that the short- and long-term KD feeding increases liver total antioxidant capacity, and long-term KD feeding increases Gpx protein expression levels and decreases liver protein carbonyl levels (numerically versus SC-fed rats and significantly versus SC + KS-fed rats) indicate a potential protective effect of the KD feeding on liver tissue. Several studies have observed that KD feeding increase inflammation and fibroblast growth factor 21-mediated hepatocyte steatosis in rodents [3,39,40]. However, we have previously reported decreased liver inflammatory signaling and triglyceride accumulation with KD feeding in rats whereby the KD was similar to the one utilized in the present study [24]. Milder et al. [18] have also reported an increase in nuclear Nrf2 and a decrease in liver mitochondrial glutathione levels following 3 weeks of KD feeding in rats. While we did not assess nuclear Nrf2 levels in the current study and report that the long-term KD feeding does not alter liver glutathione levels, the contention by Milder et al. that short-term KD feeding leads to long-term increases in oxidative stress defenses may be a potential explanation as to why protein carbonyls were lowest in long-term KD fed animals while Gpx protein and total antioxidant levels were relatively higher in these same animals. It is also plausible that lower liver inflammatory signaling and/or lower triglyceride accumulation, as reported in our past study, could be a contributing mechanism to increased liver oxidative stress protection given the link between ectopic tissue fat accumulation, inflammation and oxidative stress [41].

Interestingly, we also observed that long-term ketone salt supplementation significantly increased liver protein carbonyl levels. While there is currently no literature to date documenting such effects, this suggests that the daily low-dose consumption of ketone salts may induce repetitive bouts of oxidative stress to the liver and this finding warrants further mechanistic investigation. It should also be noted that unaltered brain and liver citrate synthase activity (a well-validated surrogate of mitochondrial density [42]) in KD-fed rats is also discordant with previous literature given that increases in citrate synthase activity have been reported to occur in the hippocampus of KD-fed rodents after a 3-week period [10,43]. Again, however, discrepancies in our findings versus the aforementioned literature may be explained by the long-term nature of our study as well as tissue-specific differences that occur in response to ketogenic diet feedings. Beneficial KD-induced mitochondrial adaptations that occur in the liver or brain, if they do occur, may be transient once the rodents are given an extended period time for dietary adaptation.

4.4. Long-Term Ketogenic Diet Feeding Negatively Impacts Skeletal Muscle Mitochondrial Physiology

While long-term KD feeding did not alter markers of brain and liver mitochondrial quality or quantity, we observed that long-term KD feeding robustly impacts several of these variables in skeletal muscle. Specifically, long-term KD-fed rats expressed lower gastrocnemius glutathione levels, lower gastrocnemius maximal citrate synthase activity, higher gastrocnemius mitochondrial ROS production, and an impairment in gastrocnemius pyruvate-malate respiratory control. Our laboratory has previously investigated skeletal muscle mitochondrial density and mitochondrial quality in 3 months old KD-fed rats following a 6 week dietary intervention [9]. We reported that succinate respiratory control increased in KD-fed rats compared to rats fed a Western diet, suggesting that shorter-term ketogenic dieting may improve lipid oxidation. However, the present study indicates that this mitochondrial adaptation does not persist after 8 months of KD feeding. Regarding the pyruvate-malate respiratory control ratio in skeletal muscle, the depression observed in KD-fed rats may be related to increased concentrations of uncoupling proteins in the gastrocnemius mitochondria. This phenomenon has been observed by others who reported KD-induced uncoupling protein expression increases in brown adipose and neural tissue, respectively [13,44]. Likewise, the KD-induced decrease in gastrocnemius mitochondrial pyruvate-malate respiratory control ratio may also be related to the increased ROS production observed given that ROS-induced mitochondrial damage is implicated in the reduction of mitochondrial function [45–47]. The KD-induced decrease in gastrocnemius citrate synthase activity observed may also be indicative of decrements in skeletal muscle mitochondrial quality or quantity. In this regard, past literature suggests that ketonemia, high fat diet feeding and/or fasting in humans and rodents down-regulates oxidative phosphorylation, lowers skeletal muscle mitochondrial density, or lowers the efficiency mitochondrial respiration [48–50]. The KD-induced increase in gastrocnemius mitochondrial ROS and/or decrease in mitochondrial function and density may have also been due to mitochondrial glutathione levels being low in KD-fed animals. In support of this hypothesis, others have reported that low mitochondrial glutathione levels are associated with increased ROS production and decrements in mitochondrial function [51,52]. Collectively, our results along with other observations in past studies suggest that while short-term ketogenic dieting may enhance mitochondrial biogenesis and/or alter mitochondrial physiology in certain tissues, skeletal muscle may be particularly susceptible to KD-induced alterations in mitochondrial physiology following long-term feeding, and this phenomenon may be due to KD-induced decrements in mitochondrial glutathione levels and/or another unidentified mechanism.

It is notable that, despite the aforementioned observations in the gastrocnemius, KD-fed rats exhibited higher rotarod performance throughout the long-term intervention relative to the other two groups. While rotarod testing does not test maximal aerobic capacity, it does provide a reliable measure of coordination, muscle strength and endurance [23]. Therefore, our rotarod performance data suggest that long-term ketogenic dieting may result in increased skeletal muscle metabolic efficiency (i.e., similar or improved work output with lowered mitochondrial function and/or density in muscle). However, it is notable that rotarod performance robustly decreased with age in KD-fed rats, whereas this did not occur in the SC and SC + KS groups; this being suggestive that long-term KD-induced decrements in mitochondrial function may be a contributor to age-related declines in rotarod performance. In this regard, further research is needed in order to examine: (a) the KD-associated mechanisms that affect mitochondrial function and/or density; and/or (b) if KD feeding affects metabolic efficiency in skeletal muscle.

4.5. Experimental Considerations

Certain limitations are evident in the current study. First, due to resource constraints, short-term tissue analyses herein were only performed at one sacrificial time point which differs from Milder et al. [18] who examined the effects of ketogenic dieting at 1-day, 3-day, 1-week and 3-week sacrificial time points. Second, and as mentioned prior, our brain assays were performed on cortical brain lysates; this is an experimental consideration which likely masked any potential benefit of KD feeding and/or ketone salt supplementation on the hippocampus or other brain regions. Third, it is notable that total antioxidant capacity levels were assayed in different tissues rather than individual antioxidant enzyme activities due to resource constraints. Thus, given that liver antioxidant capacity increases in response to long-term ketogenic dieting, studying the effects of ketogenic dieting on single antioxidant activity levels are warranted. Lastly, it is noteworthy that sedentary rats were studied herein, and this may be a potential reason as to why skeletal muscle mitochondria is negatively impacted with long-term KD feeding. In this regard, we have previously reported that increases in skeletal muscle mitochondrial quality and quantity are not impaired in KD-fed rats that exercised via resistance-loaded voluntary wheel running over a 6-week period [9]. Moreover, elite ultra-endurance athletes that chronically engage in ketogenic-like dieting have been reported to perform equally as well in laboratory-based exercise tests compared to athletes that consume high-carbohydrate diets [53]; this is suggestive that long-term KD athletes do not present functional impairments in skeletal muscle mitochondrial integrity. Hence, again, a future research aim would be to examine if skeletal muscle mitochondrial deficits observed in 8 months KD-fed rats are mitigated with exercise training. One final experimental consideration is the rapid weight loss observed in one long-term SC + KS rat as well as the tumor observed during dissections in one long-term SC-fed rat. While the mechanistic causes of both events were not more thoroughly explored, it is notable that no adverse events were observed in KD-fed rats. Hence, long-term KD feeding in rodents appears to be safe and, with regard to the latter observation, future research is needed in order to determine if KD feeding has anti-tumorigenic effects.

5. Conclusions

In spite of the aforementioned limitations, we posit that these data provide a comprehensive evaluation as to how short-term and long-term KD feeding affect oxidative stress markers, endogenous antioxidant gene and protein expression and mitochondrial function in multiple tissues in rats. Longer-term human studies examining muscle biopsy specimens are needed in order to validate our findings, suggesting that long-term KD feeding appreciably alters mitochondrial physiology.

Acknowledgments: We would like to thank Patricia Rynders, Bettina Schemera and Timothy Knox for their critical assistance with animal husbandry.

Author Contributions: W.C.K. performed most animal husbandry and performed or oversaw most experiments. P.W.M., X.M., M.A.R., H.W.H., Y.Z., C.B.M., J.C.Q., K.C.Y., D.T.B., J.S.M. and D.J.M. critically assisted with experiments, statistical analyses and/or data interpretation. R.P.L., J.M.W., A.N.K. and M.D.R. conceived experimental design and dictated all analytical procedures. W.C.K. and M.D.R. primarily drafted the manuscript and all co-authors edited the manuscript prior to submission.

Conflicts of Interest: No authors declare competing interests regarding the publication of these data. Funding for all rats, per diems and reagents were provided though a contract from Applied Sports Performance Institute to MDR.

References

1. Paoli, A.; Rubini, A.; Volek, J.S.; Grimaldi, K.A. Beyond weight loss: A review of the therapeutic uses of very-low-carbohydrate (ketogenic) diets. *Eur. J. Clin. Nutr.* **2013**, *67*, 789–796. [CrossRef] [PubMed]

2. Kulak, D.; Polotsky, A.J. Should the ketogenic diet be considered for enhancing fertility? *Maturitas* **2013**, *74*, 10–13. [CrossRef] [PubMed]

3. Douris, N.; Melman, T.; Pecherer, J.M.; Pissios, P.; Flier, J.S.; Cantley, L.C.; Locasale, J.W.; Maratos-Flier, E. Adaptive changes in amino acid metabolism permit normal longevity in mice consuming a low-carbohydrate ketogenic diet. *Biochim. Biophys. Acta* **2015**, *1852*, 2056–2065. [CrossRef] [PubMed]

4. Simeone, K.A.; Matthews, S.A.; Rho, J.M.; Simeone, T.A. Ketogenic diet treatment increases longevity in kcna1-null mice, a model of sudden unexpected death in epilepsy. *Epilepsia* **2016**, *57*, e178–e182. [CrossRef] [PubMed]

5. Martin, B.; Ji, S.; White, C.M.; Maudsley, S.; Mattson, M.P. Dietary energy intake, hormesis, and health. In *Hormesis*; Springer: Berlin, Germany, 2010; pp. 123–137.

6. Stadtman, E. *The Status of Oxidatively Modified Proteins as a Marker of Aging*; Life Science Research Reports; John Wiley & Sons Ltd.: Hoboken, NJ, USA, 1995; p. 129.

7. Ames, B.N.; Shigenaga, M.K.; Hagen, T.M. Oxidants, antioxidants, and the degenerative diseases of aging. *Proc. Natl. Acad. Sci. USA* **1993**, *90*, 7915–7922. [CrossRef] [PubMed]

8. Milder, J.; Patel, M. Modulation of oxidative stress and mitochondrial function by the ketogenic diet. *Epilepsy Res.* **2012**, *100*, 295–303. [CrossRef] [PubMed]

9. Hyatt, H.W.; Kephart, W.C.; Holland, A.M.; Mumford, P.; Mobley, C.B.; Lowery, R.P.; Roberts, M.D.; Wilson, J.M.; Kavazis, A.N. A ketogenic diet in rodents elicits improved mitochondrial adaptations in response to resistance exercise training compared to an isocaloric western diet. *Front. Physiol.* **2016**, *7*, 533. [CrossRef] [PubMed]

10. Bough, K.J.; Wetherington, J.; Hassel, B.; Pare, J.F.; Gawryluk, J.W.; Greene, J.G.; Shaw, R.; Smith, Y.; Geiger, J.D.; Dingledine, R.J. Mitochondrial biogenesis in the anticonvulsant mechanism of the ketogenic diet. *Ann. Neurol.* **2006**, *60*, 223–235. [CrossRef] [PubMed]

11. McDaniel, S.S.; Rensing, N.R.; Thio, L.L.; Yamada, K.A.; Wong, M. The ketogenic diet inhibits the mammalian target of rapamycin (mtor) pathway. *Epilepsia* **2011**, *52*, e7–e11. [CrossRef] [PubMed]

12. Kennedy, A.R.; Pissios, P.; Otu, H.; Roberson, R.; Xue, B.; Asakura, K.; Furukawa, N.; Marino, F.E.; Liu, F.F.; Kahn, B.B.; et al. A high-fat, ketogenic diet induces a unique metabolic state in mice. *Am. J. Physiol. Endocrinol. Metab.* **2007**, *292*, E1724–E1739. [CrossRef] [PubMed]

13. Sullivan, P.G.; Rippy, N.A.; Dorenbos, K.; Concepcion, R.C.; Agarwal, A.K.; Rho, J.M. The ketogenic diet increases mitochondrial uncoupling protein levels and activity. *Ann. Neurol.* **2004**, *55*, 576–580. [CrossRef] [PubMed]

14. Jarrett, S.G.; Milder, J.B.; Liang, L.P.; Patel, M. The ketogenic diet increases mitochondrial glutathione levels. *J. Neurochem.* **2008**, *106*, 1044–1051. [CrossRef] [PubMed]

15. Ziegler, D.R.; Ribeiro, L.C.; Hagenn, M.; Siqueira, I.R.; Araujo, E.; Torres, I.L.; Gottfried, C.; Netto, C.A.; Goncalves, C.A. Ketogenic diet increases glutathione peroxidase activity in rat hippocampus. *Neurochem. Res.* **2003**, *28*, 1793–1797. [CrossRef] [PubMed]

16. Erickson, A.M.; Nevarea, Z.; Gipp, J.J.; Mulcahy, R.T. Identification of a variant antioxidant response element in the promoter of the human glutamate-cysteine ligase modifier subunit gene. Revision of the are consensus sequence. *J. Biol. Chem.* **2002**, *277*, 30730–30737. [CrossRef] [PubMed]

17. Mulcahy, R.T.; Wartman, M.A.; Bailey, H.H.; Gipp, J.J. Constitutive and beta-naphthoflavone-induced expression of the human gamma-glutamylcysteine synthetase heavy subunit gene is regulated by a distal antioxidant response element/tre sequence. *J. Biol. Chem.* **1997**, *272*, 7445–7454. [CrossRef] [PubMed]

18. Milder, J.B.; Liang, L.P.; Patel, M. Acute oxidative stress and systemic nrf2 activation by the ketogenic diet. *Neurobiol. Dis.* **2010**, *40*, 238–244. [CrossRef] [PubMed]

19. Maalouf, M.; Sullivan, P.G.; Davis, L.; Kim, D.Y.; Rho, J.M. Ketones inhibit mitochondrial production of reactive oxygen species production following glutamate excitotoxicity by increasing nadh oxidation. *Neuroscience* **2007**, *145*, 256–264. [CrossRef] [PubMed]

20. Kim, D.Y.; Davis, L.M.; Sullivan, P.G.; Maalouf, M.; Simeone, T.A.; van Brederode, J.; Rho, J.M. Ketone bodies are protective against oxidative stress in neocortical neurons. *J. Neurochem.* **2007**, *101*, 1316–1326. [CrossRef] [PubMed]

21. Roberts, M.D.; Holland, A.M.; Kephart, W.C.; Mobley, C.B.; Mumford, P.W.; Lowery, R.P.; Fox, C.D.; McCloskey, A.E.; Shake, J.J.; Mesquita, P.; et al. A putative low-carbohydrate ketogenic diet elicits mild nutritional ketosis but does not impair the acute or chronic hypertrophic responses to resistance exercise in rodents. *J. Appl. Physiol.* **2016**, *120*, 1173–1185. [CrossRef] [PubMed]

22. Reagan-Shaw, S.; Nihal, M.; Ahmad, N. Dose translation from animal to human studies revisited. *FASEB J.* **2008**, *22*, 659–661. [CrossRef] [PubMed]

23. Hamm, R.J.; Pike, B.R.; O'Dell, D.M.; Lyeth, B.G.; Jenkins, L.W. The rotarod test: An evaluation of its effectiveness in assessing motor deficits following traumatic brain injury. *J. Neurotrauma* **1994**, *11*, 187–196. [CrossRef] [PubMed]

24. Holland, A.M.; Kephart, W.C.; Mumford, P.W.; Mobley, C.B.; Lowery, R.P.; Shake, J.J.; Patel, R.K.; Healy, J.C.; McCullough, D.J.; Kluess, H.A.; et al. Effects of a ketogenic diet on adipose tissue, liver, and serum biomarkers in sedentary rats and rats that exercised via resisted voluntary wheel running. *Am. J. Physiol. Regul. Integr. Comp. Physiol.* **2016**, *311*, R337–351. [CrossRef] [PubMed]

25. Ghosh, R.; Gilda, J.E.; Gomes, A.V. The necessity of and strategies for improving confidence in the accuracy of western blots. *Expert Rev. Proteomics* **2014**, *11*, 549–560. [CrossRef] [PubMed]

26. Thacker, J.S.; Yeung, D.H.; Staines, W.R.; Mielke, J.G. Total protein or high-abundance protein: Which offers the best loading control for western blotting? *Anal. Biochem.* **2016**, *496*, 76–78. [CrossRef] [PubMed]

27. Gilda, J.E.; Gomes, A.V. Stain-free total protein staining is a superior loading control to beta-actin for western blots. *Anal. Biochem.* **2013**, *440*, 186–188. [CrossRef] [PubMed]

28. Kavazis, A.N.; Smuder, A.J.; Min, K.; Tümer, N.; Powers, S.K. Short-term exercise training protects against doxorubicin-induced cardiac mitochondrial damage independent of HSP72. *Am. J. Physiol. Heart. Circ. Physiol.* **2010**, *299*, H1515–H1524. [CrossRef] [PubMed]

29. Mowry, A.V.; Kavazis, A.N.; Sirman, A.E.; Potts, W.K.; Hood, W.R. Reproduction does not adversely affect liver mitochondrial respiratory function but results in lipid peroxidation and increased antioxidants in house mice. *PLoS ONE* **2016**, *11*, e0160883. [CrossRef] [PubMed]

30. Messer, J.I.; Jackman, M.R.; Willis, W.T. Pyruvate and citric acid cycle carbon requirements in isolated skeletal muscle mitochondria. *Am. J. Physiol. Cell Physiol.* **2004**, *286*, C565–C572. [CrossRef] [PubMed]

31. Kurz, L.C.; Shah, S.; Frieden, C.; Nakra, T.; Stein, R.E.; Drysdale, G.R.; Evans, C.T.; Srere, P.A. Catalytic strategy of citrate synthase: Subunit interactions revealed as a consequence of a single amino acid change in the oxaloacetate binding site. *Biochemistry* **1995**, *34*, 13278–13288. [CrossRef] [PubMed]

32. Bielohuby, M.; Menhofer, D.; Kirchner, H.; Stoehr, B.J.; Muller, T.D.; Stock, P.; Hempel, M.; Stemmer, K.; Pfluger, P.T.; Kienzle, E.; et al. Induction of ketosis in rats fed low-carbohydrate, high-fat diets depends on the relative abundance of dietary fat and protein. *Am. J. Physiol. Endocrinol. Metab.* **2011**, *300*, E65–E76. [CrossRef] [PubMed]

33. Frommelt, L.; Bielohuby, M.; Menhofer, D.; Stoehr, B.J.; Bidlingmaier, M.; Kienzle, E. Effects of low carbohydrate diets on energy and nitrogen balance and body composition in rats depend on dietary protein-to-energy ratio. *Nutrition* **2014**, *30*, 863–868. [CrossRef] [PubMed]

34. Morens, C.; Keijzer, M.; de Vries, K.; Scheurink, A.; van Dijk, G. Effects of high-fat diets with different carbohydrate-to-protein ratios on energy homeostasis in rats with impaired brain melanocortin receptor activity. *Am. J. Physiol. Regul. Integr. Comp. Physiol.* **2005**, *289*, R156–R163. [CrossRef] [PubMed]

35. Hall, K.D.; Chen, K.Y.; Guo, J.; Lam, Y.Y.; Leibel, R.L.; Mayer, L.E.; Reitman, M.L.; Rosenbaum, M.; Smith, S.R.; Walsh, B.T.; et al. Energy expenditure and body composition changes after an isocaloric ketogenic diet in overweight and obese men. *Am. J. Clin. Nutr.* **2016**, *104*, 324–333. [CrossRef] [PubMed]

36. Srivastava, S.; Kashiwaya, Y.; King, M.T.; Baxa, U.; Tam, J.; Niu, G.; Chen, X.; Clarke, K.; Veech, R.L. Mitochondrial biogenesis and increased uncoupling protein 1 in brown adipose tissue of mice fed a ketone ester diet. *FASEB J.* **2012**, *26*, 2351–2362. [CrossRef] [PubMed]

37. Kesl, S.L.; Poff, A.M.; Ward, N.P.; Fiorelli, T.N.; Ari, C.; Van Putten, A.J.; Sherwood, J.W.; Arnold, P.; D'Agostino, D.P. Effects of exogenous ketone supplementation on blood ketone, glucose, triglyceride, and lipoprotein levels in sprague-dawley rats. *Nutr. Metab. (Lond.)* **2016**, *13*, 9. [CrossRef] [PubMed]

38. Shimazu, T.; Hirschey, M.D.; Newman, J.; He, W.; Shirakawa, K.; Le Moan, N.; Grueter, C.A.; Lim, H.; Saunders, L.R.; Stevens, R.D.; et al. Suppression of oxidative stress by beta-hydroxybutyrate, an endogenous histone deacetylase inhibitor. *Science* **2013**, *339*, 211–214. [CrossRef] [PubMed]

39. Dushay, J.; Chui, P.C.; Gopalakrishnan, G.S.; Varela-Rey, M.; Crawley, M.; Fisher, F.M.; Badman, M.K.; Martinez-Chantar, M.L.; Maratos-Flier, E. Increased fibroblast growth factor 21 in obesity and nonalcoholic fatty liver disease. *Gastroenterology* **2010**, *139*, 456–463. [CrossRef] [PubMed]

40. Tendler, D.; Lin, S.; Yancy, W.S., Jr.; Mavropoulos, J.; Sylvestre, P.; Rockey, D.C.; Westman, E.C. The effect of a low-carbohydrate, ketogenic diet on nonalcoholic fatty liver disease: A pilot study. *Dig. Dis. Sci.* **2007**, *52*, 589–593. [CrossRef] [PubMed]

41. Reuter, S.; Gupta, S.C.; Chaturvedi, M.M.; Aggarwal, B.B. Oxidative stress, inflammation, and cancer: How are they linked? *Free Radic. Biol. Med.* **2010**, *49*, 1603–1616. [CrossRef] [PubMed]

42. Larsen, S.; Nielsen, J.; Hansen, C.N.; Nielsen, L.B.; Wibrand, F.; Stride, N.; Schroder, H.D.; Boushel, R.; Helge, J.W.; Dela, F.; et al. Biomarkers of mitochondrial content in skeletal muscle of healthy young human subjects. *J. Physiol.* **2012**, *590*, 3349–3360. [CrossRef] [PubMed]

43. Hughes, S.D.; Kanabus, M.; Anderson, G.; Hargreaves, I.P.; Rutherford, T.; O'Donnell, M.; Cross, J.H.; Rahman, S.; Eaton, S.; Heales, S.J. The ketogenic diet component decanoic acid increases mitochondrial citrate synthase and complex i activity in neuronal cells. *J. Neurochem.* **2014**, *129*, 426–433. [CrossRef] [PubMed]

44. Srivastava, S.; Baxa, U.; Niu, G.; Chen, X.; Veech, R.L. A ketogenic diet increases brown adipose tissue mitochondrial proteins and ucp1 levels in mice. *IUBMB Life* **2013**, *65*, 58–66. [CrossRef] [PubMed]

45. Lee, H.Y.; Choi, C.S.; Birkenfeld, A.L.; Alves, T.C.; Jornayvaz, F.R.; Jurczak, M.J.; Zhang, D.; Woo, D.K.; Shadel, G.S.; Ladiges, W.; et al. Targeted expression of catalase to mitochondria prevents age-associated reductions in mitochondrial function and insulin resistance. *Cell. Metabolism* **2010**, *12*, 668–674. [CrossRef] [PubMed]

46. Anderson, E.J.; Lustig, M.E.; Boyle, K.E.; Woodlief, T.L.; Kane, D.A.; Lin, C.T.; Price, J.W., 3rd; Kang, L.; Rabinovitch, P.S.; Szeto, H.H.; et al. Mitochondrial H_2O_2 emission and cellular redox state link excess fat intake to insulin resistance in both rodents and humans. *J. Clin. Investig.* **2009**, *119*, 573–581. [CrossRef] [PubMed]

47. Sakellariou, G.K.; Pearson, T.; Lightfoot, A.P.; Nye, G.A.; Wells, N.; Giakoumaki, I.I.; Vasilaki, A.; Griffiths, R.D.; Jackson, M.J.; McArdle, A. Mitochondrial ros regulate oxidative damage and mitophagy but not age-related muscle fiber atrophy. *Sci. Rep.* **2016**, *6*, 33944. [CrossRef] [PubMed]

48. Lecker, S.H.; Jagoe, R.T.; Gilbert, A.; Gomes, M.; Baracos, V.; Bailey, J.; Price, S.R.; Mitch, W.E.; Goldberg, A.L. Multiple types of skeletal muscle atrophy involve a common program of changes in gene expression. *Faseb. J.* **2004**, *18*, 39–51. [CrossRef] [PubMed]

49. Sparks, L.M.; Xie, H.; Koza, R.A.; Mynatt, R.; Hulver, M.W.; Bray, G.A.; Smith, S.R. A high-fat diet coordinately downregulates genes required for mitochondrial oxidative phosphorylation in skeletal muscle. *Diabetes* **2005**, *54*, 1926–1933. [CrossRef] [PubMed]

50. Iossa, S.; Lionetti, L.; Mollica, M.P.; Crescenzo, R.; Botta, M.; Barletta, A.; Liverini, G. Effect of high-fat feeding on metabolic efficiency and mitochondrial oxidative capacity in adult rats. *Br. J. Nutr.* **2003**, *90*, 953–960. [CrossRef] [PubMed]

51. Fernandez-Checa, J.C.; Garcia-Ruiz, C.; Ookhtens, M.; Kaplowitz, N. Impaired uptake of glutathione by hepatic mitochondria from chronic ethanol-fed rats. Tracer kinetic studies in vitro and in vivo and susceptibility to oxidant stress. *J. Clin. Investig.* **1991**, *87*, 397–405. [CrossRef] [PubMed]

52. Wilkins, H.M.; Kirchhof, D.; Manning, E.; Joseph, J.W.; Linseman, D.A. Mitochondrial glutathione transport is a key determinant of neuronal susceptibility to oxidative and nitrosative stress. *J. Biol. Chem.* **2013**, *288*, 5091–5101. [CrossRef] [PubMed]

53. Volek, J.S.; Freidenreich, D.J.; Saenz, C.; Kunces, L.J.; Creighton, B.C.; Bartley, J.M.; Davitt, P.M.; Munoz, C.X.; Anderson, J.M.; Maresh, C.M.; et al. Metabolic characteristics of keto-adapted ultra-endurance runners. *Metabolism* **2016**, *65*, 100–110. [CrossRef] [PubMed]

nutrients

MDPI

Article

Kiwifruit Non-Sugar Components Reduce Glycaemic Response to Co-Ingested Cereal in Humans

Suman Mishra [1], Haley Edwards [2], Duncan Hedderley [1], John Podd [2] and John Monro [1,3,*]

[1]   New Zealand Institute for Plant and Food Research, Palmerston North 4442, New Zealand;
     suman.mishra@plantandfood.co.nz (S.M.); Duncan.hedderley@plantandfood.co.nz (D.H.)
[2]   Department of Psychology, Massey University, Palmerston North 4442, New Zealand;
     haley.edwards@massey.ac.nz (H.E.); john.podd@massey.ac.nz (J.P.)
[3]   Riddet Institute, Massey University, Palmerston North 4442, New Zealand
*   Correspondence: john.monro@plantandfood.co.nz; Tel.: +64-(06)-355-6137

Received: 30 August 2017; Accepted: 25 October 2017; Published: 30 October 2017

Abstract: Kiwifruit (KF) effects on the human glycaemic response to co-ingested wheat cereal were determined. Participants ($n = 20$) consumed four meals in random order, all being made to 40 g of the same available carbohydrate, by adding kiwifruit sugars (KF sug; glucose, fructose, sucrose 2:2:1) to meals not containing KF. The meals were flaked wheat biscuit (WB)+KFsug, WB+KF, WB+guar gum+KFsug, WB+guar gum+KF, that was ingested after fasting overnight. Blood glucose was monitored 3 h and hunger measured at 180 min post-meal using a visual analogue scale. KF and guar reduced postprandial blood glucose response amplitude, and prevented subsequent hypoglycaemia that occurred with WB+KFsug. The area between the blood glucose response curve and baseline from 0 to 180 min was not significantly different between meals, 0–120 min areas were significantly reduced by KF and/or guar. Area from 120 to 180 min was positive for KF, guar, and KF+guar, while the area for the WB meal was negative. Hunger at 180 min was significantly reduced by KF and/or guar when compared with WB. We conclude that KF components other than available carbohydrate may improve the glycaemic response profile to co-ingested cereal food.

Keywords: kiwifruit; postprandial glycaemia; hypoglycaemia; carbohydrates

1. Introduction

Glucose intolerance is growing in prevalence globally [1–3]. It is an important condition to address because of the multiple illnesses that arise as complications of long-term exposure to elevated blood glucose concentrations and of repeated acute postprandial blood glucose excursions [4].

The rate of portal loading of sugars released from the foods by digestion is a dominant net process governing the relative blood glucose response to foods. It is governed by a number of factors that are operating at the gut level. Transfer of digestion products from the digesta to the brush border by mixing, diffusive transfer at the brush border, and absorption, are all rate-limiting steps that modulate the influence of a food on glycaemic response [5,6]. In fruit, factors that may determine the glycaemic impact of sugars include the effect of intactness of fruit tissue particles on sugar release [7], the influence of the fruit tissue and cell wall remnants, including soluble and insoluble cell wall polysaccharides, on gut rheology [8], inhibition of sugar transport at the gut wall by phenolics [9–11], and organic acid-induced delay in gastric emptying [12,13].

We have found that ripe kiwifruit disintegrate into a dispersion of cell wall remnants during in vitro digestion. The settled volume occupied by the dispersion is several times greater than the volume of the original fruit; two fruit (200 mL) would yield a dispersion of pulp (800 mL) occupying most of the gastric volume. Within the dispersion, the rate of glucose diffusion and the rate of mixing were each reduced by about 40%, and particles of the dispersion greatly amplified the viscosity of hydrocolloid that was present in the same suspension, with a further reduction in mixing [14].

The ability of kiwifruit cell wall remnants to reduce diffusion and mixing, as well as to interact with soluble fibres in a meal, and thereby alter gut processes involved in the glycaemic response, raised the possibility that kiwifruit consumed with a highly glycaemic food might be able to beneficially alter the glycaemic response to the food. Furthermore, the physical interaction of guar gum with kiwifruit remnants demonstrated in vitro [14] suggested that kiwifruit consumed with hydrocolloids such as guar gum might accentuate the anti-glycaemic influence of hydrocolloids in the diet [6]. Guar has been shown to substantially reduce the amplitude and extend the duration of the postprandial glycaemic response to available carbohydrate. Either a reduced amplitude of the immediate postprandial response and/or a change in the response profile could indicate an improved carbohydrate management by the body, with consequences for downstream states such as hunger and cognition.

Being able to demonstrate that kiwifruit may lower the glycaemic impact of a starchy meal by more than the simple substitution of fructose for starch is important in the context of health claims. Fructose, either free of in sucrose, constitutes about half of the available carbohydrate in kiwifruit, as in many other fruit, and it is disposed of in the liver in obesogenic metabolic pathways. Therefore, although fructose confers low glycaemic potency, there is concern about its possible negative metabolic effects [15], so any health claims for a lowering of glycaemic response by kiwifruit would be likely to be rejected if the effect were due solely to substitution of glucose-yielding carbohydrate by fructose. For instance, a low glycaemic index (GI) claim for a fruit juice was recently rejected because the low GI could not be attributed to more than the fructose content of the drink [15].

To test for a role for non-sugar (non-fructose) components in any glycaemia-moderating effect of kiwifruit we measured changes in blood glucose concentrations for 180 min in response to four meals that contained:

- breakfast cereal plus the sugar loading of kiwifruit,
- breakfast cereal with complete kiwifruit,
- breakfast cereal with guar gum (a hydrocolloid) plus the sugar loading of kiwifruit, and
- and breakfast cereal with guar gum plus complete kiwifruit.

Hunger experienced at 180 min was also measured. The meals were formulated to contain exactly the same type and quantity of available carbohydrate, so that any differences could be attributed to the non-available carbohydrate components of the kiwifruit, to guar gum, or to the interaction of kiwifruit with guar gum. We hypothesised that kiwifruit would cause beneficial changes in the blood glucose response to co-ingested glycaemic cereal product by mechanisms unrelated to the intake and type of sugars in the kiwifruit.

2. Materials and Methods

2.1. Meal Components

Green kiwifruit (var 'Hayward') were obtained from the Plant & Food Research orchard at Te Puke, New Zealand. They were harvested in good condition, and firm but close to ripe. The fruit were stored at room temperature and processed as they ripened, with ripeness being assessed by hand-felt firmness and confirmed by eating. They were peeled and the hard apical core removed, halved, and stored frozen ($-20\,^\circ$C) until enough fruit for the trial had been accumulated over a period of about two weeks. The frozen fruit were allowed to partially thaw and were then crushed to a coarse pulp by briefly (10 s) chopping in a Halde food processor. The pulp was then divided accurately into individual 200 g portions, each stored frozen within a plastic, capped, freezer-proof sundae container.

The breakfast cereal was Weet-Bix™ (WB; Sanitarium, Auckland, New Zealand), which is a whole wheat biscuit that is commonly consumed in Australia and New Zealand and made of cooked wheat flakes compressed into a biscuit that disintegrates rapidly on wetting, and consists mainly of rapidly digestible starch with little sugar (Nutrient information: Protein 12%, Fat 1.4%, Carbohydrate 67% of which sugars 2.8%).

Food-grade guar gum in the form of a fine powder (ISGUAR.5, Ingredient Stop Guar), and fructose were obtained from Hawkins Watts Limited, Penrose, Auckland, New Zealand.

The glucose used was dextrose monohydrate (Davis Food Ingredients, Palmerston North, New Zealand), which contains 91% glucose. It is henceforth referred to as glucose, and allowance was made for its water content in all calculations and weights.

2.2. Analysis of Kiwifruit

Subsamples (15 mL) of the kiwifruit pulp were measured accurately into 70 mL specimen pottles in triplicate. They were adjusted to pH 6.5 by titration with 1 M NaOH solution, 50 mL 0.2 M Na maleate buffer (5 mL, pH 6.5) was added, and made to 50 mL with 1% NaCl solution. The pottles were closed, heated to 80 °C for 10 min to gelatinise starch, and cooled to 37 °C before adding 1 mL of 2% pancreatin (Sigma P1750) (Sigma, St. Louis, MO, USA) and 0.1 mL of invertase concentrate (Megazyme E-INVERT) (Megazyme, Bray, Ireland) and the tubes were incubated for 1 h at 30 °C to hydrolyse starch and sucrose to glucose and fructose for analysis. A 1 mL aliquot was removed to a tube containing 4 mL absolute ethanol, mixed and allowed to stand at least 30 min at room temperature before centrifuging (2000 rpm) prior to an analysis of sugar in the supernatant. An invertase-free digestion was also conducted and the difference in reducing sugar between the digestions with and without invertase was used to estimate the sucrose content of the kiwifruit by difference in reducing sugars.

The available carbohydrate content of the digested pulp was measured using a reduced scale modification of the dinitrosalicylic acid (DNS) method [16], and the fructose content was measured by the thiobarbituric acid procedure [17].

2.3. Available Carbohydrate Analysis of the Meals

The available carbohydrate content of the meals was measured by a validated in vitro digestion procedure [16], using exactly one tenth of each of the meals in a volume of 50 mL. The samples were moistened with 10 mL of 1% NaCl solution and adjusted to pH 2.5 with 1 M HCl. One millilitre of 10% pepsin (Sigma P-7125) (Sigma, St Louis, MO, USA) solution was added and the pottle incubated at 37 °C for 45 min to simulate gastric digestion. Maleate buffer (5 mL, 0.2 M) was added and the pH adjusted to 6.5 with 0.1 M NaOH. The volume was accurately made to the 53 mL mark and pancreatic digestion commenced by adding 1 mL of 2% pancreatin (Sigma P-1750) solution. Samples (1 mL) were removed at 0, 20, 60, and 120 min into tubes containing 4 mL ethanol to stop the digestion, and the tubes were mixed before storing cold until the sugar analysis. Sugar analyses were conducted as outline above and the results plotted to provide a digestion curve for each meal.

A 2 mL aliquot of the remaining digest was removed from the invertase-containing digests to a 12 mL screw capped tube, 5 mL acetate buffer pH 5.2 added and the tube heated for 20 min in a boiling water bath. Heat-stable amylase (50 µL, Megazyme E-BLAAM) (Megazyme, Bray, Ireland) was added and the tubes mixed and allowed to cool to 40 °C when 50 µL of amyloglucosidase (Megazyme E-AMG) (Megazyme, Bray, Ireland) was added. The tubes were mixed and allowed to stand for 30 min, after which a 1 mL aliquot was removed to 4 mL of ethanol for analysis of total "available" carbohydrate digestion by reducing sugar analysis.

2.4. Buffering Capacity of WB+S and WB+KF Meals

The buffering capacity of kiwifruit in the WB+S and WB+KF meals was tested in light of the difference in glycaemic response to the two meals obtained in the present study, in a post hoc titration. The quantities of the meals ingested in a glycaemic response test were each titrated with 0.1 M HCl from their initial pH to pH 2.5, and the titration then continued using 0.1 M NaOH to pH 7.0. The quantities of acid and alkali used were expressed as milliequivalents (mEq).

2.5. Formulation of the Meals

Four meals were used in the intervention study, all containing the same quantity of WB and the same quantity and type of sugars as in kiwifruit, added in kiwifruit or as free sugars (S) (Table 1). The meals were formulated to contain 40 g available carbohydrate based on the analysis of the available carbohydrate in the kiwifruit pulp, and on the available carbohydrate value for WB as determined by in vitro digestive analysis.

In meals not containing 200 g of kiwifruit, the kiwifruit was substituted by the amount of sugars that were present in the kiwifruit, so that all of the meals had the same and equal amounts of available carbohydrate. The ratio of the sugars in the kiwifruit was: glucose:fructose:sucrose (2:2:1), so the 18.3 g of available carbohydrate added to the non-kiwifruit meals was made up of 7.32 g glucose plus 7.32 g fructose plus 3.66 g sucrose (Table 2).

The dose of guar gum (10 g) was based on that used by Nilsson, et al. [18] to obtain a reduced glycaemic response by feeding subjects 179 g bread made from flour that had been 15% substituted by guar gum. To ensure that all of the meals in the preset study were of almost equal volume 180 mL of water was consumed with meals not containing kiwifruit.

Table 1. Weights of meal components used (g).

Meal Component	Meal 1	Meal 2	Meal 3	Meal 4
	WB+S	WB+KF	WB+GG+S	WB+KF+GG
Wheat biscuit (WB)	47.3	47.3	47.3	47.3
Kiwifruit	-	200	-	200
Guar gum	-	-	10	10
Sugar mix [1]	19	-	19	-
Water [2]	180 mL [2]	-	180 mL	-

[1] Glucose:fructose:sucrose, 2:2:1; [2] Approximate water content of 200 g kiwifruit. "–": absent from meal.

Table 2. Available carbohydrate in meals by formulation (g).

Available Carbohydrate Source	Meal 1	Meal 2	Meal 3	Meal 4
	WB+S	WB+KF	WB+GG+S	WB+KF+GG
WB	21.7	21.7	21.7	21.7
Kiwifruit	-	18.3	-	18.3
Glucose	7.32	-	7.32	-
Fructose	7.32	-	7.32	-
Sucrose	3.66	-	3.66	-
Total	40	40	40	40

"–": absent from meal.

2.6. Human Intervention Study

The human intervention study was approved by the Human and Disabilities Ethics Committee of the New Zealand Ministry of Health (Ethics number 14/STH/77), and the trial was registered with the Australia New Zealand Clinical Trials Registry (Trial ID: ACTRN12615001259538). The participant flow chart shows ethical approval, recruitment, and intervention processes for the trial, as shown in Figure 1. The CONSORT checklist is attached as Supplementary Table S1. A written informed consent was obtained from participants.

The trial was run as a non-blinded randomised repeated measures study. It was not possible to blind the subjects to the meals they were consuming. However, the data and statistical analysis was performed by an analyst who was blinded to the treatments.

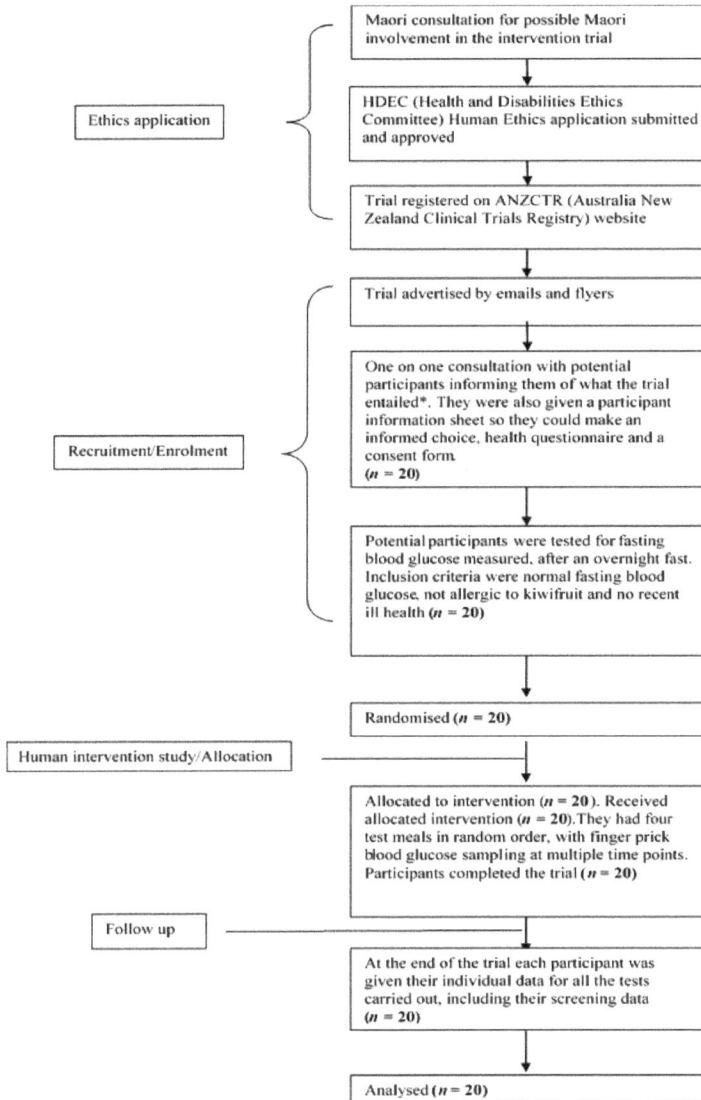

Figure 1. Participant flow chart shows ethical approval, recruitment and intervention processes for the trial. * The participants were allowed to bring a family member of their "Whanau" (support person, family or friend) as family support is highly valued in Maori culture.

2.7. Subjects

Twenty subjects, 6 male and 14 female between the ages of 26 and 66, with a mean age of 36 were recruited by flyer and email. Subject numbers were based on similar published trials, comparing glycaemic responses to foods, and the number was confirmed as adequate by the results of this study. Respondents were interviewed and given an information pack including a description of the study and a consent form. Prospective participants were asked to complete a health questionnaire. Exclusion criteria included known intolerance of kiwifruit, glucose intolerance, and recent ill health.

2.8. Preparation of Meals

The dry ingredients for each meal were thoroughly mixed. The moist component (kiwifruit pulp or 180 mL water) was then added with rapid mixing to prevent the formation of lumps. The meals were consumed with 200 mL water in addition to that within the meal.

2.9. Glycaemic Response

Subjects were allowed to consume their customary diet but were asked to fast overnight for at least 12 h and present themselves at 8.30 a.m. for the dietary intervention. They were asked to consume the meals within a 10 min period and to avoid physical exertion for three hours afterward, during which time blood glucose determinations were made. Blood glucose concentrations were measured by finger-prick analysis of capillary blood using a HemoCue (HemoCue, Ängelholm, Sweden) blood glucose analyser. Blood samplings were made immediately before consuming the meals (duplicate, baseline), and at 20, 40, 60, 90, 120, 150, and 180 min after the start of food consumption. At 180 min the subjects were asked to rate their hunger, using a visual analogue scale, along the dimension "Not at all" to "Extremely" in response to the question "How hungry are you?", based on published research on VAS scales for assessing appetite [19].

2.10. Data Analysis

Incremental blood glucose responses were calculated by subtracting each individual's baseline value from subsequent measurements and were then used to determine the incremental area under the curve (IAUC) for each individual by trapezoid summation. The highest postprandial blood glucose peak for each individual, irrespective of the time of occurrence (nearly all were at either 30 or 40 min) was used to determine the mean peak height for each meal (Table 3). Data were entered into an Excel spreadsheet for preliminary analysis. For statistical comparison of means, analysis of variance (ANOVA) was used (GenStat version 11.1, VSNi Ltd., Hemel Hempstead, UK). Breakfasts were described by two factors, kiwifruit (present/absent) and guar gum (present/absent), and these two factors (and their interaction) were tested in the ANOVAs. Participant and week were fitted as blocking factors. Residuals were checked to ensure that the assumptions of ANOVA were met. Least significant differences (LSDs) were used to compare means. A power calculation based on the IAUCs obtained in a comparison of breakfast cereal and kiwifruit-substituted breakfast, showed a sample size of $n = 17$ would be required ($p = 0.05$) in a cross-over design with a power of 80% to detect a difference.

Table 3. Peak incremental blood glucose concentrations (mmol L^{-1}) in response to consuming meals containing wheat biscuit (WB), kiwifruit (KF) and guar gum (GG).

	Meals				LSD	*p*
	WB+S	WB+KF	WB+GG+S	WB+KF+GG		
Mean	2.72	2.17	1.26	1.26	0.346	<0.001
Sem	0.18	0.21	0.15	0.15		

Sem: Means and standard error of mean.

3. Results

3.1. Analysis of Kiwifruit

Digestive analysis of the kiwifruit provided a value of 9.15 ± 0.34 g/100 mL for total sugars. Consistent with previous analyses of six varieties of kiwifruit, the sugars consisted of approximately equal proportions of glucose and fructose with a minor sucrose component. Therefore the sugar mixture made to substitute for kiwifruit sugars in the kiwifruit-free meals was approximated by a 2:2:1 mixture of glucose, fructose and sucrose.

3.2. Digestive Analysis of the Meals

The digestion profiles of the meals during simulated gastrointestinal digestion in vitro confirmed that all four of the meals to be consumed contained the same quantity of available carbohydrate (Figure 2).

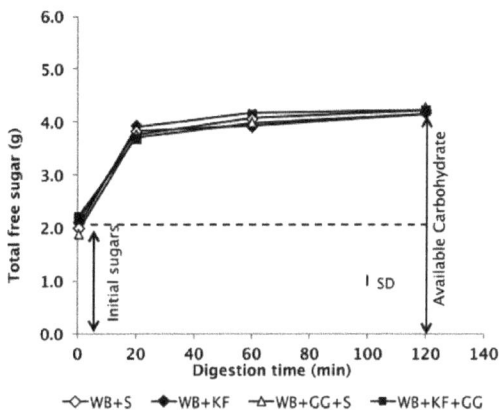

Figure 2. Available carbohydrate content of the meals measured by in vitro digestion. The meals had been formulated to deliver the same quantities of cereal starch and sugars at time of ingestion (S = sugar, WB = wheat biscuit, KF = kiwifruit, GG = guar gum) SD = Mean standard deviation of analyses.

3.3. Post Hoc Titration of WB+S and WB+KF Meals

The 200 g of kiwifruit ingested by participants in the present study had a pH of 3.3 which was raised to only pH 3.7 by adding 47.3 g of wheat biscuit in the WB+KF meal. It took 22.5 milliequivalents of acid (0.5 M HCl) to reduce the pH of the KF+WB meaL to pH 2.5, as compared with 8.7 mEq to reduce the pH of WB+S from 5.3 to 2.5. Alkali (0.5 M NaOH) required to subsequently raise the pH from 2.5 to 7.0 were 66 mEq for KF+WB and only 11 mEq for WB+S (Figure 3).

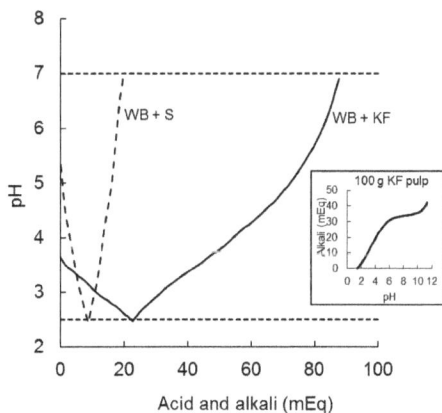

Figure 3. Acid and alkali (milliequivilants) required to reduce the pH of meals ingested from their initial pH to pH 2.5 (with 0.5 M HCl) and to subsequently raise it to pH 7.0 (with 0.5 M NaOH). WB+S (dashed line) was 47.3 g wheat biscuit +20 g sugars; KF+WB (solid line) was 200 g kiwifruit + 47.3 g wheat biscuit. The inset graph shows the typical titration curve of a weak acid with a strong alkali for 100 g kiwifruit pulp, indicating buffering capacity of kiwifruit flesh.

3.4. Blood Glucose Responses

3.4.1. Response Amplitude

The baseline values for the different groups were very similar, and the means were not significantly different (Means: 4.5, 4.5, 4.4, 4.6 mmol/L. $p = 0.471$), so the responses are presented as increments over baseline. The different meals induced blood glucose responses that were clearly distinctive (Figure 4).

Peak height showed a very significant difference between the four breakfasts. When comparing the means by LSD (Table 3), the WB+KF induced a significantly lower peak than the WB+S breakfast, but the peaks induced by breakfasts containing guar gum were very significantly lower than induced by both the WB+KF and WB+S breakfasts. That is, GG had a large effect in reducing peak height, KF had a smaller effect on its own, and no extra effect beyond GG when served with GG (GG, $p < 0.001$; KF, $p = 0.028$).

The high amplitude postprandial response to WB+S was followed by an overshoot to below the baseline starting at 120 min. In the presence of kiwifruit, the amplitude of the average incremental response was reduced by about 18% (from 7.03 to 6.46 mmol/L). The rate of post-peak decline (mainly blood glucose clearance) was reduced by kiwifruit and/or guar gum, and did not reach the baseline during the three hours of blood glucose measurements. In contrast to the breakfast cereal alone, the meals containing kiwifruit and/or guar were able to sustain a small elevation of blood glucose throughout the entire three-hour measurement period (Figures 4 and 5).

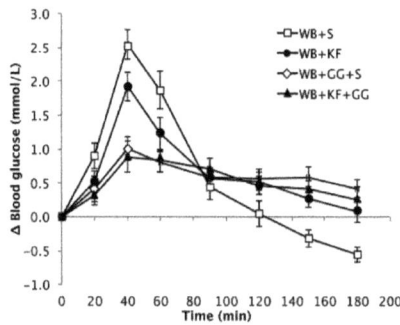

Figure 4. Increase over baseline in blood glucose concentration induced by the meals wheat biscuit (WB)+sugars (S) (WB+S), WB+kiwifruit (WB+KF), WB+guar gum+sugars (WB+GG+S), and WB+KF+GG (Means ± sem).

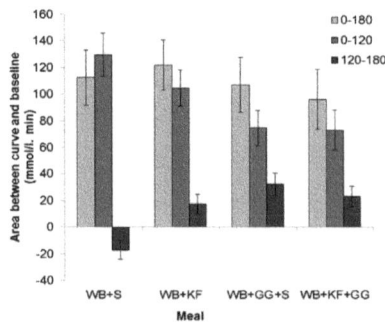

Figure 5. Distribution of incremental areas under the blood glucose response curve in the meals consumed: Wheat bix+sugars (WB+S), WB+kiwifruit (WB+KF), WB+guar gum+sugars (WB+GG+S), and WB+KF+GG. (Means ± sem). LSDs: 0–180 min, 38.9 ($p = 0.61$); 0–120 min, 28.0 ($p < 0.001$); 120–180 min, 16.0 ($p < 0.001$).

3.4.2. Area between the Blood Glucose Response Curves and Baseline

The IAUC for the meals was calculated for the full 180 min sampling period and for the 0–120 min and 120–180 min sub-periods.

The IAUC over the period 0–180 min showed no significant difference between the four breakfasts based on LSDs (Figure 5) and ANOVA (GG, $p = 0.266$; KF, $p = 0.969$, GG × KF interaction, $p = 0.462$).

The IAUC for the first 120 min, which is the time period recommended for glycaemic index measurements [20] and stipulated in Australian Standard AS 4694-2007 "Glycaemic index of foods", showed a very significant difference between the four meals ($p < 0.001$). It was greatest for WB+S (129 mmol/L/min), less for KF+WB (105 mmol/L/min), and significantly lower for the meals containing GG (WB+KF+GG, 75 mmol/L/min; WB+GG+S, 73 mmol/L/min). ANOVA, however, confirmed that the only statistically significant effect could be attributed to the GG (GG, $p < 0.001$, KF, $p = 0.189$, GG × KF interaction, $p = 0.462$).

In the period 120–180 min, WB+KF (17 mmol/L/min), WB+GG+S (32 mmol/L/min) and WB+KF+GG (23 mmol/L/min) all gave positive areas between the curve and baseline that differed significantly from the area for the WB+S meal (-17 mmol/L/min), which alone was negative (Figure 5). ANOVA revealed significant effects of both KF and GG, and an interaction (GG, $p < 0.001$, KF, $p = 0.030$, GG × KF, $p < 0.001$)

3.4.3. Blood Glucose at 180 min

Differences from baseline in blood glucose at 180 min followed a similar pattern to the 120–180 min IAUC. Comparing the means, treatments containing GG gave the highest incremental blood glucose values (WB+GG+S, 0.42 mmol/L; WB+KF+GG, 0.25 mmol/L), the WB+KF was just positive (0.08 mmol/L) but significantly greater than the WB+S breakfast, which was negative (-0.56 mmol/L) (Figure 6). ANOVA revealed significant effects of KF and GG and an interaction (GG, $p < 0.001$, KF, $p = 0.028$, KF × GG, $p = 0.026$).

Whether blood glucose at 180 min was significantly different from baseline was tested using a one-sample t-test comparing the means to zero. After the WB+S breakfast blood glucose at 180 min was significantly lower than baseline ($p < 0.001$). After the WB+GG+S breakfast, it was significantly higher ($p = 0.005$), while for the WB+KF and WB+KF+GG blood glucose was not significantly different from baseline ($p = 0.629$, and $p = 0.222$).

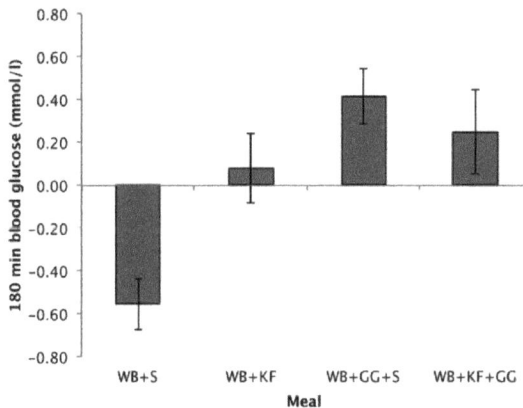

Figure 6. Blood glucose concentrations at 180 min after consuming wheat biscuit+sugars (WB+S), WB+kiwifruit (WB+KF), WB+guar gum+sugars (WB+GG+S), and WB+KF+GG. (Means ± sem; LSD 0.347, $p < 0.001$).

3.5. Hunger Experienced

The hunger scores showed a significant difference between the four breakfasts ($p < 0.001$). Factorial ANOVA revealed that GG ($p < 0.001$) and KF ($p < 0.001$) both lowered the hunger scores and that there was also a further lowering with the interaction of both factors ($p = 0.089$). Both WB+KF and WB+GG+S breakfasts had significantly lower hunger scores than WB alone, and the WB+KF+GG had a lower hunger score than the WB+KF alone (Figure 7).

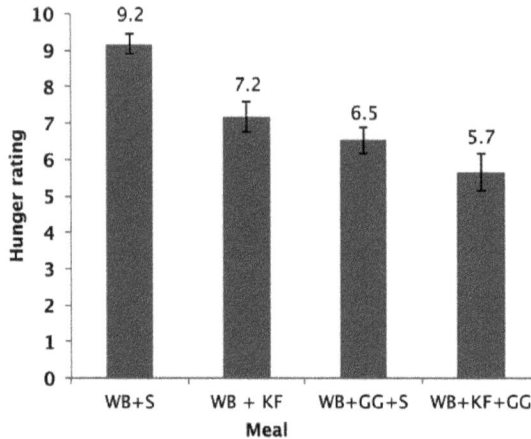

Figure 7. Hunger rating at 180 min after consuming WB+sugars (WB+S), WB+kiwifruit (WB+KF), WB+guar gum+sugars (WB+GG+S), and WB+kiwifruit+guar gum (WB+KF+GG). The subjects were asked to respond to the question "How hungry do you feel?" on a visual analogue scale extending from "Not at all" to "Extremely", scored on a scale of 1–10. (Means ± sem; LSD 0.96, $p < 0.001$).

4. Discussion

As the primary aim of the research reported here was to measure the role of non-carbohydrate factors in any differences between the meals in glycaemic response, it was important to ensure that the meals contained the same quantities of available carbohydrate after formulation. The digestive analysis confirmed that all meals were nearly equal in available carbohydrate and that the available carbohydrate in them did not differ between treatments in intrinsic digestibility, based on the in vitro digestograms (Figure 2).

A secondary objective was to see whether an interaction of kiwifruit with co-consumed hydrocolloid could augment any modulatory effect of kiwifruit on the glycaemic response. The study suggested that kiwifruit components other than sugars might be acting to modulate glycaemic response to co-consumed cereal, with significant effects on response amplitude and response distribution over the 3 h testing period. However, no substantial interactions between kiwifruit and guar were demonstrated, and the response curves for the meals containing guar gum, with or without kiwifruit, were very similar. Therefore, the interaction between kiwifruit remnants and guar gum that resulted in a marked reduction in mixing in vitro [14] was not evident in the present study, probably because of the high concentration of guar used.

The meals containing kiwifruit and/or guar gum improved the glycaemic response profile in two ways—by reducing peak height and by averting the overshoot to below the baseline seen in the WB+S meal. Delays in gastric processing of WB in the WB+KF, WB+GG+S and WB+KF+GG meals may have contributed to the reduced amplitude with a correspondingly less intense glucose disposal so that the glycaemic response was extended and the postprandial overshoot to below baseline avoided. As digestion continues well beyond the immediate postprandial period [21], meals that reduce the rate of digestion may lead to the prolonged slight elevation of blood glucose above the baseline demonstrated in this study, indicating an adequate and sustained supply of glucose from the gut.

Guar gum consumed at the dose used in the present study has been shown to improve cognitive performance, while maintaining blood glucose slightly above baseline. Like guar, kiwifruit alone was able to prevent blood glucose from dropping below the baseline and to significantly reduce hunger at 180 min (Figure 6). This effect may be a consequence of differences in blood glucose combined with physical effects in the gut. Irrespective of the mechanisms involved, the results suggest that if kiwifruit contributes to maintenance of satiety, its effects on appetite are worth investigating further with the fully validated appetite scale in the context of obesity management. Similarly, assuming that satiety reflects physiological state, and given the results obtained with guar [18], the effects of kiwifruit on cognition should be explored in further research. Although the hunger results were possibly limited by experienced hunger not being a complete measure of appetite [19], it correlates with appetite. A possible limitation of the hunger data is that it is a subjective rating, and it was not possible to blind the subjects to the taste of kiwifruit or to the sensation of guar gum in the meals.

The 10 g dose of guar gum used in this study was selected because it had been shown to be effective in suppressing postprandial glycaemic response and to maintain blood glucose concentrations above the baseline for 180 min [18], so can be regarded as a positive control. The concentration of guar in the stomach and intestine would depend on the volume of liquid consumed and on the volume of digestive juices that were produced. A final concentration of 1%, if the guar was dispersed in a volume of 1 L, would be quite viscous [14]. So, it is likely that effects on gastric emptying as well as retardation of digestive processes such as diffusion and mixing contributed to the observed reduction in glycaemic response in the guar-containing treatments. However, a lower concentration may have been required to detect an interaction with kiwifruit in vivo.

A number of specific factors that may have contributed to the overall effect of kiwifruit could be the subject of further research. Remnants of kiwifruit after in vitro digestion, mainly cell wall residues, have been shown to substantially retard gut-level processes involved in the glycaemic response; at physiological concentrations kiwifruit residues reduced glucose diffusion and simulated luminal mixing each by about 50% [6]. Phenolic compounds present in kiwifruit may have inhibited glucose uptake from the gut and/or stimulated glucose disposal in the body [9]. It is also possible that the organic acids of the kiwifruit may have delayed gastric emptying [22] and ileal digestion through the combined effect of them lowering pH of the kiwifruit-containing meal when ingested, and delaying gastric and duodenal pH adjustment through the considerable buffering capacity of kiwifruit demonstrated in the *post hoc* analysis reported here. Therefore, delays in gastrointestinal pH adjustment may have contributed to the retardation of digestive processes involved in the glycaemic response.

The results of the present study have implications for the inclusion of fruit in the diets of people who suffer from glucose intolerance. There are numerous well-established nutritional benefits from consuming fruit, and most nutritional recommendations include the regular daily consumption of fruit, while those with glucose intolerance commonly assume that fruit should be avoided because of their high sugar content. However, the present study has shown that if kiwifruit were to be included in a diet on an equal carbohydrate exchange basis, involving the substitution of readily digested starchy food by kiwifruit, there would be a glycaemic response lowering effect of the non-carbohydrate components of the fruit, as well as an effect of partial replacement of glucose by fructose.

5. Conclusions

The present study was designed to show the influence of kiwifruit components other than the available carbohydrates on the glycaemic response, and used meals containing the same quantities of the same carbohydrates to do so. The effects that were noted were worthwhile in terms of improving the glycaemic response profile. Maintaining blood glucose at near baseline and avoiding consequent hunger also has important health implications given the well-established links between energy intake, obesity, and ill-health in a range of forms. However, in terms of incorporating kiwifruit, for its health benefits, into the meals for those who are glucose intolerant and who need to monitor their energy intakes for the control of obesity, it may be necessary to consider using kiwifruit within a carbohydrate exchange regimen to avoid increasing energy intake.

If kiwifruit were to be included in a meal by equi-carbohydrate (approximately iso-energetic) substitution of breakfast cereal, such as WB, one might expect even greater reductions in glycaemic response than were observed in the present study, because the fructose (GI = 22) portion of the carbohydrate in kiwifruit would substitute digestible starch (GI = 70) in a flaked breakfast cereal.

The present results have potentially important implications for health claims related to a role for kiwifruit in diets of populations at risk of diabetes. With a growing concern for the negative impact of increased fructose intakes in fruit and fruit products, it is important to demonstrate that the lowering of glycaemic response by equi-carbohydrate partial exchange of kiwifruit for readily digested starchy foods is due to more than simply substitution of low GI fructose for high GI starch. The design of the present study allowed us to demonstrate that kiwifruit components other than available carbohydrate contribute to the glycaemic response-lowering effect of equal carbohydrate exchange of kiwifruit for starchy staple.

Supplementary Materials: The following are available online at http://www.mdpi.com/2072-6643/9/11/1195/s1, Supplementary Table S1: CONSORT checklist.

Acknowledgments: The study was funded as part of the "Kiwi, fruity and friendly" subcontract (UOAX1421) of the New Zealand National Science Challenge, with contribution from Zespri International Limited and by the Kiwifruit Royalty Investment Programme of The New Zealand Institute for Plant & Food Research Limited, New Zealand.

Author Contributions: J.M. designed the study, processed the kiwifruit, formulated and prepared the meals, analysed the blood glucose results and wrote the report. H.E. obtained ethics approval, recruited the subjects and carried out the blood glucose measurements. S.M. carried out the available carbohydrate analyses and helped with kiwifruit processing and meal preparation. J.P. advised on experimental design. D.H. conducted statistical analysis of the results.

Conflicts of Interest: The authors declare no conflict of interest.

References

1. Nanditha, A.; Ma, R.C.W.; Ramachandran, A.; Snehalatha, C.; Chan, J.C.N.; Chia, K.S.; Shaw, J.E.; Zimmet, P.Z. Diabetes in asia and the pacific: Implications for the global epidemic. *Diabetes Care* **2016**, *39*, 472–485. [CrossRef] [PubMed]
2. Danaei, G.; Finucane, M.M.; Lu, Y.; Singh, G.M.; Cowan, M.J.; Paciorek, C.J.; Lin, J.K.; Farzadfar, F.; Khang, Y.H.; Stevens, G.A.; et al. National, regional, and global trends in fasting plasma glucose and diabetes prevalence since 1980: Systematic analysis of health examination surveys and epidemiological studies with 370 country-years and 2.7 million participants. *Lancet* **2011**, *378*, 31–40. [CrossRef]
3. Shaw, J.E.; Sicree, R.A.; Zimmet, P.Z. Global estimates of the prevalence of diabetes for 2010 and 2030. *Diabetes Res. Clin. Pract.* **2010**, *87*, 4–14. [CrossRef] [PubMed]
4. Brownlee, M. Biochemistry and molecular cell biology of diabetic complications. *Nature* **2001**, *414*, 813–820. [CrossRef] [PubMed]
5. Parada, J.; Aguilera, J.M. Review: Starch matrices and the glycemic response. *Food Sci. Technol. Int.* **2011**, *17*, 187–204. [CrossRef] [PubMed]
6. Gidley, M.J. Hydrocolloids in the digestive tract and related health implications. *Curr. Opin. Colloid Interface Sci.* **2013**, *18*, 371–378. [CrossRef]
7. Ha, M.A.; Mann, J.I.; Melton, L.D.; Lewisbarned, N.J. Relationship between the glycemic index and sugar content of fruits. *Diabetes Nutr. Metab.* **1992**, *5*, 199–203.
8. Lentle, R.G.; Janssen, P.W.M. *The Physical Processes of Digestion*; Springer: New York, NY, USA, 2011.
9. Hanhineva, K.; Torronen, R.; Bondia-Pons, I.; Pekkinen, J.; Kolehmainen, M.; Mykkanan, H.; Poutanen, K. Impact of dietary polyphenols on carbohydrate metabolism. *Int. J. Mol. Sci.* **2010**, *11*, 1365–1402. [CrossRef] [PubMed]

10. Manzano, S.; Williamson, G. Polyphenols and phenolic acids from strawberry and apple decrease glucose uptake and transport by human intestinal caco-2 cells. *Mol. Nutr. Food Res.* **2010**, *54*, 1773–1780. [CrossRef] [PubMed]

11. Sequeira, I.R.; Popitt, S.D. Unfolding novel mechanisms of polyphenol flavonoids for better glycaemic control: Targetting pancreatic islet amyloid polypeptide (iapp). *Nutrients* **2017**, *9*, 788. [CrossRef] [PubMed]

12. Ostman, E.; Granfeldt, Y.; Persson, L.; Bjorck, I. Vinegar supplementation lowers glucose and insulin responses and increases satiety after a bread meal in healthy subjects. *Eur. J. Clin. Nutr.* **2005**, *59*, 983–988. [CrossRef] [PubMed]

13. Fardet, A.; Leenhardt, F.; Lioger, D.; Scalbert, A.; Remesy, C. Parameters controlling the glycaemic response to breads. *Nutr. Res. Rev.* **2006**, *19*, 18–25. [CrossRef] [PubMed]

14. Mishra, S.; Monro, J. Kiwifruit remnants from digestion in vitro have functional attributes of potential importance to health. *Food Chem.* **2012**, *135*, 2188–2194. [CrossRef] [PubMed]

15. Herman, M.A.; Samuel, V.T. The sweet path to metabolic demise: Fructose and lipid synthesis. *Trends Endocrinol. Metab.* **2016**, *27*, 719–730. [CrossRef] [PubMed]

16. Monro, J.A.; Mishra, S.; Venn, B. Baselines representing blood glucose clearance improve in vitro prediction of the glycemic impact of customarily consumed food quantities. *Br. J. Nutr.* **2010**, *103*, 295–305. [CrossRef] [PubMed]

17. Blakeney, A.; Mutton, L. A simple colourimetric method for determination of sugars in fruit and vegetables. *J. Sci. Food Agric.* **1980**, *31*, 889–897. [CrossRef]

18. Nilsson, A.; Radeborg, K.; Bjorck, I. Effects on cognitive performance of modulating the postprandial blood glucose profile at breakfast. *Eur. J. Clin. Nutr.* **2012**, *66*, 1039–1043. [CrossRef] [PubMed]

19. Flint, A.; Raben, A.; Blundell, J.E.; Astrup, A. Reproducibility, power and validity of visual analogue scares in assessment of appetite sensations in single test meal studies. *Int. J. Obes.* **2000**, *24*, 38–48. [CrossRef]

20. Brouns, F.; Bjorck, I.; Frayn, K.; Gibbs, A.; Lang, V.; Slama, G.; Wolever, T. Glycaemic index methodology. *Nutr. Res. Rev.* **2005**, *18*, 1–28. [CrossRef] [PubMed]

21. Lentle, R.G.; Sequeira, I.R.; Hardacre, A.K.; Reynolds, G. A method for assessing real time rates of dissolution and absorption of carbohydrate and other food matrices in human subjects. *Food Funct.* **2016**, *7*, 2820–2832. [CrossRef] [PubMed]

22. Leeman, M.; Ostman, E.; Bjorck, I. Vinegar dressing and cold storage of potatoes lowers postprandial glycaemic and insulinaemic responses in healthy subjects. *Eur. J. Clin. Nutr.* **2005**, *59*, 1266–1271. [CrossRef] [PubMed]

nutrients

MDPI

Article

Slowly Digestible Carbohydrate for Balanced Energy: In Vitro and In Vivo Evidence

Vishnupriya Gourineni *, Maria L. Stewart, Rob Skorge and Bernard C. Sekula

Global Nutrition R & D, Ingredion Incorporated, 10 Finderne Ave, Bridgewater, NJ 08807, USA;
maria.stewart@ingredion.com (M.L.S.), rob.skorge@ingredion.com (R.S.) bernie.sekula@ingredion.com (B.C.S.)
* Correspondence: vishnupriya.gourineni@ingredion.com; Tel.: +1-908-575-6169

Received: 19 October 2017; Accepted: 7 November 2017; Published: 10 November 2017

Abstract: There is growing interest among consumers in foods for sustained energy management, and an increasing number of ingredients are emerging to address this demand. The SUSTRA™ 2434 slowly digestible carbohydrate is a blend of tapioca flour and corn starch, with the potential to provide balanced energy after a meal. The aim of the study was to characterize this starch's digestion profile in vitro (modified Englyst assay) and in vivo (intact and cecectomized rooster study), and to determine its effects on available energy, by measuring post-prandial glycemia in healthy adults ($n = 14$), in a randomized, double-blind, placebo-controlled, cross-over study, with two food forms: cold-pressed bar and pudding. The in vitro starch digestion yielded a high slowly digestible fraction (51%) compared to maltodextrin (9%). In the rooster digestibility model, the starch was highly digestible (94%). Consumption of slowly digestible starch (SDS), in an instant pudding or bar, yielded a significantly lower glycemic index compared to a control. At individual time points, the SDS bar and pudding yielded blood glucose levels with significantly lower values at 30–60 min and significantly higher values at 120–240 min, demonstrating a balanced energy release. This is the first study to comprehensively characterize the physiological responses to slowly digestible starch (tapioca and corn blend) in in vitro and in vivo studies.

Keywords: slowly digestible carbohydrates (SDC); slowly digestible starch (SDS); sustained energy; sustained blood glucose

---

1. Introduction

According to the World Health Organization (WHO), diabetes will be one of the leading causes of deaths by 2030 [1,2]. Elevated post-prandial glycemia may be one of many risk factors contributing to chronic diseases, such as diabetes [3]. Improvement of the blood glucose response to carbohydrate is pivotal for diabetes management. Carbohydrates influence blood glucose and are ranked using a standardized measurement, the glycemic index (GI). In a meta-analysis of 12 randomized trials, low GI diets, compared to high GI diets, were shown to improve the glycemic response by lowering glycated albumin (HbA1c), reaching levels of clinical significance in diabetes [4]. Low GI diets (GI < 55) release glucose at a sustainable rate, thus providing balanced available energy, and health benefits, such as diabetes management and improved heart health [4,5].

Carbohydrates are a commonly consumed staple food and represent 45–55% of daily energy intake in a Western diet [6]. Carbohydrates play an important role as preferred substrates for brain and red blood cells; thus, the Institute of Medicine (IOM) established a recommended dietary allowance (RDA) for carbohydrates, of 130 g/day for adults and children at least 1 year of age. Food starches derived from cereals, legumes, tubers have semi-crystalline granules, which differ in size and shape, amylose and amylopectin content, chain lengths, degree of branching and X-ray diffraction patterns. Earlier studies reported the influence of starch structure on its digestibility [7]. The digestion rate of starch is measured in vitro by the Englyst method, which classifies starch into three fractions: rapidly digestible starch (RDS), slowly digestible starch (SDS) and non-digestible resistant starch (RS) [8].

Some native starches have inherently high SDS content (A-type X-ray diffraction pattern granule), such as in waxy corn starch or RS (B-type X-ray diffraction pattern granule), such as in potato starch [9]. During processing in the presence of heat and moisture, SDS/RS content is replaced by RDS due to changes in the structural order of starch. This leads to increased accessibility to digestive enzymes and rapid release of glucose. This phenomenon is described as starch gelatinization [10]. Starch gelatinization influences SDS content. Cereal products with high SDS contents and limited starch gelatinization, such as breakfast biscuits, have been shown to have a low GI [11,12].

Starch digestibility is associated with GI. Slowly digestible carbohydrates, commonly abbreviated as SDC, are digested steadily, resulting in prolonged glucose release from the lumen of the small intestine into the blood stream, with blunted glycemia and therefore a lower insulin requirement [13]. Because SDCs provide a high amount of available carbohydrate and are typically low in dietary fiber, they can be formulated into products that have a low GI. Low GI carbohydrates, such as SDCs, include certain types of sugars and starch [11]. There are a few commercial SDCs for sustained energy management, including isomaltulose (disaccharide), trehalose (disaccharide), sucromalt (alternan oligosaccharide) and pullulans (maltotriose-based polysaccharide) [14]. The aforementioned SDCs may have application challenges due to their inherent albeit low sweetness and inability to provide textural functionality. Starch-based SDCs have the advantage of allowing easy replacement of RDS with SDS, in bakery, snack and beverage applications.

The investigational material in the present study is a new, commercially available product, the SUSTRA™ 2434 slowly digestible carbohydrate. This product will be referred to as "slowly digestible starch" or SDS for the remainder of the report. This SDS is a blend of corn starch and tapioca flour, designed to deliver balanced energy and a reduced glycemic response when added to non-thermal applications. The aims of this study were to (1) characterize the in vitro digestibility profile of SDS; (2) confirm in vivo digestibility using a rooster model and (3) assess its glycemic response in a healthy population.

2. Materials and Methods

2.1. In Vitro Testing

The in vitro digestion method compared maltodextrin (Globe® Plus 10 DE maltodextrin, Ingredion Incorporated, Bridgewater, NJ, USA), and SDS (SUSTRA™ 2434 slowly digestible carbohydrate, Ingredion Incorporated, Bridgewater, NJ, USA). A modified version of the Englyst method [8] was used to analyze the starches for RDS, SDS and RS contents. Three minor modifications were made to the assay: (1) a commercial source of amyloglucosidase (AMG) was used to replace the original AMG 400, which is no longer available; 1 mL AMG enzyme (Sigma A7095, from *Aspergillus niger*, ≥260 U/mL, St. Louis, MO, USA) was added to centrifuged 12 g pancreatin in solution; (2) Invertase (40 mg) (Sigma I4505, Grade VII from baker's yeast, ≥300 U/mg) was added, replacing 4 mL of the original liquid enzyme (3000 EU/mL and (3) Guar gum (Sigma G4129) was prehydrated in 0.05 M HCl solution (0.5% *w/v*) with 0.5% pepsin before being added to the digestion tubes. Released glucose was determined using glucose oxidase/peroxidase (GOPOD) reagent. Slowly available glucose (SAG) was calculated using the following equation: SAG = (slowly digestible starch/glucose amount release at 120 min) × 100. The total dietary fiber of test starch was measured using the AOAC 991.43 method [15].

2.2. In Vivo Rooster Study

The aforementioned SDS and maltodextrin were evaluated in roosters. Nutrient digestibility is determined by measuring nitrogen-corrected true metabolizable energy (TME_n) content of a starch and maltodextrin, using both conventional and cecectomized roosters, as previously described [16,17]. Briefly, two precision-fed rooster assays, utilizing conventional Single Comb White Leghorn roosters and cecectomized Single Comb White Leghorn roosters, were conducted. All animal housing, handling, and surgical procedures were approved by the University of Illinois Animal Care and Use Committee. Aliquots of the test carbohydrates were mixed with either 20 or 33 mL of water prior to dosing. After 26 h of feed withdrawal, 10 conventional roosters (5 roosters per treatment) and 10 cecectomized roosters (5 roosters per treatment) were tube-fed an average of 26.7 g (dry matter basis) of the test starches. Following crop intubation, excreta (urine and feces) were collected for 48 h on plastic trays placed under each individual cage. Excreta samples then were lyophilized, weighed, and ground prior to analysis. The two test carbohydrates were analyzed for dry matter (DM; 105 °C; AOAC, 2006; method 934.01) [18], and both the two test carbohydrates and the excreta samples were analyzed for N or crude protein (CP; TruMac® N, LECO Corporation, St. Joseph, MI, USA; AOAC, 2006), and gross energy (GE) using a bomb calorimeter (Parr Instruments, Moline, IL, USA). The TME$_n$ values, corrected for endogenous energy excretion based on previous data, were calculated using the following equation:

$$TME_n(\text{kcal/g}) = \frac{EI_{fed} - (EE_{fed} + 8.22 \times N_{fed}) + \left(EE_{fasted} + 8.22 \times N_{fasted}\right)}{FI}$$

where EI_{fed} equals the gross energy intake of the test substrate consumed; EE_{fed} equals the energy excreted by the fed birds; 8.22 is the correction factor for uric acid; N_{fed} equals the g nitrogen retained by the fed birds; EE_{fasted} equals the energy excreted by the fasted birds; N_{fasted} equals the g nitrogen retained by the fasted birds; and FI equals the grams of dry test substrate consumed. The database with conventional and cecectomized birds indicates that the N-corrected endogenous energy excretion by fasted birds was 9.25 kcal.

2.3. Clinical Study

This study was conducted in accordance with the ethical principles outlined in the Declaration of Helsinki and the protocol was approved by the Western Institutional Review Board (Vancouver, BC, Canada). All subjects provided written informed consent prior to starting the study. The clinical study was conducted at GI Labs (Toronto, ON, Canada).

2.3.1. Subject Screening

Inclusion criteria: Participants were healthy male or non-pregnant females, 18–75 years of age, with a body mass index (BMI) of $\geq$20 and $\leq$40 kg/m^2. Participants were required to maintain their regular diet, supplement intake, physical activity and body weight throughout the study duration and refrain from smoking prior to each visit. On test days, subjects were not allowed to take any dietary supplements, until dismissal from the GI labs. Subjects were required to have normal fasting serum glucose (<7.0 mmol/L capillary corresponding to whole blood glucose <6.3 mmol/L), abstain from alcohol consumption and to avoid vigorous physical activity for 24 h prior to all test visits. Subjects had to have an understanding of the study procedures and be willing to provide informed consent to participate in the study and authorization to release relevant protected health information to the investigator.

Exclusion criteria: Subjects were excluded if they failed to meet inclusion criteria, had a history of chronic disease, such Type 1 or 2 Diabetes, cardiovascular disease, cancer, gastrointestinal disorders; used medications within four weeks of the screening, had surgery within 3 months of screening, had an intolerance or allergy to test ingredients, had extreme dietary habits, had drastic body weight changes >3.5 kg within four weeks of screening duration, had the presence of any symptoms of an active infection during screening or study visits, had a history of alcohol or substance abuse, or had an unwillingness or inability to comply with the experimental procedures and to follow GI Labs safety guidelines.

2.3.2. Study Design and Subjects

The study was a randomized, double-blinded, placebo controlled, cross-over design, with 14 healthy adults (age 18–75 years, body mass index (BMI) $\geq$20.0 and <40.0 kg/m^2). Eligible participants were studied on separate days over a period of 2 to 6 weeks. The interval between successive tests was no less than 48 h and no more than 2 weeks. Subjects completed six study visits in a random order, during which they consumed one of the following treatments: SDS bar, control bar, SDS pudding, control pudding, dextrose beverage 1 (50 g dextrose in 250 mL water) or dextrose beverage 2 (50 g dextrose in 250 mL water). The dextrose beverage was administered twice for glycemic index calculations. The study subject flow is shown in Figure 1. Descriptions of the study foods are provided in Section 2.3.3.

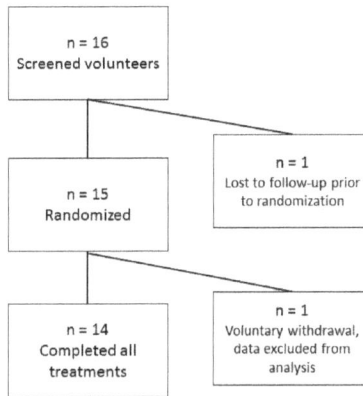

Figure 1. Subject flow through study.

2.3.3. Study Foods

Cold-pressed bar: The SDS bar and control bar were similar in appearance and were packaged in an opaque envelope with an alpha-numeric code for identification. Neither the study subjects nor the investigators knew the identity of the study foods. Both the SDS bar and control bar were matched for available carbohydrates (50 g). The SDS bar contained 24.9 g SDS (SUSTRA^{TM} 2434 slowly digestible carbohydrate, Ingredion Incorporated, Bridgewater, NJ, USA) and 18.9 g corn syrup (Globe$^{®}$ Plus 63 DE maltodextrin, Ingredion Incorporated, Bridgewater, NJ, USA) while the control bar contained 22.1 g maltodextrin (Globe$^{®}$ Plus 10 DE maltodextrin, Ingredion Incorporated, Bridgewater, NJ, USA) and 21.8 g corn syrup (Globe$^{®}$ Plus 63 DE maltodextrin, Ingredion Incorporated, Bridgewater, NJ, USA). The portion sizes and nutrient composition of the study bars are shown in Table 1.

Table 1. Nutrient composition of cold-pressed bars and instant pudding. SDS: slowly digestible starch.

Nutrient Content (g)	Bar SDS	Bar Control	Pudding SDS	Pudding Control
Serving size	72	70	200	200
Total carbohydratess	52.9	52.8	52.6	52.4
Available carbohydratess	50.1	51.0	50.0	50.7
Sugars	19.0	20.7	17.0	43.4
Dietary fiber	2.8	1.9	2.6	1.7
Protein	4.8	4.4	5.8	5.7
Fat	4.4	4.2	5.6	5.6

Pudding: The SDS pudding and control pudding were identical in appearance and were packaged in an opaque envelope with an alpha-numeric code for identification. Neither the study subjects nor the investigators knew the identity of the study foods. Both the SDS pudding and control pudding were matched for available carbohydrate (50 g). The test pudding contained 30.4 g SDS (SUSTRA^{TM} 2434 slowly digestible carbohydrate, Ingredion Incorporated, Bridgewater, NJ USA) and 10.9 g dextrose (CERELOSE® dextrose, Ingredion Incorporated, Bridgewater, NJ, USA), and the control pudding contained 39.6 g dextrose (CERELOSE® dextrose, Ingredion Incorporated, Bridgewater, NJ, USA). The remainder of available carbohydrate in both puddings was provided by whole (full-fat) milk. The puddings were prepared by mixing an entire 300 g package of powder with 840 g whole milk, whisking for 2 min and refrigerating overnight. The portion sizes and nutrient composition of the study puddings are shown in Table 1.

2.3.4. Study Visit Procedures

Participants were asked to maintain stable dietary and activity habits throughout the study. Prior to each study visit, participants refrained from drinking alcohol and from unusual levels of food intake or physical activity for 24 h. On each test occasion, subjects arrived at the clinical site after fasting for 10 to 12 h. Two fasting blood samples for glucose analysis (2–3 drops into a fluoro-oxalate tube) were obtained by fingerprick, 5 min apart and after the second sample, the subject started to consume a test meal. Each test meal was served with a drink of 1 or 2 cups of coffee or tea with 30 mL of 2% milk if desired, or water. At the first visit, each subject selected the type and volume of drink desired and the same type and volume of drink was consumed on subsequent visits. Subjects consumed the entire test meal within 10 min. At the first bite, a timer was started and additional blood samples for glucose analysis (2–3 drops into a fluoro-oxalate tube) were taken at 15, 30, 45, 60, 90, 120, 150, 180, 210 and 240 min after starting to eat. Blood samples were obtained from hands warmed with an electric heating pad for 3–5 min prior to each sample.

2.4. Biochemical Analysis

After blood collection the tubes containing blood for glucose analysis were rotated to mix the blood with an anti-coagulant and then placed in a refrigerator until the last blood sample in the set had been collected. After all tubes were collected from one subject, the tubes were stored in a −20 °C freezer until analysis. Analysis was performed within 3 days of the study visit, using a YSI model 2300 STAT analyzer (Yellow Springs, OH, USA).

2.5. Data Analysis and Statistics

2.5.1. Sample Size Calculation for Clinical Study and Randomization Method

The main effect of treatment was expected to have a SD of 23%/sqrt (2) because each subject tested each treatment in 2 types of food (bar and pudding). We wished to be able to detect a difference in glycemic response of 20%; n = 14 subjects provided 80% of the power required to detect a difference of 19%.

The study was conducted in two blocks of three treatments; the two blocks contained either the bar or the pudding and within each block the tests consisted of one dextrose test and the control and test food. The six possible orders of three treatments were listed twice, one for each block. The six orders in each block were randomized using the Rand () function on an Excel spreadsheet. The order of the blocks was randomized in the same way and the process was repeated three times.

The order of testing was assigned to the 14 subjects in the order they attended for the first visit; the extra orders were included in case of drop-outs which needed to be replaced.

2.5.2. Data Analysis

Glycemic index values were calculated based on previously published methods [19]. The net incremental area under the curve was calculated using the trapezoidal rule. Values below the baseline were treated as negative values.

2.5.3. Statistical Analysis

Paired t-tests were conducted on blood glucose values at individual time points, net incremental area-under-curve (iAUC), and glycemic index using GraphPad Prism 7 (v 7.03, GraphPad Software, Inc., La Jolla, CA, USA). p values < 0.05 were deemed statistically significant.

3. Results

3.1. In Vitro Testing

The digestibility profile of SDS and maltodextrin (control) was determined by measuring glucose release, as shown in Figure 2. Glucose released in the first 20 min reflects RDS, while the SDS fraction was obtained by subtracting the amount of glucose hydrolyzed at 120 min with glucose released at 20 min (120 min–20 min). Glucose release beyond 120 min reflects the resistant starch fraction. The tested SDS showed a 13% rapidly digestible fraction and 51% was slowly digestible, while in the control, 88% was rapidly digestible and 9% was slowly digestible. At the end of two hours digestion, slowly available glucose (SAG) was higher for SDS (79.4%), as compared to the control (9.3%). Total dietary fiber of SDS measured by AOAC 991.43 method was 10%, as-is.

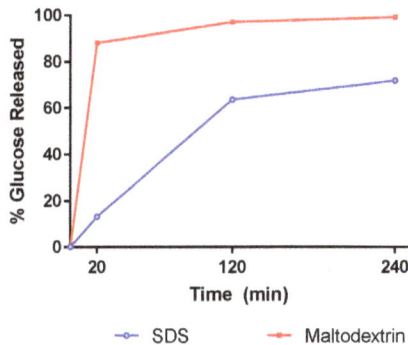

Figure 2. In Vitro digestibility profile of slowly digestible starch (SDS) and maltodextrin.

3.2. In Vivo Rooster Study

Roosters provide a robust model to determine digestibility and/or fermentability of an ingredient. The true metabolizable energy (TMEn) for both SDS and maltodextrin (control) was high, indicating their complete digestibility (Table 2). However, the energy value of SDS was slightly lower than the maltodextrin in roosters, suggesting that around 6% is non-digestible, aligning with the fiber fraction.

Table 2. True metabolizable energy (TMEn) evaluation of slowly digestible starch (SDS) in roosters.

Ingredients	Gross Energy [1]	TMEn	Digested (%)
SDS	3.986	3.756	94.2
Maltodextrin	4.008	4.014	100

[1] Bomb calorimetry, all values are kcal/g, based on dry-basis.

3.3. Clinical Study

Study demographics are shown in Table 3. All participants were healthy.

Table 3. Subjects demographics.

Mean ± Standard Deviation (SD)	Participants (n = 14)
Age (years)	38.3 ± 13.3
Gender (male/female)	10/4
Weight (kg)	78.6 ± 10.4
Body mass index (kg/m^2)	26.8 ± 2.7
Fasting blood glucose (mg/dL)	82.5 ± 8.8

The inclusion of SDS in a cold-pressed bar and instant pudding yielded a significantly lower glycemic index compared to the control (Table 4).

Table 4. Glycemic index and net incremental area-under-curve (iAUC) for SDS and control.

	Bars			Puddings		
Ingredient	SDS	Control	p-Value	SDS	Control	p-Value
Glycemic Index [1],*	49.9 ± 4.6	93.0 ± 8.1	<0.0001	45.9 ± 4.5	92.6 ± 9.2	<0.0001
Net iAUC (0–2 h) *	91.8 ± 16.9	183.4 ± 23.1	0.0001	83.7± 14.8	191.1 ± 33.5	0.014

[1] Glycemic Index (GI) Scale: Low GI: ≤55; Medium GI: 56–69; High GI: ≥70. * Values are presented as mean ± standard error mean (SEM), control vs SDS are statistically different in the t-test if $p < 0.05$.

SDS inclusion in the cold-pressed bar yielded a significantly different mean blood glucose concentration (Figure 3a) and a 50% lower net iAUC (0–2 h) (Table 4), compared to control bars. The SDS bar resulted in significantly lower blood glucose values at 30, 45, 60 and 90 min, and significantly higher blood glucose values at 120 and 150 min, compared to the control (Figure 3a). Similarly, SDS inclusion in pudding yielded significantly different mean blood glucose concentrations (Figure 3b) and a 56% lower net iAUC (0–2 h), compared to the control pudding. The SDS pudding resulted in significantly lower blood glucose values at 30, 45, and 60 min, and significantly higher blood glucose values at 150, 180, 210, and 240 min, compared to the control (Figure 3b). The blood glucose concentrations in response to SDS in the bar and pudding after two hours remained above the baseline for an average of 15 min longer compared to the control foods, and the drop below baseline was less substantial for SDS incorporated foods.

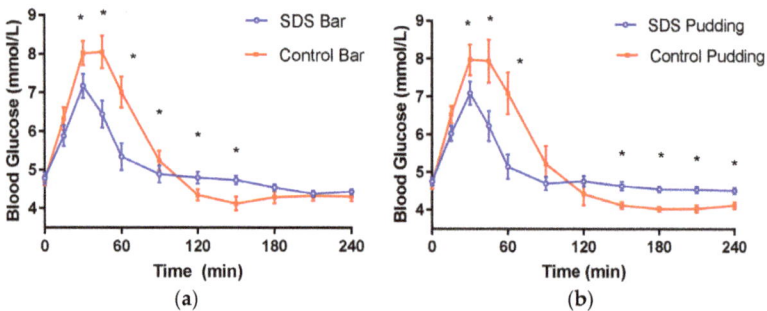

Figure 3. Post-prandial glycemic response of SDS in healthy adults (a) cold-pressed bars (b) instant pudding. Data are mean ± standard error mean (SEM); * indicates treatments were significantly different at specific time points in the paired t-test ($p < 0.05$).

4. Discussion

Global health concerns, such as diabetes, can be addressed with lifestyle changes, including dietary modifications by replacing rapidly digestible carbohydrates with slowly digestible carbohydrates. Natural SDCs, such as whole grains and pulses, comprise slowly digestible starch and resistant starch/fiber fractions. Dietary fiber has profound health benefits, but can lead to gastrointestinal discomfort. Thus, there is a heightened interest in ingredients with high amounts of SDS and low RDS content, in order to deliver sustained glucose energy.

Carbohydrate quality or prolonged digestion is an emerging area and studies to evaluate the effect of SDCs are currently relying on glycemic response and glycemic index measurements to define "sustained energy" [20]. However, "sustained energy" may have multiple consumer interpretations [21]. Not all carbohydrates are the same, with respect to digestion and absorption. Some carbohydrates, such as maltodextrin, are rapidly digested and absorbed, which results in a rapid blood glucose spike and crash. This places greater demands on the pancreas and the liver to manage the carbohydrate load. In contrast, SDS results in a steadier blood glucose rise and a slower blood glucose drop [13].

This is the first study to characterize a SDS by adopting a translational approach, which involved in vitro screening of the ingredient and in vivo validation of slow but near complete digestibility and prolonged glucose release. In the in vitro assay, SDS had an over 50% slowly digestible fraction and showed a steady glucose release over 120 min of digestion which simulates digestion in the small intestine. SDS had low fiber and RDS fractions. Carbohydrate digestibility in humans is commonly determined in ileostomates [22]. Alternately, the rooster model provides a quick and reliable tool to measure true metabolizable energy of ingredients [23]. SDS in roosters showed high digestibility similar to control, but the energy value was slightly lower than the control, indicating slower glucose release. In Vitro and rat studies on hydrogenated isomaltulose also showed slower hydrolysis and resulted in lower glycemic response [24].

A recently published meta-analysis [25], with three clinical trials, including 79 subjects, showed a direct association of SDS content on glucose metabolism. A high-SDS containing cereal had a 15-fold higher chance of having a low rate of appearance of exogenous glucose (RaE). In the current study, SDS formulated in cold-pressed bars and pudding showed a significantly lower glucose response, iAUC and glycemic index, compared to the control foods with high RDS. The low glycemic response may be attributed to the high content of SDS and slowly available glucose, beyond 2 h of digestion. A similar study with two native granular uncooked starches, which differed in digestion rate, was evaluated for glycemic and insulin responses in healthy and diabetic populations [26]. Slowly hydrolyzed starch significantly lowered glycemic and insulin responses over 6 h and reduced 3-hydroxybutyrate, an indicator for sustained use of glucose as oxidizable fuel.

Another study with starches differing in hydrolysis rates, indicated a relation between starch in vitro digestibility and glycemic response in healthy men [27]. Subjects consumed five soups containing 50 g maltodextrin (SDS = 1 g), whole-grain (SDS = 4 g), high-amylose starch (SDS = 6 g), regular cornstarch (SDS = 16 g) or no added starch at 1-week intervals. Ad libitum food intake was measured at 30 min or 120 min, which were the estimated times of digestion of a rapidly digestible starch (RDS) and slowly digestible starch, respectively. Blood glucose concentrations and appetite were measured pre- and post-meal. Food intake was significantly reduced by maltodextrin at 30 min and by whole-grain, high-amylose starch and corn starch at 120 min. Blood glucose AUC was significantly lower for all starches at 30 min, while at 120 min blood glucose AUC was significantly reduced for corn starch. This study showed that ingredients with high SDS contents impact digestibility, glycemic response and satiety.

Glycemic responses of other sugar-based SDCs, such as isomaltulose and sucromalt, were tested in low GI formulated meal replacement drinks in diabetics [28]. Four enteral formulas were tested. Two drinks had SDCs, which were compared to a standard formula and high-fat drink. SDC-enriched and high-fat beverages significantly attenuated blood glucose and blood insulin concentrations, while the high-fat formula had a greater reduction effect on triglycerides. However, the SDC content of these slowly hydrolyzed carbohydrates has not been reported.

The current study has few limitations. The modified in vitro method is a screening tool to characterize digestibility profile and is less dynamic in terms of measuring interactions with other nutrients and absorption uptake and does not completely reflect physiological responses. The human ileostomate model, which has been utilized to measure ingredient digestibility in some studies, was replaced with the cost-effective rooster model. Other limitations included lack of insulin response measurements, rate of glucose appearance and indicators of carbohydrate and fat oxidation.

5. Conclusions

The SUSTRA™ 2434 slowly digestible carbohydrate is a blend of tapioca flour and corn starch, with the potential to provide balanced energy. The new SDS has been shown to be highly digestible in vivo and yielded a lower GI, compared to a rapidly digestible control. Incorporating SDS into a cold-pressed bar and instant pudding resulted in a lower glucose response in the first 60–90 min and a higher glucose response at two hours and beyond, suggesting steadily available energy. The reduced blood glucose response to SDS is due to the presence of a significant slowly digestible starch fraction and low rapidly digestible and resistant fractions, as documented in the in vitro study. The balanced energy from the SUSTRA™ 2434 slowly digestible carbohydrate is a solution to improve carbohydrate quality in non-thermal applications.

Acknowledgments: We would like to acknowledge Qin Zhao and Matthew Park for their technical support. Both are employees of Ingredion Incorporated.

Author Contributions: V.G.; R.S. and B.C.S. conceived and designed the experiments; R.S. performed the in vitro experiment; V.G.; B.C.S. and M.L.S. analyzed the data; V.G. and M.L.S. wrote the paper.

Conflicts of Interest: The authors are employees of Ingredion Incorporated. The studies were financially supported by Ingredion Incorporated, 10 Finderne Avenue, Bridgewater, NJ, USA.

References

1. World Health Organization. *Diabetics Key Facts*; World Health Organization: Geneva, Switherland, 2017.
2. Rowley, W.R.; Bezold, C.; Arikan, Y.; Byrne, E.; Krohe, S. Diabetes 2030: Insights from Yesterday, Today, and Future Trends. *Popul. Health Manag.* **2017**, *20*, 6–12. [CrossRef] [PubMed]
3. Blaak, E.E.; Antoine, J.M.; Benton, D.; Bjorck, I.; Bozzetto, L.; Brouns, F.; Diamant, M.; Dye, L.; Hulshof, T.; Holst, J.J.; et al. Impact of postprandial glycaemia on health and prevention of disease. *Obes. Rev.* **2012**, *13*, 923–984. [CrossRef] [PubMed]
4. Thomas, D.E.; Elliott, E.J. The use of low-glycaemic index diets in diabetes control. *Br. J. Nutr.* **2010**, *104*, 797–802. [CrossRef] [PubMed]
5. Goff, L.M.; Cowland, D.E.; Hooper, L.; Frost, G.S. Low glycaemic index diets and blood lipids: A systematic review and meta-analysis of randomised controlled trials. *Nutr. Metab. Cardiovasc. Dis.* **2013**, *23*, 1–10. [CrossRef] [PubMed]
6. Trumbo, P.; Schlicker, S.; Yates, A.A.; Poos, M. Food and Nutrition Board of the Institute of Medicine; The National Academies. Dietary reference intakes for energy, carbohydrate, fiber, fat, fatty acids, cholesterol, protein and amino acids. *J. Am. Diet. Assoc.* **2002**, *102*, 1621–1630. [CrossRef]
7. Magallanes-Cruz, P.A.; Flores-Silva, P.C.; Bello-Perez, L.A. Starch Structure Influences Its Digestibility: A Review. *J. Food Sci.* **2017**, *82*, 2016–2023. [CrossRef] [PubMed]
8. Englyst, H.N.; Kingman, S.M.; Cummings, J.H. Classification and measurement of nutritionally important starch fractions. *Eur. J. Clin. Nutr.* **1992**, *46* (Suppl. S2), S33–S50. [PubMed]
9. Zhang, G.; Hamaker, B.R. Slowly digestible starch: Concept, mechanism, and proposed extended glycemic index. *Crit. Rev. Food Sci. Nutr.* **2009**, *49*, 852–867. [CrossRef] [PubMed]
10. Zhang, G.; Ao, Z.; Hamaker, B.R. Slow digestion property of native cereal starches. *Biomacromolecules* **2006**, *7*, 3252–3258. [CrossRef] [PubMed]

11. Englyst, K.N.; Vinoy, S.; Englyst, H.N.; Lang, V. Glycaemic index of cereal products explained by their content of rapidly and slowly available glucose. *Br. J. Nutr.* **2003**, *89*, 329–340. [CrossRef] [PubMed]

12. Garsetti, M.V.; Vinoy, S.; Lang, V.; Holt, S.; Loyer, S.; Brand-Miller, J.C. The glycemic and insulinemic index of plain sweet biscuits: Relationships to in vitro starch digestibility. *J. Am. Coll. Nutr.* **2005**, *24*, 441–447. [CrossRef] [PubMed]

13. Glycemic Index Foundation. GLYCEMIC INDEX & DIABETES Making Healthy Choices Easy. Available online: http://www.gisymbol.com/cms/wp-content/uploads/2013/10/GIF_HP_Diabetes.pdf (accessed on 19 October 2017).

14. Miao, M.; Jiang, B.; Cui, S.W.; Zhang, T.; Jin, Z. Slowly digestible starch—A review. *Crit. Rev. Food Sci. Nutr.* **2015**, *55*, 1642–1657. [CrossRef] [PubMed]

15. AOAC International. AOAC Official Method 991.43 Total, Soluble, and Insoluble Dietary Fiber in Foods. In *Enzymatic-Gravimetric Method, MES-TRIS Buffer*, 19th ed.; Official Methods of Analysis of AOAC International; AOAC International: Rockville, MD, USA, 2012.

16. Parsons, C. Influence of caecectomy on digestibility of amino acids by roosters fed distillers' dried grains with solubles. *J. Agric. Sci.* **1985**, *104*, 469–472. [CrossRef]

17. Parsons, C.M.; Potter, L.M.; Bliss, B.A. True Metabolizable Energy Corrected to Nitrogen Equilibrium. *Poultry Sci.* **1982**, *61*, 2241–2246. [CrossRef]

18. AOAC International. *Official Methods of Analysis*, 17th ed.; Association of Official Analytical Chemists: Gaithersburg, MD, USA, 2006.

19. Wolever, T.M.; Jenkins, D.J.; Jenkins, A.L.; Josse, R.G. The glycemic index: Methodology and clinical implications. *Am. J. Clin. Nutr.* **1991**, *54*, 846–854. [CrossRef] [PubMed]

20. Vinoy, S.; Laville, M.; Feskens, E.J.M. Slow-release carbohydrates: Growing evidence on metabolic responses and public health interest. Summary of the symposium held at the 12th European Nutrition Conference (FENS 2015). *Food Nutr. Res.* **2016**, *60*, 31662. [CrossRef] [PubMed]

21. Marinangeli, C.P.; Harding, S.V. Health claims using the term 'sustained energy' are trending but glycaemic response data are being used to support: Is this misleading without context? *J. Hum. Nutr. Diet.* **2016**, *29*, 401–404. [CrossRef] [PubMed]

22. Jenkins, D.J.; Cuff, D.; Wolever, T.M.; Knowland, D.; Thompson, L.; Cohen, Z.; Prokipchuk, E. Digestibility of carbohydrate foods in an ileostomate: Relationship to dietary fiber, in vitro digestibility, and glycemic response. *Am. J. Gastroenterol.* **1987**, *82*, 709–717. [PubMed]

23. Van der Klis, J.D.; Kwakernaak, C. Proving a concept: An in vitro approach[1]. *J. Appl. Poultry Res.* **2014**, *23*, 301–305. [CrossRef]

24. Grupp, U.; Siebert, G. Metabolism of hydrogenated palatinose, an equimolar mixture of alpha-D-glucopyranosido-1,6-sorbitol and alpha-D-glucopyranosido-1,6-mannitol. *Res. Exp. Med.* **1978**, *173*, 261–278. [CrossRef]

25. Vinoy, S.; Meynier, A.; Goux, A.; Jourdan-Salloum, N.; Normand, S.; Rabasa-Lhoret, R.; Brack, O.; Nazare, J.A.; Peronnet, F.; Laville, M. The Effect of a Breakfast Rich in Slowly Digestible Starch on Glucose Metabolism: A Statistical Meta-Analysis of Randomized Controlled Trials. *Nutrients* **2017**, *9*, 318. [CrossRef] [PubMed]

26. Seal, C.J.; Daly, M.E.; Thomas, L.C.; Bal, W.; Birkett, A.M.; Jeffcoat, R.; Mathers, J.C. Postprandial carbohydrate metabolism in healthy subjects and those with type 2 diabetes fed starches with slow and rapid hydrolysis rates determined in vitro. *Br. J. Nutr.* **2003**, *90*, 853–864. [CrossRef] [PubMed]

27. Anderson, G.H.; Cho, C.E.; Akhavan, T.; Mollard, R.C.; Luhovyy, B.L.; Finocchiaro, E.T. Relation between estimates of cornstarch digestibility by the Englyst in vitro method and glycemic response, subjective appetite, and short-term food intake in young men. *Am. J. Clin. Nutr.* **2010**, *91*, 932–939. [CrossRef] [PubMed]

28. Vanschoonbeek, K.; Lansink, M.; van Laere, K.M.; Senden, J.M.; Verdijk, L.B.; van Loon, L.J. Slowly digestible carbohydrate sources can be used to attenuate the postprandial glycemic response to the ingestion of diabetes-specific enteral formulas. *Diabetes Educ.* **2009**, *35*, 631–640. [CrossRef] [PubMed]

nutrients

MDPI

Article

Prebiotic Dietary Fiber and Gut Health: Comparing the in Vitro Fermentations of Beta-Glucan, Inulin and Xylooligosaccharide

Justin L. Carlson [1,†], Jennifer M. Erickson [1,†], Julie M. Hess [1], Trevor J. Gould [2] and Joanne L. Slavin [1,*]

[1] Department of Food Science and Nutrition, University of Minnesota, 1334 Eckles Ave, St. Paul, MN 55108, USA; carl2814@umn.edu (J.L.C.); eric2472@umn.edu (J.M.E.); jmhess@umn.edu (J.M.H.)
[2] Informatics Institute, University of Minnesota, 101 Pleasant St., Minneapolis, MN 55455, USA; goul0109@umn.edu
*   Correspondence: jslavin@umn.edu; Tel.: +1-612-624-7234
†   Authors contributed equally to this work.

Received: 27 October 2017; Accepted: 13 December 2017; Published: 15 December 2017

Abstract: Prebiotic dietary fiber supplements are commonly consumed to help meet fiber recommendations and improve gastrointestinal health by stimulating beneficial bacteria and the production of short-chain fatty acids (SCFAs), molecules beneficial to host health. The objective of this research project was to compare potential prebiotic effects and fermentability of five commonly consumed fibers using an in vitro fermentation system measuring changes in fecal microbiota, total gas production and formation of common SCFAs. Fecal donations were collected from three healthy volunteers. Materials analyzed included: pure beta-glucan, Oatwell (commercially available oat-bran containing 22% oat β-glucan), xylooligosaccharides (XOS), WholeFiber (dried chicory root containing inulin, pectin, and hemi/celluloses), and pure inulin. Oatwell had the highest production of propionate at 12 h ($4.76 \ \mu mol/mL$) compared to inulin, WholeFiber and XOS samples ($p < 0.03$). Oatwell's effect was similar to those of the pure beta-glucan samples, both samples promoted the highest mean propionate production at 24 h. XOS resulted in a significant increase in the genus *Bifidobacterium* after 24 h of fermentation (0 h:0.67 OTUs (operational taxonomic unit); 24 h:5.22 OTUs; $p = 0.038$). Inulin and WholeFiber increased the beneficial genus *Collinsella*, consistent with findings in clinical studies. All analyzed compounds were fermentable and promoted the formation of beneficial SCFAs.

Keywords: prebiotic; microbiota; fermentation; dietary fiber; microbiome

---

1. Introduction

Prebiotic definitions vary among different scientific and political arenas across the world [1]. Depending on the local definition, nearly all prebiotics can be classified as dietary fiber, but not all fibers are considered prebiotics [2]. The most recent definition describes a prebiotic as "a substrate that is selectively utilized by host microorganisms conferring a health benefit" [3]. Functional characteristics of prebiotics include the ability to: resist the low pH of the stomach, resist hydrolysis by mammalian enzymes, resist absorption in the upper gastrointestinal tract, the ability to be fermented by intestinal microbiota and selectively stimulate the growth and/or activity of intestinal bacteria associated with host health and overall well-being [4,5]. Inulin, beta-glucans, and xylooligosaccharides all provide health benefits to consumers that are related to the fermentation of these compounds in the distal gastrointestinal tract, and are also considered functional fibers with many other benefits [6]. As the definition of "prebiotic" broadens to include the overall impact from the metabolism from these compounds, the category of prebiotics will expand [7]. The importance of displaying direct health benefits due to bacterial fermentation is still the driving mechanism for all prebiotics.

As our awareness and understanding of the importance of the gut microbiome and gut microbiota increases, it is imperative for consumers to understand the key differences between different forms of prebiotics, and where they can be found in various foods and food products. XOS is an emerging prebiotic with well-displayed, consistent health benefits [8] and is composed of sugar oligomers composed of xylose units [9], found naturally in fruits, vegetables, milk, honey and bamboo shoots. XOS is commonly produced from xylan containing lignocellulosic materials through various chemical methods, direct enzymatic hydrolysis, or a combination of both treatments [10–14]. Inulin is a heterogeneous blend of fructose polymers (degree of polymerization, DP < 10) [15] which occurs naturally in thousands of plant species, including wheat, onion, bananas, garlic and chicory [16]. Beta-glucan is a polysaccharide composed of D-glucose monomers with beta-glycosidic linkages, present in either linear chains in grains, such as oat and barley (up to 7%), or in branched structures in fungi, yeast and certain bacteria [17]. These prebiotics, or prebiotic mixtures, each provide a unique carbon source for selective stimulation of different bacterial taxa and are important microbiota-shaping compounds.

Because no analytical method currently exists to measure the prebiotic capacity of foods in terms of their influence on gastrointestinal taxa, this field relies heavily on fecalbiotics (living or once living fecal microbial populations) to quantify the effects of these compounds. In vitro fermentation models allow for quantitative analysis of specific materials and are semi-representative models of colonic fermentation [18]. Although not a complete substitute for human studies, when paired with in vivo models, in vitro analysis can be an accurate systematic approach to analyzing different parameters and end points in colonic fermentation [19].

With the recent release of the International Scientific Association for Probiotics and Prebiotics consensus statement, XOS has been categorized as a prebiotic or prebiotic candidate [3]. The prebiotic effects of XOS have previously been summarized by Broekaert et al. [20]. While there is less evidence supporting the prebiotic effects of XOS compared to other types of prebiotics, studies have shown that XOS supplementation in humans can increase SCFA and bifidobacteria, as well as improve stool consistency and frequency [21–23]. This paper compares the fermentation effects of XOS to previously established prebiotics (inulin and beta-glucans) in a controlled in vitro model. To the authors' knowledge, this is the first controlled in vitro study comparing the effects of XOS to these known prebiotics. The objective of this project was to compare currently available prebiotics by their ability to change specific taxa as well as compare differences in the production of gas and common short chain fatty acids (SCFA) between these products. Inulin, XOS and beta-glucan based products were chosen for this experiment because they are established and emerging prebiotics that are commonly consumed, and offer well-demonstrated health benefits to their consumers.

2. Materials and Methods

2.1. Prebiotic Dietary Fibers Analyzed

Five common prebiotic dietary fibers were chosen for this study (Table 1), including different types of beta-glucans, inulin and xylooligosaccharide supplements.

Table 1. Comparison Prebiotic Dietary Fibers Analyzed with in vitro Fermentation System.

Prebiotic Dietary Fibers	Supplier Information
OatWell (Oatbran containing 28% beta-glucan)	DSM Nutritional Products, Ltd. (Kaiseraugust, Switzerland)
WholeFiber (A dried chicory root blend containing: inulin, pectin, hemi/cellulose)	WholeFiber, Inc. (Pennington, NJ, USA)
Xylooligosaccharide (XOS)	AIDP, Inc. (Industry, CA, USA)
Pure Inulin	Cargill, Inc. (Wayzata, MN, USA)
Pure Beta-glucan	Megazyme, Inc. (Bray, Wicklow, Ireland)

2.2. Donor Information

Fecal samples were collected from three healthy volunteers (2 males, 1 female) under anaerobic conditions. Donors included individuals (ages 22–28) consuming non-specific Western diets, who do not consume any supplements, including fiber supplements. Donors were non-smokers, did not receive any antibiotic treatments in the last year, and were not affected by any known gastrointestinal diseases (Table 2).

Table 2. Demographic Characteristics of Three Fecal Donors.

Demographic characteristics	Donor 1	Donor 2	Donor 3
Age	26	25	22
Sex	Female	Male	Male
Body Mass Index (kg/m^2)	28.1	26.3	23.0

2.3. Fecal Collection

Fecal samples were anaerobically collected within 5 min of the start of the fermentation (Medline Specimen Collection Kit, Medline, Inc., Rogers, MN, USA), and homogenized immediately upon collection. All data and samples collected were done in accordance with University of Minnesota policies and procedures.

2.4. Fermentation

The fiber samples were fermented using in vitro methods to mimic the environment of the distal colon. These methods have been used in previous in vitro studies, including Koecher et al., who found complementary results between these in vitro methods and a human intervention study of the same fibers [24]. Fiber samples (0.5 g) were hydrated in 40 mL of prepared sterile trypticase peptone fermentation media in 100 mL serum bottles, capped, and incubated for 12 h at 4 °C to limit possibility of microbial growth [25]. Following incubation, serum bottles were transferred to a circulating water bath at 37 °C for 2 h to allow the samples to reach body temperature. Post-collection, fecal samples were mixed using a 6:1 ratio of phosphate buffer solution to fecal sample. After mixing, obtained fecal slurry was combined with prepared reducing solution (2.52 g cysteine hydrochloride, 16 mL 1 N NaOH, 2.56 g sodium sulfide nonanhydride, 380 mL DD H_2O) at a 2:15 ratio. 10 mL of the prepared fecal inoculum was added to each of the serum bottles, 0.8 mL Oxyrase® was added, flushed with CO_2, sealed, and then immediately placed in a 37 °C circulating water bath. Fecal inoculum control samples with no fiber added were prepared for SCFA and gas production comparison. Baseline pH of the fermentation media was measured, with a mean of 6.83 ± 0.04, to mimic the environment of the distal colon. Samples were prepared in triplicate and analyzed at 0, 12 and 24 h. Upon removal at each time point, total gas volume was measured. Then samples were divided into aliquots for analysis and 1 mL of copper sulfate (200 g/L) was added to cease fermentation. All samples were immediately frozen and stored at −80 °C for further analysis.

2.5. SCFA Analysis

SCFA samples were extracted according to Schneider et al. [26] with minor modifications, and analyzed with previously described methods [27].

2.6. DNA Extractions

Fecal bacteria DNA from the in vitro system were extracted using a PowerSoil DNA Isolation Kit (Mo Bio Laboratories, Inc., Carlsbad, CA, USA) following the provided operating instruction, including bead beating for 20 min.

2.6.1. Primary/Secondary Amplification

The V1-V3 region of the 16S rRNA was amplified using a two-step PCR protocol. The primary amplification was done using an ABI7900 qPCR machine (Applied Biosystems, Foster City, CA, USA). The following recipe was used: 3 µL template DNA, 0.48 µL nuclease-free water, 1.2 µL 5× KAPA HiFi buffer (Kapa Biosystems, Woburn, MA, USA), 0.18 µL 10 mM dNTPs (Kapa Biosystems, Woburn, MA, USA), 0.3 µL DMSO (Fisher Scientific, Waltham, MA, USA), 0.12 µL ROX (25 µM) (Life Technologies, Carlsbad, CA, USA), 0.003 µL 1000× SYBR Green, 0.12 µL KAPA HiFi Polymerase (Kapa Biosystems, Woburn, MA, USA), 0.3 µL forward primer (10 µM), 0.3 µL reverse primer (10 µM). Cycling conditions were: 95 °C for 5 min, followed by 20 cycles of 98 °C for 20 s, 55 °C for 15 s, and 72 °C for 1 min. The primers for the primary amplification contained both 16S-specific primers (V1_27F and V3_V34R), as well as adapter tails for adding indices and Illumina flow cell adapters in a secondary amplification. The following primers were used (16S-specific sequences in bold): Meta_V1_27F (TCGTCGGCAGCGTCAGATGTGTATAAGAGACAG**AGAGTTTGATCMTGGCTCAG**) and Meta_V3_534R (GTCTCGTGGGCTCGGAGATGTGTATAAGAGACAG**ATTACCGCGGCTGCTGG**).

The amplicons from the primary PCR were diluted 1:100 in sterile, nuclease-free water, and a second PCR reaction was set up to add the Illumina flow cell adapters (Illumina Inc., San Diego, CA, USA) and indices. The secondary amplification was done on a fixed block BioRad Tetrad PCR machine (Bio-Rad Laboratories, Inc., Hercules, CA, USA) using the following recipe: 5 µL template DNA, 1 µL nuclease-free water, 2 µL 5× KAPA HiFi buffer (Kapa Biosystems, Woburn, MA, USA), 0.3 µL 10 mM dNTPs (Kapa Biosystems, Woburn, MA, USA), 0.5 µL DMSO (Fisher Scientific, Waltham, MA, USA) 0.2 µL KAPA HiFi Polymerase (Kapa Biosystems, Woburn, MA, USA), 0.5 µL forward primer (10 µM), 0.5 µL reverse primer (10 µM). Cycling conditions were: 95 °C for 5 min, followed by 10 cycles of 98 °C for 20 s, 55 °C for 15 s, 72 °C for 1 min, followed by a final extension at 72 °C for 10 min. The following indexing primers were used (X indicates the positions of the 8 bp indices): Forward indexing primer: AATGATACGGCGACCACCGAGATCTACACXXXXXXXXTCGTCGGCAGCGTC and Reverse indexing primer: CAAGCAGAAGACGGCATACGAGATXXXXXXXXGTCTCGTGGGCTCGG.

2.6.2. Normalization and Sequencing

The samples were normalized using a SequalPrep capture-resin bead plate (Life Technologies, Carlsbad, CA, USA) and pooled using equal volume. The final pools were quantified via PicoGreen dsDNA assay (Life Technologies, Carlsbad, CA, USA) and diluted to 2 nM. 10 µL of the 2 nM pool was denatured with 10 µL of 0.2 N NaOH, diluted to 8 pM in Illumina's HT1 buffer, spiked with 15% phiX, heat denatured at 96 °C for 2 min, and sequenced using a MiSeq 600 cycle v3 kit (Illumina, San Diego, CA, USA).

2.6.3. Sequence Processing and Analysis

Generated sequence data was processed and analyzed using QIIME [28]. Fastq sequence data was processed with the University of Minnesota's gopher-pipeline for metagenomics [29]. Sequence data had adapters removed and sliding quality trimming window by Trimmomatic [30]; primers removed and overlapping reads merged by Pandaseq [31]. Within QIIME, chimera checking done by chimera slayer, Open reference OTU picking completed with Usearch61, taxonomic identification using GreenGenes (Version 13.8) reference database, rarefied to 14,393 sequences per sample. Analysis was performed using *R* (R Development Core Team, Vienna, Austria, 2012).

2.7. Statistical Analysis

All statistical analysis was performed using *R* software (Version 3.2.2, *R* Development Core Team, Vienna, Austria, 2012). Differences in means were determined using the Kruskal-Wallis ANOVA test, testing the null hypothesis that the location parameter of the groups of abundancies for a given OTU is the same. Multiple comparisons were corrected using the Benjamini-Hochberg

FDR (false discovery rate) procedure for multiple comparisons. For gas and SCFA data, ANOVA with Tukey HSD was used to compare means. Significance was set for p-values < 0.05 for all statistical tests.

2.8. Consent Ethics Approval Code

Voluntary informed consent was obtained from all fecal donors prior to this study according to University of Minnesota policies and procedures.

3. Results

3.1. Gas Production

At 12 h, the OatWell and the pure beta-glucan samples produced similar amounts of total gas (Figure 1). The XOS samples produced significantly more gas than the pure beta glucan samples ($p < 0.01$) or the OatWell samples ($p < 0.01$). The WholeFiber and pure inulin samples produced similar amounts of total gas ($p = 0.102$), and the total gas production for both of these prebiotic dietary fibers was significantly higher than the XOS samples, ($p < 0.01$ and $p = 0.045$), respectively. At 24 h, the OatWell samples had the lowest gas production (46.2 mL) and were similar to the pure beta-glucan samples (63.7 mL; $p = 0.498$). The 24 h XOS samples (74.0 mL) were also similar to the beta-glucan samples ($p = 0.926$). However, the 24 h WholeFiber (109.6 mL) and pure inulin (107.1 mL) samples produced significantly more gas than XOS, beta-glucan and Oatwell samples ($p < 0.01$). Individual variation in gas production can be seen in Table S1.

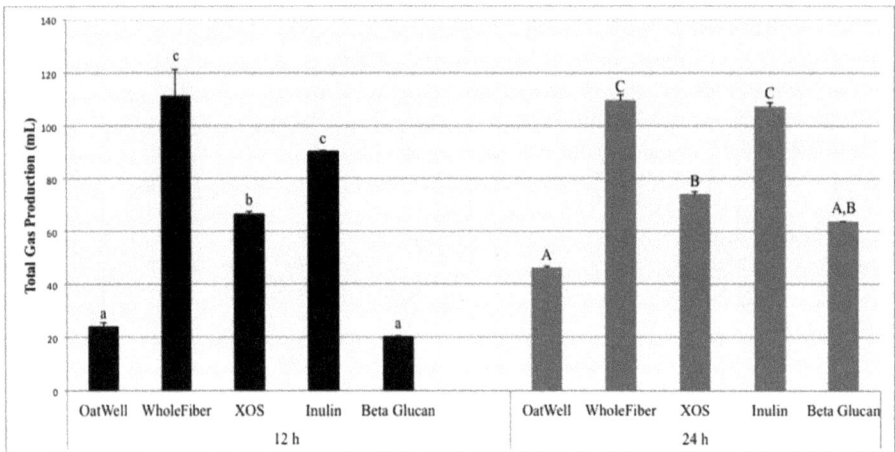

Figure 1. Total gas production comparing fermentation differences among five prebiotic dietary fibers for three individuals at 12 h and 24 h post-exposure to fecal microbiota in an in vitro fermentation system. Data displayed are means (3 donors × 3 replicates = 9) for each prebiotic dietary fiber ± SD. Columns with different letters are significantly different from one another within each time measurement (lowercase: 12 h; uppercase: 24 h). Data were analyzed using ANOVA with Tukey HSD ($p < 0.05$).

3.2. SCFA Production

For all SCFA analysis, analysis at 12 and 24 h shows production only, from baseline corrected samples. Acetate, propionate and butyrate production is shown as μmol/mL of fermentation media. Individual variation in SCFA production can be seen in Table S1.

Acetate production at 12 h was similar for the Oatwell, WholeFiber and beta-glucan samples (Figure 2). The XOS samples produced significantly more acetate at 12 h than the Oatwell, WholeFiber or beta-glucan samples ($p < 0.05$). The inulin samples had similar amounts of acetate compared to the WholeFiber and XOS samples, and significantly more than the Oatwell ($p = 0.024$) and beta-glucan ($p = 0.013$) samples at 12 h. After 24 h, the inulin samples contained less acetate than the XOS samples ($p = 0.038$), while the Oatwell, WholeFiber and beta-glucan samples were similar to both the XOS and inulin samples.

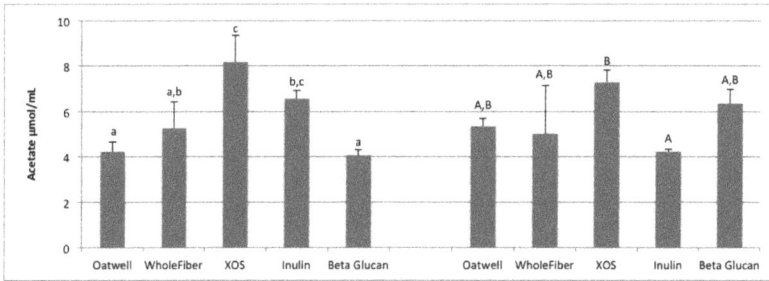

Figure 2. Acetate production at 12 h and 24 h of fermentation for five prebiotic dietary fibers displayed as μmol/mL of fermentation inoculum. Data displayed are means (3 donors × 3 replicates = 9) for each prebiotic dietary fiber ± SD. Columns with different letters are significantly different from one another (lowercase: 12 h; uppercase: 24 h). Data were analyzed using ANOVA with Tukey HSD ($p < 0.05$).

Propionate production at 12 h of fermentation was highest for the OatWell samples (4.76 μmol/mL) and was significantly greater than the WholeFiber ($p = 0.029$), XOS ($p = 0.005$) and inulin samples ($p = 0.004$), and similar to the beta-glucan samples (Figure 3). At 24 h of fermentation, the Oatwell samples had the highest mean production 5.05 μmol/mL, which was significantly greater than the XOS samples (2.58 μmol/mL; $p = 0.021$), and similar to WholeFiber, inulin and beta-glucan samples.

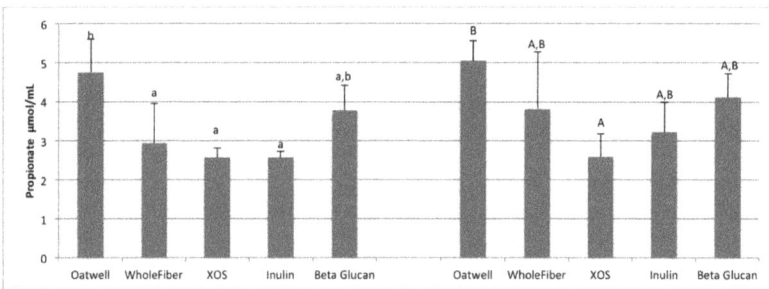

Figure 3. Propionate production at 12 h and 24 h of fermentation for five prebiotic dietary fibers displayed as μmol/mL of fermentation inoculum. Data displayed are means (3 donors × 3 replicates = 9) for each prebiotic dietary fiber ± SD. Columns with different letters are significantly different from one another (lowercase: 12 h; uppercase: 24 h). Data were analyzed using ANOVA with Tukey HSD ($p < 0.05$).

Butyrate production after 12 h of fermentation ranged from 7.30 μmol/mL for the beta-glucan samples to 16.76 μmol/mL for the inulin samples (Figure 4). The inulin samples had the highest average production, and were similar to the XOS (16.38 μmol/mL) and WholeFiber samples (12.89 μmol/mL). The XOS samples were significantly higher than the Oatwell ($p = 0.035$) and beta-glucan samples ($p = 0.014$). At 24 h of fermentation, all five prebiotic dietary fibers were statistically similar to one another, ranging from 7.93–14.08 μmol/mL due to a wide ranges in response differences between the three fecal donors used in this study.

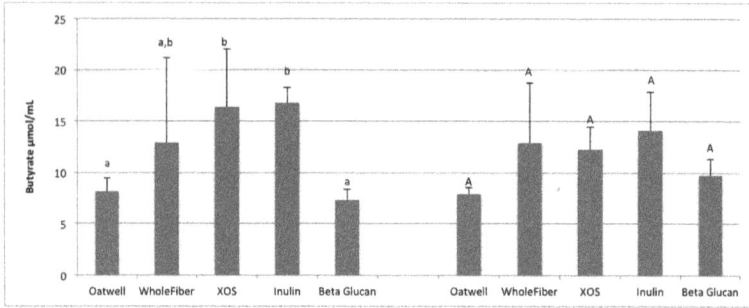

Figure 4. Butyrate production at 12 h and 24 h of fermentation for five prebiotic dietary fibers displayed as μmol/mL of fermentation inoculum. Data displayed are means (3 donors × 3 replicates = 9) for each prebiotic dietary fiber ± SD. Columns with different letters are significantly different from one another (lowercase: 24 h; uppercase: 24 h). Data were analyzed using ANOVA with Tukey HSD ($p < 0.05$).

3.3. Microbiota Analysis

Extracted DNA from in vitro samples were sequenced using the MiSeq Illumina platforms (Illumina Inc., San Diego, CA, USA) generating a total of 31,591,899 sequence reads. Sequencing parameters identified reads belonging to 11 bacterial phyla, 61 families and 97 genera.

For all three donors, the phyla Bacteroidetes and Firmicutes represented > 80% of all sequence reads (Figure S1) across 24 h of fermentation. At the family level, 13 families represented 85% of all sequence reads (Figure S2), while 11 genera represented >75% of all sequence reads (Figure S3). Six metrics measuring α-diversity for all donors showed various degrees of similarity by donors (Figure S4), and by treatment (Figure S5). Both Unifrac and Bray-Curtis β-diversity metrics (measuring pairwise dissimilarity between samples), showed similarity among technical replicates of treatment groups for each donor (Figure S6) as well as for all treatment groups for each respective donor (Figure S7).

After 24 h of fermentation, the Oatwell samples significantly decreased the population of SMB53 (0 h:9.11 OTUs; 24 h:2.11 OTUs; $p = 0.008$), Lachnospira and Faecalibacterium (0 h:26.56 OTUs; 24 h:4.44 OTUs; $p = 0.008$ and 0 h 136.44 OTUs; 24 h:66 OTUs; $p = 0.022$, respectively) (Table 3). No genera analyzed showed significant increases in 24 h for the Oatwell samples measured for the three fecal donors in this study. The WholeFiber samples (Table 4) significantly increased the genus Collinsella at 24 h compared to 0 h (0 h:68 OTUs; 24 h:299.78 OTUs; $p = 0.011$). Bifidobacterium populations were only significantly increased at 24 h compared to 0 h for the XOS samples (0 h:0.67 OTUs; 24 h:5.22 OTUs; $p = 0.038$), while the same samples showed a significant decrease in Lachnospira and Faecalibacterium ($p = 0.038$ and $p = 0.03$) (Table 5). The inulin samples (Table 6) increased Collinsella (0 h:55.11 OTUs; 24 h:291.44 OTUs; $p = 0.016$). The pure beta glucan samples significantly decreased Lachnospira and Faecalibacterium ($p = 0.008$) (Table 7).

Table 3. Combined changes across 24 h of fermentation for Oatwell samples of identified genera [1].

Genera	0 h	24 h	*p*-Value
Actinobacteria			
Bifidobacterium	1.22	0.89	0.660
Adlercreutzia	1.44	3.00	0.470
Collinsella	48.44	140.56	0.089
Bacteroidetes			
Alistipes	2.56	1.33	0.674
Parabacteroides	135.00	155.89	0.952
Bacteroides	755.67	865.00	0.192

Table 3. *Cont.*

Genera	0 h	24 h	p-Value
Firmicutes			
Eubacterium	0.44	0.56	0.817
Veillonella	1.11	1.22	0.980
Dorea	2.33	3.56	0.516
Acidaminococcus	3.22	10.44	0.817
Clostridium	7.67	8.33	0.769
Anaerostipes	8.11	6.00	0.674
Turicibacter	8.67	1.22	0.286
SMB53	9.11	2.11	0.008 *
Ruminococcus	11.22	23.22	0.263
Lactococcus	11.67	10.67	0.980
Streptococcus	15.22	8.11	0.511
Roseburia	20.22	22.33	0.980
Oscillospira	21.78	36.67	0.121
Lachnospira	26.56	4.44	0.008 *
Phascolarctobacterium	27.78	173.33	0.263
Dialister	39.56	43.00	0.560
Blautia	41.89	53.11	0.470
Coprococcus	49.89	39.00	0.396
Ruminococcus	61.33	40.67	0.289
Faecalibacterium	136.44	66.00	0.022 *
Proteobacteria			
Escherichia	0.44	1.44	0.325
Haemophilus	10.22	0.67	0.286
Sutterella	10.78	14.44	0.980
Bilophila	13.67	14.78	0.788
Verrucomicrobia			
Akkermansia	5.00	12.00	0.980

[1] Replicate samples were pooled among donor at each respective time point (3 donors × 3 replicated = 9). Samples were analyzed between differentially represented OTUs for significant changes after 24 h of fermentation compared to 0 h samples. Values are the number of OTUs after rarefaction to 3668 sequences per sample. Data were analyzed using the Kruskal-Wallis ANOVA test, with the FDR (false discovery rate) multiple comparisons correction. * Indicates significance at $p \leq 0.05$.

Table 4. Combined changes across 24 h of fermentation for WholeFiber samples of identified genera [1].

Genera	0 h	24 h	p-Value
Actinobacteria			
Adlercreutzia	0.89	3.89	0.239
Bifidobacterium	1.11	1.11	0.785
Collinsella	68.00	299.78	0.011 *
Bacteroidetes			
Alistipes	1.11	0.56	0.894
Parabacteroides	131.44	142.00	0.913
Bacteroides	743.56	776.56	0.785
Firmicutes			
Eubacterium	1.11	0.78	0.799
Veillonella	1.22	1.00	0.960
Dorea	2.00	5.00	0.785
Acidaminococcus	2.67	11.33	0.894
SMB53	5.67	4.00	0.239
Clostridium	7.33	13.22	0.896
Anaerostipes	10.22	1.22	0.239
Ruminococcus	10.89	19.67	0.943

Table 4. *Cont.*

Genera	0 h	24 h	p-Value
Firmicutes			
Streptococcus	12.67	8.78	0.785
Turicibacter	14.22	2.44	0.960
Lachnospira	14.78	72.00	0.237
Oscillospira	17.22	14.89	0.647
Lactococcus	20.44	9.22	0.896
Phascolarctobacterium	24.67	60.44	0.501
Dialister	26.11	58.22	0.943
Roseburia	28.56	6.00	0.674
Blautia	32.44	49.44	0.156
Coprococcus	45.78	66.44	0.501
Ruminococcus	54.22	39.33	0.261
Faecalibacterium	154.89	93.11	0.080
Proteobacteria			
Escherichia	0.78	1.44	0.960
Sutterella	4.00	32.44	0.894
Haemophilus	10.67	0.56	0.107
Bilophila	10.67	7.67	0.896
Verrucomicrobia			
Akkermansia	17.00	3.33	0.501

[1] Replicate samples were pooled among donor at each respective time point (3 donors × 3 replicated = 9). Samples were analyzed between differentially represented OTUs for significant changes after 24 h of fermentation compared to 0 h samples. Values are the number of OTUs after rarefaction to 3668 sequences per sample. Data were analyzed using the Kruskal-Wallis ANOVA test, with the FDR multiple comparisons correction. * Indicates significance at $p \leq 0.05$.

Table 5. Combined changes across 24 h of fermentation for xylooligosaccharide samples of identified genera [1].

Genera	0 h	24 h	p-Value
Actinobacteria			
Bifidobacterium	0.67	5.22	0.038 *
Adlercreutzia	1.33	1.78	0.972
Collinsella	58.44	154.00	0.413
Bacteroidetes			
Alistipes	1.44	0.56	0.413
Parabacteroides	147.33	133.33	0.972
Bacteroides	770.89	870.44	0.189
Firmicutes			
Eubacterium	0.33	1.67	0.364
Veillonella	0.67	0.00	0.162
Acidaminococcus	1.33	2.33	0.972
Dorea	2.11	3.67	0.423
SMB53	7.33	5.44	0.558
Anaerostipes	7.44	3.44	0.447
Turicibacter	8.00	8.56	0.972
Clostridium	8.44	4.00	0.087
Ruminococcus	12.78	26.11	0.087
Streptococcus	14.11	4.67	0.367
Lachnospira	21.11	5.33	0.038 *
Oscillospira	21.33	21.78	0.972
Phascolarctobacterium	23.44	16.33	0.972
Lactococcus	23.89	21.00	0.982
Roseburia	28.89	35.33	0.972
Dialister	33.89	41.56	0.831

Table 5. *Cont.*

Genera	0 h	24 h	p-Value
Firmicutes			
Blautia	39.22	65.00	0.087
Ruminococcus	45.11	37.33	0.385
Coprococcus	47.11	48.67	0.705
Faecalibacterium	148.56	79.56	0.030 *
Proteobacteria			
Escherichia	0.89	0.44	0.972
Haemophilus	6.44	3.11	0.972
Bilophila	17.78	6.22	0.107
Sutterella	25.78	40.89	0.831
Verrucomicrobia			
Akkermansia	2.78	5.00	0.841

[1] Replicate samples were pooled among donor at each respective time point (3 donors × 3 replicated = 9). Samples were analyzed between differentially represented OTUs for significant changes after 24 h of fermentation compared to 0 h samples. Values are the number of OTUs after rarefaction to 3668 sequences per sample. Data were analyzed using the Kruskal-Wallis ANOVA test, with the FDR multiple comparisons correction. * Indicates significance at $p \leq 0.05$.

Table 6. Combined changes across 24 h of fermentation for pure inulin samples of identified genera [1].

Genera	0 h	24 h	p-Value
Actinobacteria			
Bifidobacterium	1.33	5.44	0.304
Adlercreutzia	1.33	2.00	0.845
Collinsella	55.11	291.44	0.016 *
Bacteroidetes			
Alistipes	1.56	0.89	0.878
Parabacteroides	147.44	164.78	0.887
Bacteroides	726.78	644.44	0.652
Firmicutes			
Veillonella	0.78	0.56	0.908
Eubacterium	0.89	1.56	0.908
Dorea	1.78	7.00	0.640
Acidaminococcus	3.11	18.67	0.887
SMB53	7.44	9.11	0.965
Turicibacter	7.78	4.89	0.652
Clostridium	8.22	7.11	0.845
Ruminococcus	9.56	34.11	0.309
Anaerostipes	11.22	4.67	0.652
Streptococcus	13.00	12.44	0.887
Lactococcus	19.11	9.67	0.908
Lachnospira	21.00	4.89	0.022 *
Phascolarctobacterium	26.22	21.00	0.887
Oscillospira	26.33	10.11	0.034 *
Roseburia	26.78	14.11	0.887
Dialister	32.67	95.11	0.887
Blautia	38.22	50.22	0.690
Coprococcus	48.11	60.89	0.640
Ruminococcus	52.33	43.00	0.908
Faecalibacterium	148.11	187.33	0.652
Proteobacteria			
Escherichia	1.00	1.22	0.908
Haemophilus	9.11	2.67	0.652
Sutterella	14.00	31.22	0.908
Bilophila	16.89	7.78	0.309

Table 6. *Cont.*

Genera	0 h	24 h	*p*-Value
Verrucomicrobia			
Akkermansia	7.78	7.44	0.304

[1] Replicate samples were pooled among donor at each respective time point (3 donors × 3 replicated = 9). Samples were analyzed between differentially represented OTUs for significant changes after 24 h of fermentation compared to 0 h samples. Values are the number of OTUs after rarefaction to 3668 sequences per sample. Data were analyzed using the Kruskal-Wallis ANOVA test, with the FDR multiple comparisons correction. * Indicates significance at $p \le 0.05$.

Table 7. Combined changes across 24 h of fermentation for pure beta-glucan samples of identified genera [1].

Genera	0 h	24 h	*p*-Value
Actinobacteria			
Bifidobacterium	0.33	0.33	1.000
Adlercreutzia	2.00	1.89	0.843
Collinsella	69.22	85.11	0.723
Bacteroidetes			
Alistipes	0.78	0.89	0.778
Parabacteroides	119.56	179.78	0.778
Bacteroides	776.11	854.33	0.664
Firmicutes			
Eubacterium	0.11	0.44	0.778
Veillonella	0.56	0.22	0.778
Dorea	0.89	3.11	0.110
Acidaminococcus	2.33	15.11	0.778
SMB53	6.11	4.89	0.778
Lactococcus	6.11	0.67	0.778
Anaerostipes	7.44	5.22	0.778
Turicibacter	8.11	3.00	0.803
Ruminococcus	9.44	18.67	0.166
Clostridium	10.11	3.33	0.110
Streptococcus	14.89	6.44	0.256
Roseburia	16.11	54.33	0.510
Lachnospira	21.22	3.89	0.008 *
Oscillospira	24.33	35.11	0.389
Phascolarctobacterium	29.00	125.33	0.283
Dialister	30.56	43.67	0.819
Coprococcus	44.11	20.78	0.211
Blautia	45.11	68.11	0.408
Ruminococcus	59.67	44.44	0.500
Faecalibacterium	152.11	62.67	0.008 *
Proteobacteria			
Escherichia	0.89	0.56	0.778
Haemophilus	11.00	0.78	0.110
Sutterella	14.00	35.44	0.778
Bilophila	14.44	13.89	0.778
Verrucomicrobia			
Akkermansia	9.11	15.89	0.778

[1] Replicate samples were pooled among donor at each respective time point (3 donors × 3 replicated = 9). Samples were analyzed between differentially represented OTUs for significant changes after 24 h of fermentation compared to 0 h samples. Values are the number of OTUs after rarefaction to 3668 sequences per sample. Data were analyzed using the Kruskal-Wallis ANOVA test, with the FDR multiple comparisons correction. * Indicates significance at $p \le 0.05$.

4. Discussion

The aim of this study was to investigate the beneficial effects of commonly consumed prebiotic dietary fibers, including their ability to influence the growth of identified bacterial populations, form beneficial SCFAs, and the amount of gas they produce due to fermentation. Total gas production due to fiber fermentation depends on a wide range of factors. The inulin samples and the WholeFiber samples (mixture of dried chicory root inulin, pectin and hemi/cellulose) resulted in the highest gas production at both 12 and 24 h. These results are consistent with results from both clinical feeding studies and other in vitro experiments, in which fermentation of inulin products resulted in high amounts of gas production, sometimes resulting in mild negative gastrointestinal symptoms, depending on the dosage [32,33]. Similar in vitro studies have found inulin to be much more fermentable than beta-glucan products, for both barley and oat-derived beta-glucans [34]. XOS fermentation results in less gas production than the inulin products, and more gas than beta-glucan products. Because of these findings, previous studies based on digestive tolerance and parameters have established a tolerated daily dosage for XOS of approximately 12 g/day [35].

SCFA production due to the fermentation of prebiotic dietary fibers promotes many beneficial health outcomes to the host. SCFA production may contribute to up to 10% of the host's metabolizable energy daily, with production of total SCFAs usually between 100–200 mmol/day, but is highly dependent on the donor and availability of substrates for fermentation [36,37]. At 12 h of fermentation, the OatWell and beta-glucan samples had significantly higher concentrations of propionate, and the highest mean concentration at 24 h, compared to the other prebiotic dietary fibers analyzed. Similar in vitro studies with beta-glucan based products have also shown similar preference for these products to result in propionate formation [34]. Although no mechanism has been identified, and studies show conflicting results [38], elevated serum propionate concentrations have been shown to have a hypocholesterolaemic effect [39]. Propionate may also play an influential role in satiety, although mechanisms still remain unclear [40,41]. Cholesterol-lowering properties of beta-glucans may be limited to effects from the upper-GI, although many propionate-producing bacteria have a preference to fermenting various types of beta-glucans (*Bacteroides*, *Prevotella*, *Clostridium*) based on the presence of genes responsible for endo-β-glucanase enzyme production [42].

Microbial diversity among fecal donors complicates the identification of trends among the five treatment groups (Supplemental Figures S8 and S9). In terms of taxonomic shift, the inulin-based products were fermented nearly identically by all three fecal donors. Both pure inulin and WholeFiber promoted the growth of *Collinsella* comparing the 24 h samples to the 0 h samples. Inulin-type fructans have been shown in clinical studies to promote substantial growth of *Collinsella*, paralleled with increased urinary hippurate levels [43]. Hippurate is a metabolite derived from various fermentation processes in the gut that has been found in decreased concentrations in obese individuals compared to lean individuals, and also between diabetics and non-diabetics [44–46]. The genus *Collinsella* has been found in lower concentrations in individuals with IBD compared to healthy controls [47], while *Collinsella aerofaciens* has been associated with low risk of colorectal cancer [48]. Increases in *Collinsella* and increased urinary hippurate levels are considered a beneficial effect of inulin consumption due to its prebiotic capacity [43]. In vivo studies with inulin, scFOS and resistant starch supplementation have found decreases in the SMB53 genus, consistent with the OatWell treatment in the present study [30,31].

A significant increase in the genus *Bifidobacterium* was observed only with the XOS treatment. Rycroft et al. found a similar affinity of *Bifidobacterium* toward XOS [49]. However, previous studies have shown inulin to also stimulate the growth of *Bifidobacterium* [49–52]. While there was an overall rise in *Bifidobacterium* with the inulin treatment group between 0 h and 24 h (1.33, 5.44 OTUs respectively), this study did not find that increase to be significant ($p = 0.304$). The small sample size and individual microbiome variability likely played a role in this result. Increases in *Bifidobacterium* have been heavily studied and reviewed, and are considered a beneficial effect due to their correlation with many positive health outcomes [1]. *Bifidobacteria* reside naturally in the gastrointestinal tract of healthy human adults and have a strong affinity to ferment oligosaccharides, making them a common marker for prebiotic capacity. *Bifidobacterium* is a unique genus of bacteria in that no gas is formed as an end product of metabolism [53]. Like *Lactobacillus*, these bacteria are saccharolytic, often considered a beneficial trait [54]. *Bifidobacteria* also do not produce any known carcinogenic substances in vivo. *Bifidobacteria* concentrations have been negatively associated with obesity and weight gain [55–58]. Increases in *Bifidobacteria* have also been correlated with a decrease in blood lipopolysaccharides (LPS), inflammatory reagents that play a role in the development of inflammatory metabolic disorders and conditions, and are primarily found in gram-negative bacteria [59]. LPS induce the activation of Toll-like receptor 4 (TLR4), which leads to inflammation due to release of pro-inflammatory cytokines and chemokines [60].

In vitro fermentations are semi-representative models of colonic fermentation, but have limitations [18]. This study did not include an in vitro digestion process, which would remove digestible contents from the samples prior to fermentation, and would be a more representative model. However, because the test substrates are primarily fiber, which is non-digestible, this should have minimal impact on the results of this study. In vivo, formed gases are continually absorbed and colonic absorption is rapid. Because SCFAs are rapidly absorbed and difficult to measure, in vitro models help to understand the kinetics of colonic fermentation. However, in vitro models must be paired with similar in vivo models to better understand the full mechanisms of action resulting from colonic fermentation of prebiotic dietary fibers. Because the SCFA are not absorbed in in vitro models, the SCFA produced can alter the pH of the fermentation media. While the fermentation media was designed to mimic the pH of the distal colon at baseline, the pH of the media in this study was not further controlled throughout the experiment. This is another limitation of this in vitro model. An additional limitation of this study was the small sample size. The present study was conducted using fecal inoculum from only three donors. Due to the individual variability between the donor's microbiota (Figure S7), a larger sample size may be needed to achieve a more representative view of the effects of each of the fibers.

5. Conclusions

All five prebiotics measured in this study display fermentability and SCFA production that could have potential health benefits. Depending on their structure, each compound offers a specific carbon source for fermentation by different bacterial populations, yielding changes in beneficial taxa and production of various amounts of SCFAs and gas in vitro. For instance, while OatWell and beta-glucans promoted propionate production, XOS increased concentrations of *Bifodobacterium*, and WholeFiber and pure inulin promoted *Collinsella* growth. Findings in this study are consistent with other in vitro studies with similar prebiotic dietary fibers, as well as clinical feeding studies [34,43,49,61–63].

Supplementary Materials: The following are available online at www.mdpi.com/2072-6643/9/12/1361/s1. Table S1: Gas and SCFA Production Comparing Fermentation Differences of Five Prebiotic Dietary Fibers at 12 h and 24 h Post-Exposure to Fecal Microbiota from Three Individual Donors in an In Vitro Fermentation System, Figure S1: Identified phyla from three fecal donors microbiota at 0, 12 and 24 h of fermentation for five prebiotic dietary fibers analyzed based on percent of sequence reads, Figure S2: Identified abundant families for three fecal donors at 0, 12 and 24 h of fermentation for five prebiotic dietary fibers analyzed based on percent of sequence reads, Figure S3: Identified abundant genera for three fecal donors at 0, 12 and 24 h of fermentation for five prebiotic dietary fibers analyzed based on percent of sequence reads, Figure S4: Six metrics of analysis for alpha-diversity among samples at 0, 12 and 24 h of analysis, grouped by donor for all five prebiotic dietary fibers analyzed, Figure S5: Six metrics of analysis for alpha-diversity among samples at 0, 12 and 24 h of analysis, grouped by treatment for all three fecal donors, Figure S6: Bray-Curtis β-diversity principal component analysis of technical replicates among each treatment group between microbiota analysis of three fecal donors, Figure S7: Bray-Curtis β-diversity principal component analysis among microbiota of three fecal donors at 0, 12 and 24 h of analysis, Figure S8: Variations in abundant phyla among three donors analyzed, Figure S9: Variations in treatment groups and pooled donors analyzed.

Acknowledgments: This study was funded by DSM. The authors would like to thank Hannah Paruzynski for the laboratory assistance, the University of Minnesota Genomics Center for the analysis of the samples, The University of Minnesota Doctoral Dissertation Fellowship for financial research support, and the companies who donated time and resources to this project.

Author Contributions: Joanne L. Slavin and Justin L. Carlson conceived and designed the experiment; Justin L. Carlson, Jennifer M. Erickson and Julie M. Hess performed the experiments. Justin L. Carlson analyzed the data, Trevor J. Gould was responsible for analysis of sequencing data. Manuscript was prepared by Justin L. Carlson and Jennifer M. Erickson, and read and approved by all authors.

Conflicts of Interest: The authors declare no conflict of interest. The funding sponsors had no role in the design of the study; in the collection, analyses, or interpretation of data; in the writing of the manuscript, and in the decision to publish the results.

References

1. Carlson, J.; Slavin, J. Health benefits of fibre, prebiotics and probiotics: A review of intestinal health and related health claims. *Qual. Assur. Saf. Crop. Foods* **2016**, *6*, 1–16. [CrossRef]
2. Brownawell, A.M.; Caers, W.; Gibson, G.R.; Kendall, C.W.C.; Lewis, K.D.; Ringel, Y.; Slavin, J.L. Prebiotics and the Health Benefits of Fiber: Current Regulatory Status, Future Research, and Goals. *J. Nutr.* **2012**, *142*, 962–974. [CrossRef] [PubMed]
3. Gibson, G.R.; Hutkins, R.; Sanders, M.E.; Prescott, S.L.; Reimer, R.A.; Salminen, S.J.; Scott, K.; Stanton, C.; Swanson, K.S.; Cani, P.D.; et al. Expert consensus document: The International Scientific Association for Probiotics and Prebiotics (ISAPP) consensus statement on the definition and scope of prebiotics. *Nat. Rev. Gastroenterol. Hepatol.* **2017**. [CrossRef] [PubMed]
4. Gibson, G.R.; Roberfroid, M.B. Dietary modulation of the human colonic microbiota: Introducing the concept of prebiotics. *J. Nutr.* **1995**, *125*, 1401–1412. [CrossRef] [PubMed]
5. Gibson, G.R.; Probert, H.M.; Van Loo, J.; Rastall, R.A.; Roberfroid, M.B. Dietary modulation of the human colonic microbiota: Updating the concept of prebiotics. *Nutr. Res. Rev.* **2004**, *17*, 259–275. [CrossRef] [PubMed]
6. Slavin, J. Fiber and prebiotics: Mechanisms and health benefits. *Nutrients* **2013**, *5*, 1417–1435. [CrossRef] [PubMed]
7. Bindels, L.B.; Delzenne, N.M.; Cani, P.D.; Walter, J. Towards a more comprehensive concept for prebiotics. *Nat. Rev. Gastroenterol. Hepatol.* **2015**. [CrossRef] [PubMed]
8. Aachary, A.A.; Prapulla, S.G. Xylooligosaccharides (XOS) as an Emerging Prebiotic: Microbial Synthesis, Utilization, Structural Characterization, Bioactive Properties, and Applications. *Compr. Rev. Food Sci. Food Saf.* **2011**, *10*, 2–16. [CrossRef]
9. Vázquez, M.; Alonso, J.; Domínguez, H.; Parajó, J. Xylooligosaccharides: Manufacture and applications. *Trends Food Sci. Technol.* **2000**, *11*, 387–393. [CrossRef]
10. Christakopoulos, P.; Katapodis, P. Antimicrobial activity of acidic xylo-oligosaccharides produced by family 10 and 11 endoxylanases. *Int. J. Biol. Macromol.* **2003**, *31*, 171–175. [CrossRef]
11. Katapodis, P.; Kavarnou, A.; Kintzios, S. Production of acidic xylo-oligosaccharides by a family 10 endoxylanase from thermoascus aurantiacus and use as plant growth regulators. *Biotechnol. Lett.* **2002**, *24*, 1413–1416. [CrossRef]
12. Vardakou, M.; Katapodis, P.; Topakas, E.; Kekos, D.; Macris, B.J.; Christakopoulos, P. Synergy between enzymes involved in the degradation of insoluble wheat flour arabinoxylan. *Innov. Food Sci. Emerg. Technol.* **2004**, *5*, 107–112. [CrossRef]
13. Yang, R.; Xu, S.; Wang, Z.; Yang, W. Aqueous extraction of corncob xylan and production of xylooligosaccharides. *LWT-Food Sci. Technol.* **2005**, *38*, 677–682. [CrossRef]
14. Yuan, Q.; Zhang, H. Pilot-plant production of xylo-oligosaccharides from corncob by steaming, enzymatic hydrolysis and nanofiltration. *J. Chem. Technol. Biotechnol.* **2004**, *78*, 1073–1079. [CrossRef]
15. Niness, K.R. Inulin and Oligofructose: What Are They? *J. Nutr.* **1999**, *129*, 1402S–1406S. [PubMed]
16. Carpita, N.C.; Kanabus, J.; Housley, T.L. Linkage Structure of Fructans and Fructan Oligomers from Triticum aestivum and Festuca arundinacea Leaves. *J. Plant Physiol.* **1989**, *134*, 162–168. [CrossRef]
17. Lam, K.L.; Chi-Keung Cheung, P. Non-digestible long chain beta-glucans as novel prebiotics. *Bioact. Carbohydr. Diet. Fibre* **2013**, *2*, 45–64. [CrossRef]
18. Wisker, E.; Daniel, M. Fermentation of non-starch polysaccharides in mixed diets and single fibre sources: Comparative studies in human subjects and in vitro. *Br. J. Nutr.* **1998**, *80*, 253–261. [PubMed]

19. Payne, A.N.; Zihler, A.; Chassard, C.; Lacroix, C. Advances and perspectives in in vitro human gut fermentation modeling. *Trends Biotechnol.* **2012**, *30*, 17–25. [CrossRef] [PubMed]

20. Broekaert, W.F.; Courtin, C.M.; Verbeke, K.; Van de Wiele, T.; Verstraete, W.; Delcour, J.A. Prebiotic and Other Health-Related Effects of Cereal-Derived Arabinoxylans, Arabinoxylan-Oligosaccharides, and Xylooligosaccharides. *Crit. Rev. Food Sci. Nutr.* **2011**, *51*, 178–194. [CrossRef] [PubMed]

21. Okazaki, M.; Fufikawa, S.; Matsumoto, N. Effect of xylooligosaccharides on the growth of bifidobacteria. *Bifidobact. Microflora* **1990**, *9*, 77–86. [CrossRef]

22. Na, M.H.; Kim, W.K. Effects of Xyooligosaccharide Intake on Fecal Bifidobacteria, Lactic acid and Lipid Concentration in Korean Young Women. *Korean J. Nutr.* **2004**, *37*, 662–668.

23. Tateyama, I.; Hashii, K.; Johno, I.; Iino, T.; Hirai, K.; Suwa, Y.; Kiso, Y. Effect of xylooligosaccharide intake on severe constipation in pregnant women. *J. Nutr. Sci. Vitaminol. (Tokyo)* **2005**, *51*, 445–448. [CrossRef] [PubMed]

24. Koecher, K.J.; Noack, J.A.; Timm, D.A.; Klosterbuer, A.S.; Thomas, W.; Slavin, J.L. Estimation and Interpretation of Fermentation in the Gut: Coupling Results from a 24 h Batch in Vitro System with Fecal Measurements from a Human Intervention Feeding Study Using Fructo-oligosaccharides, Inulin, Gum Acacia, and Pea Fiber. *J. Agric. Food Chem.* **2014**, *62*, 1332–1337. [CrossRef] [PubMed]

25. McBurney, M.I.; Thompson, L.U. Effect of human faecal inoculum on in vitro fermentation variables. *Br. J. Nutr.* **1987**, *58*, 233. [CrossRef] [PubMed]

26. Schneider, S.M.; Girard-Pipau, F.; Anty, R.; van der Linde, E.G.M.; Philipsen-Geerling, B.J.; Knol, J.; Filippi, J.; Arab, K.; Hébuterne, X. Effects of total enteral nutrition supplemented with a multi-fibre mix on faecal short-chain fatty acids and microbiota. *Clin. Nutr.* **2006**, *25*, 82–90. [CrossRef] [PubMed]

27. Carlson, J.; Hospattankar, A.; Deng, P.; Swanson, K.; Slavin, J. Prebiotic Effects and Fermentation Kinetics of Wheat Dextrin and Partially Hydrolyzed Guar Gum in an In Vitro Batch ermentation System. *Foods* **2015**, *4*, 349–358. [CrossRef] [PubMed]

28. Caporaso, J.G.; Kuczynski, J.; Stombaugh, J.; Bittinger, K.; Bushman, F.D.; Costello, E.K.; Fierer, N.; Peña, A.G.; Goodrich, J.K.; Gordon, J.I.; et al. QIIME allows analysis of high-throughput community sequencing data. *Nat. Methods* **2010**, *7*, 335–336. [CrossRef] [PubMed]

29. Garbe, J.; Gould, T.; Gohl, D.; Knights, D.; Beckman, K. *Metagenomics Pipeline*, version 1; Gopher-Pipelines: Minneapolis, MN, USA, 2016.

30. Bolger, A.M.; Lohse, M.; Usadel, B. Trimmomatic: A flexible trimmer for Illumina sequence data. *Bioinformatics* **2014**, *30*, 2114–2120. [CrossRef] [PubMed]

31. Masella, A.P.; Bartram, A.K.; Truszkowski, J.M.; Brown, D.G.; Neufeld, J.D. PANDAseq: Paired-end assembler for illumina sequences. *BMC Bioinform.* **2012**, *13*, 31. [CrossRef] [PubMed]

32. Bonnema, A.L.; Kolberg, L.W.; Thomas, W.; Slavin, J.L. Gastrointestinal tolerance of chicory inulin products. *J. Am. Diet. Assoc.* **2010**, *110*, 865–868. [CrossRef] [PubMed]

33. Timm, D.A.; Stewart, M.L.; Hospattankar, A.; Slavin, J.L. Wheat Dextrin, Psyllium, and Inulin Produce Distinct Fermentation Patterns, Gas Volumes, and Short-Chain Fatty Acid Profiles In Vitro. *J. Med. Food* **2010**, *13*, 961–966. [CrossRef] [PubMed]

34. Hughes, S.A.; Shewry, P.R.; Gibson, G.R.; Mccleary, B.V.; Rastall, R.A. In vitro fermentation of oat and barley derived b-glucans by human faecal microbiota. *FEMS Microbiol. Ecol.* **2008**, *64*, 482–493. [CrossRef] [PubMed]

35. Xiao, L.; Ning, J.; Xu, G. Application of Xylo-oligosaccharide in modifying human intestinal function. *Afr. J. Microbiol. Res.* **2012**, *6*, 2116–2119. [CrossRef]

36. Livesey, G. Energy values of unavailable carbohydrate and diets: An inquiry and analysis. *Am. J. Clin. Nutr.* **1990**, *51*, 617–637. [PubMed]

37. Cook, S.I.; Sellin, J.H. Review article: Short chain fatty acids in health and disease. *Aliment. Pharmacol. Ther.* **1998**, *12*, 499–507. [CrossRef] [PubMed]

38. Malkki, Y.; Autio, K.; Hanninen, O.; Myllymaki, O.; Pelkonen, K.; Suortti, T.; Torronen, R. *Oat Bran Concentrates: Physical Properties of Beta-Glucan and Hypocholesterolemic Effects in Rats*; FAO: Rome, Italy, 1992.

39. Chen, W.-J.L.; Anderson, J.W.; Jennings, D. Propionate May Mediate the Hypocholesterolemic Effects of Certain Soluble Plant Fibers in Cholesterol-Fed Rats. *Exp. Biol. Med.* **1984**, *175*, 215–218. [CrossRef]

40. Chambers, E.S.; Morrison, D.J.; Frost, G.; Mozaffarian, D. Control of appetite and energy intake by SCFA: What are the potential underlying mechanisms? *Proc. Nutr. Soc.* **2015**, *74*, 328–336. [CrossRef] [PubMed]

41. Chambers, E.; Viardot, A.; Psichas, A.; Morrison, D. Effects of targeted delivery of propionate to the human colon on appetite regulation, body weight maintenance and adiposity in overweight adults. *Gut* **2015**, *64*, 1744–1754. [CrossRef] [PubMed]

42. Zappe, H.; Jones, W.A.; Jones, D.T.; Woods, D.R. Structure of an endo-beta-1,4-glucanase gene from *Clostridium acetobutylicum* P262 showing homology with endoglucanase genes from *Bacillus* spp. *Appl. Environ. Microbiol.* **1988**, *54*, 1289–1292. [PubMed]

43. Dewulf, E.M.; Cani, P.D.; Claus, S.P.; Fuentes, S.; Puylaert, P.G.; Neyrinck, A.M.; Bindels, L.B.; de Vos, W.M.; Gibson, G.R.; Thissen, J.-P.; et al. Insight into the prebiotic concept: Lessons from an exploratory, double blind intervention study with inulin-type fructans in obese women. *Gut* **2013**, *62*, 1112–1121. [CrossRef] [PubMed]

44. Calvani, R.; Miccheli, A.; Capuani, G.; Tomassini Miccheli, A.; Puccetti, C.; Delfini, M.; Iaconelli, A.; Nanni, G.; Mingrone, G. Gut microbiome-derived metabolites characterize a peculiar obese urinary metabotype. *Int. J. Obes.* **2010**, *34*, 1095–1098. [CrossRef] [PubMed]

45. Salek, R.M.; Maguire, M.L.; Bentley, E.; Rubtsov, D.V.; Hough, T.; Cheeseman, M.; Nunez, D.; Sweatman, B.C.; Haselden, J.N.; Cox, R.D.; et al. A metabolomic comparison of urinary changes in type 2 diabetes in mouse, rat, and human. *Physiol. Genom.* **2007**, *29*, 99–108. [CrossRef] [PubMed]

46. Waldram, A.; Holmes, E.; Wang, Y.; Rantalainen, M.; Wilson, I.D.; Tuohy, K.M.; McCartney, A.L.; Gibson, G.R.; Nicholson, J.K. Top-Down Systems Biology Modeling of Host Metabotype−Microbiome Associations in Obese Rodents. *J. Proteome Res.* **2009**, *8*, 2361–2375. [CrossRef] [PubMed]

47. Kassinen, A.; Krogius-Kurikka, L.; Mäkivuokko, H.; Rinttilä, T.; Paulin, L.; Corander, J.; Malinen, E.; Apajalahti, J.; Palva, A. The Fecal Microbiota of Irritable Bowel Syndrome Patients Differs Significantly From That of Healthy Subjects. *Gastroenterology* **2007**, *133*, 24–33. [CrossRef] [PubMed]

48. Moore, W.E.; Moore, L.H. Intestinal floras of populations that have a high risk of colon cancer. *Appl. Environ. Microbiol.* **1995**, *61*, 3202–3207. [PubMed]

49. Rycroft, C.E.; Jones, M.R.; Gibson, G.R.; Rastall, R.A. A comparative in vitro evaluation of the fermentation properties of prebiotic oligosaccharides. *J. Appl. Microbiol.* **2001**, *91*, 878–887. [CrossRef] [PubMed]

50. Wang, X.; Gibson, G.R. Effects of the in vitro fermentation of oligofructose and inulin by bacteria growing in the human large intestine. *J. Appl. Bacteriol.* **1993**, *75*, 373–380. [CrossRef] [PubMed]

51. Gibson, G.R.; Beatty, E.R.; Wang, X.; Cummings, J.H. Selective stimulation of bifidobacteria in the human colon by oligofructose and inulin. *Gastroenterology* **1995**, *108*, 975–982. [CrossRef]

52. Ramirez-Farias, C.; Slezak, K.; Fuller, Z.; Duncan, A.; Holtrop, G.; Louis, P. Effect of inulin on the human gut microbiota: Stimulation of Bifidobacterium adolescentis and Faecalibacterium prausnitzii. *Br. J. Nutr.* **2009**, *101*, 533. [CrossRef] [PubMed]

53. Buchanan, R.; Gibbons, N. *Bergey's Manual of Determinative Bacteriology*; Williams & Wilkins Co.: Baltimore, MD, USA, 1974.

54. Salyers, A. Energy sources of major intestinal fermentative anaerobes. *Am. J. Clin. Nutr.* **1979**, *32*, 158–163. [PubMed]

55. Collado, M.; Isolauri, E. Distinct composition of gut microbiota during pregnancy in overweight and normal-weight women. *Am. J. Clin. Nutr.* **2008**, *55*, 894–899.

56. Kalliomäki, M.; Collado, M. Early differences in fecal microbiota composition in children may predict overweight. *Am. J. Clin. Nutr.* **2008**, *87*, 534–538. [PubMed]

57. Santacruz, A. Gut microbiota composition is associated with body weight, weight gain and biochemical parameters in pregnant women. *Br. J. Nutr.* **2010**, *104*, 83–84. [CrossRef] [PubMed]

58. Schwiertz, A.; Taras, D.; Schäfer, K.; Beijer, S. Microbiota and SCFA in lean and overweight healthy subjects. *Obesity* **2010**, *18*, 190–195. [CrossRef] [PubMed]

59. Knaapen, M.; Kootte, R.S.; Zoetendal, E.G.; de Vos, W.M.; Dallinga-Thie, G.M.; Levi, M.; Stroes, E.S.; Nieuwdorp, M. Obesity, non-alcoholic fatty liver disease, and atherothrombosis: A role for the intestinal microbiota? *Clin. Microbiol. Infect.* **2013**, *19*, 331–337. [CrossRef] [PubMed]

60. Nakamura, Y.; Omaye, S. Metabolic diseases and pro-and prebiotics: Mechanistic insights. *Nutr. Metab. (Lond.)* **2012**, *9*, 1–11. [CrossRef] [PubMed]

61. Kedia, G.; Vázquez, J.A.; Charalampopoulos, D.; Pandiella, S.S. In Vitro Fermentation of Oat Bran Obtained by Debranning with a Mixed Culture of Human Fecal Bacteria. *Curr. Microbiol.* **2009**, *58*, 338–342. [CrossRef] [PubMed]

62. Kim, H.J.; White, P.J. In Vitro Fermentation of Oat Flours from Typical and High β-Glucan Oat Lines. *J. Agric. Food Chem.* **2009**, *57*, 7529–7536. [CrossRef] [PubMed]
63. Finegold, S.M.; Li, Z.; Summanen, P.H.; Downes, J.; Thames, G.; Corbett, K.; Dowd, S.; Krak, M.; Heber, D. Xylooligosaccharide increases bifidobacteria but not lactobacilli in human gut microbiota. *Food Funct.* **2014**, *5*, 436. [CrossRef] [PubMed]

Review

Carbohydrates for Soccer: A Focus on Skilled Actions and Half-Time Practices

Samuel P. Hills and Mark Russell *

School of Social and Health Sciences, Leeds Trinity University, Horsforth, Leeds LS18 5HD, UK;
1705952@leedstrinity.ac.uk
* Correspondence: m.russell@leedstrinity.ac.uk; Tel.: +44-(0)113-283-7100 (ext. 649)

Received: 29 November 2017; Accepted: 18 December 2017; Published: 25 December 2017

Abstract: Carbohydrate consumption is synonymous with soccer performance due to the established effects on endogenous energy store preservation, and physical capacity maintenance. For performance-enhancement purposes, exogenous energy consumption (in the form of drinks, bars, gels and snacks) is recommended on match-day; specifically, before and during match-play. Akin to the demands of soccer, limited opportunities exist to consume carbohydrates outside of scheduled breaks in competition, such as at half-time. The link between cognitive function and blood glucose availability suggests that carbohydrates may influence decision-making and technical proficiency (e.g., soccer skills). However, relatively few reviews have focused on technical, as opposed to physical, performance while also addressing the practicalities associated with carbohydrate consumption when limited in-play feeding opportunities exist. Transient physiological responses associated with reductions in activity prevalent in scheduled intra-match breaks (e.g., half-time) likely have important consequences for practitioners aiming to optimize match-day performance. Accordingly, this review evaluated novel developments in soccer literature regarding (1) the ergogenic properties of carbohydrates for skill performance; and (2) novel considerations concerning exogenous energy provision during half-time. Recommendations are made to modify half-time practices in an aim to enhance subsequent performance. Viable future research opportunities exist regarding a deeper insight into carbohydrate provision on match-day.

Keywords: glycemia; football; skill; ergogenic; blood glucose; cognition

---

1. Introduction

Soccer is the world's most popular sport [1], and is typically contested over two 45-min halves, each separated by a ~15 min half-time break. A literature-wide consensus exists that modern day players cover 10–12 km over the full 90 min [2–8]. While the predominant activities in soccer are of low-intensity, and primarily aerobic in nature [9–11], the importance of anaerobic metabolism is highlighted by observations that over 300 accelerations and decelerations may occur during each half [2], with a sprint being performed every ~90 s [3]. Despite high-intensity distance making up only 1–11% of the total distance covered during a match [3], the most decisive passages of play typically involve high-intensity actions, such as sprinting or the execution of game-specific skills [4]. Additionally, as the primary objective of soccer is to score more goals than the opposition, full-match players complete over 100 technical or skilled involvements [12]. An average of 10 shots and between 16–30 attacking plays are required for each goal scored; with dribbling and short passing being the most frequently performed technical actions [12,13]. Accordingly, interventions that improve skilled as well as physical actions are likely of interest to practitioners responsible for enhancing soccer performance.

The importance of carbohydrates in soccer has been acknowledged since the early 1970's. Seminal work interrogating muscle biopsy data, identified compromised muscle glycogen stores following soccer match-play, and observed concomitant declines in physical performance [14]. Players beginning the game with higher muscle glycogen stores (~400 mmol kg^{-1} dry weight; d.w) achieved higher movement intensities, and were better able to maintain total distance covered between-halves, than those who began with reserves of ~200 mmol kg^{-1} d.w [14]. Interestingly, Krustrup et al. [11] identified that almost 50% of individual fibers were empty, or almost empty, after 90 min of intermittent exercise; a response which was proposed to undermine force-production capabilities, even if total muscle glycogen was not substantially reduced. Given the performance benefits of attenuating glycogen depletion throughout a match, it is unsurprising that soccer players are recommended to consume 30–60 g·h^{-1} of exogenous carbohydrate during competition to maintain blood glucose concentrations and spare muscle glycogen reserves [15]. Accordingly, a 22% reduction in muscle glycogen utilization has been observed when players ingested ~1.1 L of a 6.9% carbohydrate solution throughout a simulated soccer match [16].

While the ergogenic effects of carbohydrate ingestion have been confirmed in relation to soccer performance (for reviews, see: [17,18]), studies have traditionally focused on physical actions performed during protocols, often executed on motorized treadmills (for example: [19]), that have frequently overlooked potentially important facets of the game. In a study incorporating neither skilled actions nor a scheduled half-time break, Nicholas et al. [20] observed that consuming a a 6.9% carbohydrate–electrolyte drink during 75 min of intermittent exercise, increased time to exhaustion during a subsequent running assessment. Notwithstanding the importance of maintaining physical performance, execution of soccer-specific technical skills may be a crucial determinant in the outcome of a game [21]. This narrative review summarizes pertinent research that evaluates the efficacy of carbohydrates on soccer-specific technical performance. Computerized literature searches were performed in PubMed, Google Scholar and SportDiscus databases between April 2017 and November 2017. Keywords relating to the sport (e.g., 'soccer', 'football'), outcomes of interest (e.g., 'skill', 'technical', 'passing', 'shooting', 'dribbling', 'juggling'), interventions (e.g., 'carbohydrate', 'glucose', 'maltodextrin', 'isomaltulose', 'maltose', 'sucrose', 'glycemic index') and other key terms (e.g., 'half time', 'blood glucose', 'glycogen') were used in different combinations. Articles evaluating technical proficiency in 'rugby' were excluded. All titles were scanned and relevant articles were retrieved for review. In addition, the reference lists from both original and review articles retrieved were also reviewed. Evidence highlights novel considerations for match-day carbohydrate consumption by soccer players, namely: (1) A role in the maintenance of skilled performance; and (2) possible modification of carbohydrate feeding strategies throughout match-play (with an emphasis on scheduled breaks in play, i.e., half-time).

2. Carbohydrates and Skilled Actions Performed during Soccer-Specific Exercise

Considering the disproportionate number of goals scored from 75 min onwards during soccer matches [22], and the acknowledged importance of key skills such as passing and shooting [23,24], the ability of a team to maintain technical proficiency while engaged in prolonged intermittent exercise is likely a key determinant of competitive success. Indeed, as exercise progresses into the second half, and potentially extra-time, important aspects pertaining to passing and shooting performances deteriorate [25,26]. Although the mechanisms underpinning these effects remain unclear, and environmental or match-specific factors may contribute, reductions in fuel availability (i.e., depressed muscle glycogen and blood glucose concentrations [8,10,11]), impaired cognitive function (i.e., reduced decision making skills and increased reaction times [27–29]), and dehydration (i.e., fluid losses in excess of 2% body mass, BM, increasing thermal strain [8,30]) have been implicated. It appears likely that the progressive declines in skilled performance observed during soccer matches are multifactorial in origin.

Over the past decade, research into the efficacy of nutritional ergogenic aids for soccer skills performed during fatiguing exercise has accumulated. The most common approaches have revolved around the use of carbohydrates; although the ergogenic effects of caffeine and fluid provision have also been reported [18]. In a systematic review of the literature, 75% of eligible carbohydrate papers identified that 6–8% solutions of glucose, sucrose or maltodextrin, ingested at a rate of 30–60 g·h^{-1}, enhanced at least one aspect of skilled performance over the duration of soccer-specific exercise [18].

As the brain is one of the few human organs reliant primarily on blood glucose for optimal functioning [31], it is probable that decision-making and the performance of skilled actions during soccer match-play are influenced by blood glucose concentrations. Evidence from non-exercise studies highlights that cerebral glucose uptake begins to decline when blood glucose concentrations fall below 3.6 mmol·L^{-1} [32]; a level which, although rare, is similar to those previously reported in soccer players [33]. Indeed, the relationship between glucose availability and cognitive performance has been confirmed in both healthy and diabetic populations [29,34–36]. In respect to the present discussion, Bandelow et al. [28] observed faster visual discrimination, fine motor speed, and psycho-motor speed in participants with increased blood glucose concentrations following soccer match-play in hot conditions. As cognitive processes are crucial to the skilled actions involved in team sports, and considering the role of blood glucose in the maintenance of brain function, supplementing exogenous carbohydrates that influence blood glucose concentrations could be of benefit to soccer skills performed in the latter stages of a match.

Ergogenic effects have been identified when exogenous carbohydrates are ingested before and throughout 90 [26], and 120 [25] min of soccer-specific activity. Specifically, a 6% sucrose–electrolyte drink consumed throughout exercise improved the speed of shots taken by professional youth soccer players during the second half of a soccer match simulation, whilst allowing accuracy to be maintained [26]. Similarly, carbohydrate gels (2 sachets, each containing 23 g of glucose and maltodextrin) provided to professional English Premier League youth soccer players before a simulated extra-time period, improved dribbling performance during the final 15 min of extra-time, compared with an energy-free placebo [25].

While evidence suggests that provision of exogenous carbohydrates may represent an acute strategy to prevent reductions in skilled performances throughout soccer-specific exercise, practitioners must consider the logistical implications of achieving ergogenic rates of intake (e.g., >50 g·h^{-1} [37]), without compromising gastric tolerance [38]. Although many studies have provided carbohydrates at regular intervals during exercise (i.e., every ~15 min; [26,39]), the limited number of stoppages in soccer match-play means that the potential for such frequent feeding may not exist. Indeed, the combination of high sweating rates and sporadic opportunities to drink can impair maintenance of fluid balance in soccer [40,41]; with empirical observations identifying that the warm-up and half-time are key opportunities to ingest carbohydrates. Although comparable metabolic responses (i.e., blood glucose, carbohydrate and fat oxidation) were observed when equivalent amounts of carbohydrate (~68 g ingested at ~45 g·h^{-1}) were consumed in two (i.e., before each half) or six (i.e., every 15 min) boluses, the volume of fluid ingested per feeding (>500 mL) in the two bolus trial meant that gut fullness ratings were elevated versus the high-frequency ingestion trial [42].

Accordingly, when limited opportunities exist to drink before and during match-play, consuming electrolyte beverages that contain increased (>10%) carbohydrate concentrations may provide a practical approach to enable soccer players to achieve the desired energy intake, whilst minimizing abdominal discomfort. Harper et al. [43] reported that a 12% carbohydrate–electrolyte solution, delivered in 250 mL boluses prior to the beginning of each half, improved dribbling speed and self-paced exercise performance during the later stages of a soccer match simulation, compared with either water or an electrolyte placebo. Crucially, the similar gut fullness ratings between the three conditions suggest in favor of consuming higher-concentration carbohydrate–electrolyte beverages as a practical strategy to meet the ergogenic threshold of carbohydrate ingestion, which may help attenuate declines in technical and physical performance throughout a game.

It is also important to note that the performance of skilled tasks may also be influenced by playing experience and/or physical maturity. Accordingly, it may therefore be of interest to consider whether the effects of match-day consumption of carbohydrates are the same for youth or adolescent soccer players as they are in older populations. While unrelated to soccer, improvements in basketball shooting accuracy and on-court sprinting performance have been observed when 12–15 years old players remained euhydrated with a 6% carbohydrate–electrolyte solution versus water consumption during exercise [44]. Conversely, neither shooting accuracy nor the sprinting performance of 14–15 years old basketball players were influenced by ad libitum consumption of an 8% carbohydrate–electrolyte solution compared with water [45]. As limited and equivocal literature exists on the efficacy of carbohydrate supplementation in adolescent athletes, particularly with reference to skilled actions in soccer, future research opportunities exist in this area.

3. Modification of Half-Time Practices

3.1. Current Half-Time Practice

In soccer, consecutive halves are separated by a scheduled 15 min pause in competition. This interlude may be considered a period of recovery following the first half, as well as an opportunity to prepare for the subsequent passage of play. Empirical and published evidence alike [46], highlights that upon returning to the changing room following cessation of the first half, players primarily engage with tactical debriefing alongside medical and/or nutritional practices; which include the consumption of carbohydrate–electrolyte beverages and high glycemic index (GI) foods. Personal preparation, addressing playing kit/equipment concerns, receiving video feedback, and individual player/coach interactions may also follow. Therefore, apart from potentially a short period of rewarm-up activity (which all clubs would not universally perform) in advance of the second half, the majority of half-time practices represent very low exercise intensity, and are primarily passive in nature [46].

The commercially available sports beverages that are typically consumed pre- and during-competition (including at half-time), generally contain carbohydrates in 6–10% concentrations; of which high GI carbohydrate sources, such as glucose and maltodextrin, are common constituents. Paradoxically, high GI carbohydrates consumed within an hour of commencing a single bout of exercise may elicit rebound hypoglycemia 15–30 min into subsequent activity [47]. To explain the significance of such findings to soccer, consideration should be given to the differing physiological responses when carbohydrates are consumed during exercise, such as a high-intensity warm-up or throughout a match, versus those observed following consumption in a non-exercising state [48], such as at half-time.

Briefly, carbohydrates consumed at rest promote initial elevations in blood glucose concentrations. The subsequent actions of insulin attempt to normalize glycemia via a myriad of responses including a reduction in lipolysis, and increasing the liver, skeletal muscle and fat cell uptake of glucose [48,49]. Conversely, hyperglycemic responses are observed when exogenous carbohydrates are consumed during high-intensity exercise due to the actions of catecholamines, cortisol and growth hormone [48]. An additional consideration for half-time, is that insulin-independent glucose uptake occurs following a bout of prior exercise due to translocation of type four glucose transporters (GLUT4) [48,50]. Carbohydrate supplementation, and the glycemic response to high-intensity exercise undertaken following a period of recovery from an initial bout of exercise, has received little attention to date. This is somewhat surprising given the divergent nature of the physiological responses to carbohydrate consumed at rest and during exercise, the intermittent nature of soccer match-play, and the acknowledged importance of blood glucose concentrations for augmenting soccer-specific performance [18,25,26,39].

The common notion that exogenous carbohydrate consumption before and during intermittent exercise maintains glycemia throughout the duration of match-play has recently been challenged. Notably, when blood samples are taken frequently throughout each half, transient changes in blood glucose concentrations have been reported during soccer-specific exercise; particularly at the onset of the second half [26,33,39,43]. As previously mentioned, ingesting a 6% sucrose–electrolyte drink before and during (every 15 min) a simulated soccer match, attenuated reductions in post-exercise shooting performance [26]. However, this pattern of exogenous carbohydrate provision, including consumption throughout a passive half-time period, produced blood glucose concentrations that fell by 30% in the initial stages of the second half.

This transient lowering of blood glucose concentrations, termed the 'exercise-induced rebound glycemic response', has since been confirmed [33]. Interestingly, in a soccer match simulation that included a 15 min passive half-time period, >50% of participants reported hypoglycemic values (defined as blood glucose ≤ 3.8 mmol·L^{-1}) at 60 min (i.e., 15 min into the second half) when carbohydrates were consumed before and during exercise [33]. Although the mechanisms responsible for, and the effects of, rapid reductions in blood glucose concentrations are currently unclear, optimization of match-day nutritional strategies is likely desirable for soccer players [46]. Adopting the premise of "assess then address" may provide a framework for practitioners and applied researchers to, (1) observe/audit current practices and thus establish whether key principles perceived to occur actually do so (e.g., are current carbohydrate ingestion strategies actually maintaining blood glucose concentrations for the full duration of match-play?), and (2) rationalise specific interventions (e.g., modify match-day carbohydrate intake as per the discussions below). The following discussion espouses potential modifications to current half-time practices by considering: (1) The GI of the carbohydrate consumed; (2) the timing of ingestion; (3) the amount/concentration of energy consumed; and (4) ingesting carbohydrate during a half-time rewarm-up.

3.2. Changing the Glycemic Index of Carbohydrate Consumed

Evidence, primarily from cycling studies, suggests that the GI of ingested carbohydrate may influence the rebound hypoglycemia observed at the onset of a single bout of exercise [51]; possibly due to the varying degrees of hyperinsulinemia observed post-ingestion. For example, when consumed 45 min prior to commencing activity, ingesting 75 g of trehalose (GI: 67) and galactose (GI: 20), which both exhibit a lower GI than glucose, attenuated the plasma insulin response and reduced the prevalence of hypoglycemia experienced during cycling exercise [51]. Thus, rebound hypoglycemia was substantially reduced when low to moderate GI (20–70) carbohydrates were ingested prior to a single bout of exercise, compared with high GI glucose (GI: 100).

The paucity of studies investigating the effects of consuming low GI carbohydrates in soccer, may be attributed to concerns over the risk of gastric distress associated with products offering prolonged versus expedited appearance of exogenous energy during high-intensity exercise [52]. However, work by Stevenson et al. [53] investigated the effects of an 8% solution of low GI isomaltulose (GI: 32) consumed during the warm-up and at half-time throughout a soccer match simulation that included extra-time. Isomaltulose promotes a lower insulinaemic response and slower delivery of glucose into the systemic circulation when compared to higher GI sources of carbohydrate, and participants better maintained glycemia throughout the second half of exercise than when equivalent volumes of maltodextrin were consumed (GI: 90–100). Moreover, the dampened epinephrine response in the isomaltulose condition suggests a potential sparing of muscle glycogen; depletion of which may substantially contribute to the development of peripheral fatigue during soccer [11]. Although reducing the GI of the carbohydrate consumed did not attenuate declines in either physical or skilled performance, this may be explained by the low rate of overall consumption (~20 g·h^{-1}). Crucially, abdominal comfort was not influenced by GI, therefore when limited opportunities for feeding exist (i.e., during soccer match-play), consuming low GI carbohydrates may represent an alternative to traditional forms of exogenous energy provision.

3.3. Changing the Timing of Carbohydrate Ingestion

The temporal proximity at which carbohydrates are consumed prior to a single bout of activity can influence the metabolic responses to exercise. For example, Moseley et al. [47] investigated the effects of 75 g of glucose ingested 15, 45 or 75 min before commencing cycling. Plasma glucose and insulin concentrations were significantly elevated immediately before exercise when carbohydrate had been consumed 15 min prior to the onset of activity, whereas the lowest insulin concentrations were observed when 75 min separated carbohydrate ingestion and the onset of exercise. Additionally, consuming carbohydrate–electrolyte gels (providing 0.7 ± 0.1 g·kg^{-1} BM carbohydrates) following 90 min of exercise and ~5 min prior to a simulated extra-time period, enhanced dribbling performance and elevated blood glucose concentrations by ~16% in extra-time [25]. While studies specific to the soccer half-time period are lacking, the timing of carbohydrate feeding in proximity to the start of the second half has the potential to influence subsequent responses; especially those associated with blood glucose homeostasis. It may be that ingesting carbohydrates as close to the onset of subsequent activity (i.e., the final 5 min of half-time) produces similar physiological responses to those observed when carbohydrate is consumed during high-intensity exercise, and may attenuate the decline in glycemia currently observed when exogenous energy is consumed throughout passive half-time periods [26,33,39,43]. Although the short-duration of half-time may limit the extent to which modification of nutrient timing is possible, no studies have systematically examined the physiological or performance effects of providing carbohydrates at the start, middle or end of the half-time period.

3.4. Changing the Amount/Concentration of Carbohydrate Consumed

In studies that have employed continuous exercise protocols and focused on water absorption as a priority, the detrimental effects on gastric emptying and intestinal absorption associated with high-concentration carbohydrate solutions, have led to recommendations that beverages containing between 5–10% carbohydrates are consumed during exercise [38,54]. However, limited data exists about the effects of providing carbohydrates in greater amounts (>10% solutions) when intermittent, as opposed to continuous, exercise is performed. This is somewhat surprising, given that ingesting a 20% glucose solution has been reported to enhance sprint capacity after 90 min (two 45 min periods, separated by 15 min) of intermittent cycling [55].

In recreational soccer players, elevated blood glucose concentrations have been observed from 75 min onwards when a 10% carbohydrate–electrolyte beverage was consumed before and during (including at half-time) a simulated soccer match, compared with a fluid-electrolyte placebo [33]. Interestingly, despite similarities in blood glucose concentrations (~4.0 mmol·L^{-1}) at 60 min, differences in glycemic responses were observed between conditions at 90 min. As pre-exercise ingestion of carbohydrates appear to elicit similar glycemic responses irrespective of dosage [56,57], and that the rebound hypoglycemic response appears to decay within the initial stages of exercise when high GI sources are consumed [47,55], provision of additional carbohydrate at half-time may plausibly afford ergogenic effects during the latter stages of a match. Indeed, dribbling and self-paced exercise performance were enhanced when 250 mL boluses of a 12% carbohydrate solution were consumed prior to each half of soccer specific exercise [43].

3.5. Consuming Carbohydrate during a Half-Time Rewarm-Up

It is well established that the combination of exogenous carbohydrate ingestion and high-intensity activity can elicit a hyperglycemic response in both clinical and non-clinical populations. As pancreatic beta-cell activity is inhibited by an exercise-induced catecholamine release [48], carbohydrates consumed during exercise can promote elevated blood glucose concentrations. Whilst practitioners often experience limited opportunities to engage in re-warm up activities at half-time [58], it is plausible that by combining high-intensity exercise and simultaneous carbohydrate ingestion during the half-time period, blood glucose concentrations could be better maintained thereafter.

In support, Brouns et al. [59] observed increased catecholamine concentrations, a blunted insulin response, and an increase in blood glucose concentrations at the onset of exercise, when 600 mL of a concentrated maltodextrin drink was consumed during the preceding 20 min cycle warm-up that included isolated sprint bouts. Although reductions in blood glucose concentrations were observed after 20 min of the continuous exercise that followed, these declines did not reach statistical significance. Moreover, whilst no effects on performance were observed, elevated blood glucose concentrations were reported in a subsequent exercise bout when a 16% solution containing 80 g sucrose was consumed during a previous 15 min bout of active recovery that followed a 4 km cycling time trial [60]. Although this active recovery was conducted at low intensity (i.e., <150 W), the authors speculated that the very high-intensity exercise (i.e., 4 km time trial) performed immediately before carbohydrate ingestion may have elevated blood glucose concentrations at the onset of subsequent exercise. Consequently, a half-time rewarm-up that includes a high-intensity component, combined with the ingestion of carbohydrates, may prove beneficial for soccer players who experience an exercise-induced rebound glycemic response at the onset of the second half [39]. However, the application of this approach is yet to be determined in situations where carbohydrates are provided during recovery from previous intermittent activity (i.e., first half), and when the subsequent exercise is also intermittent in nature (i.e., soccer match-play).

4. Summary

Consumption of carbohydrates both before and during competition is a nutritional strategy commonly recommended to soccer players. Such practices aim to preserve endogenous energy stores (including blood glucose concentrations and muscle glycogen), and attenuate declines in physical and skilled performance throughout the duration of match-play. Perturbations in blood glucose concentrations have been found to influence the quality of both cognitive and physical performances during soccer-specific exercise. Therefore, a role appears to exist for exogenous carbohydrate consumption to facilitate preservation of skilled performance under conditions of soccer-specific fatigue. Likewise, as evidence suggests that current practices may be sub-optimal, viable future research opportunities exist regarding strategies to maintain blood glucose concentrations for the full duration of a match. In particular, the half-time period poses unique opportunities for practitioners working in team sports such as soccer. Notably, transient reductions in blood glucose concentrations during the early stages of the second half have been observed when a variety of feeding patterns have been implemented. With these issues still unresolved, possible modifications to current match-day strategies which hold a plausible physiological rationale, include: changing the GI of the carbohydrate consumed, modifying the timing of carbohydrate provision within the half-time period, amending the dosage of carbohydrates consumed by players, and combining carbohydrate provision with a half-time rewarm-up. With this in mind, future research should seek to optimize the ergogenic potential of carbohydrate supplementation for soccer players; research which may provide commercially lucrative opportunities.

Acknowledgments: No funding was received for this review article.

Author Contributions: S.P.H. and M.R. both contributed to the structure and writing of the article. Both were also involved in reviewing the available literature, and highlighting opportunities for future research.

Conflicts of Interest: The authors declare no conflict of interest.

References

1. Palacios-Huerta, I. Structural changes during a century of the world's most popular sport. *Stat. Methods Appl.* **2004**, *13*, 241–258. [CrossRef]
2. Russell, M.; Sparkes, W.; Northeast, J.; Cook, C.J.; Love, T.D.; Bracken, R.M.; Kilduff, L.P. Changes in acceleration and deceleration capacity throughout professional soccer match-play. *J. Strength Cond. Res.* **2016**, *30*, 2839–2844. [CrossRef] [PubMed]
3. Mohr, M.; Krustrup, P.; Bangsbo, J. Match performance of high-standard soccer players with special reference to development of fatigue. *J. Sports Sci.* **2003**, *21*, 519–528. [CrossRef] [PubMed]
4. Stølen, T.; Chamari, K.; Castagna, C.; Wisløff, U. Physiology of soccer: An update. *Sports Med.* **2005**, *35*, 501–536. [CrossRef] [PubMed]
5. Di Salvo, V.; Gregson, W.; Atkinson, G.; Tordoff, P.; Drust, B. Analysis of high intensity activity in premier league soccer. *Int. J. Sports Med.* **2009**, *30*, 205–212. [CrossRef] [PubMed]

6. Russell, M.; Rees, G.; Benton, D.; Kingsley, M. An exercise protocol that replicates soccer match-play. *Int. J. Sports Med.* **2011**, *32*, 511–518. [CrossRef] [PubMed]
7. Akenhead, R.; Hayes, P.R.; Thompson, K.G.; French, D. Diminutions of acceleration and deceleration output during professional football match play. *J. Sci. Med. Sport* **2013**, *16*, 556–561. [CrossRef] [PubMed]
8. Ekblom, B. Applied physiology of soccer. *Sports Med.* **1986**, *3*, 50–60. [CrossRef] [PubMed]
9. Bangsbo, J.; Iaia, F.M.; Krustrup, P. Metabolic response and fatigue in soccer. *Int. J. Sports Physiol. Perform.* **2007**, *2*, 111–127. [CrossRef] [PubMed]
10. Bangsbo, J.; Mohr, M.; Krustrup, P. Physical and metabolic demands of training and match-play in the elite football player. *J. Sports Sci.* **2006**, *24*, 665–674. [CrossRef] [PubMed]
11. Krustrup, P.; Mohr, M.; Steensberg, A.; Bencke, J.; Kjaer, M.; Bangsbo, J. Muscle and blood metabolites during a soccer game: Implications for sprint performance. *Med. Sci. Sports Exerc.* **2006**, *38*, 1165–1174. [CrossRef] [PubMed]
12. Bloomfield, J.; Polman, R.; O'Donoghue, P. Physical demands of different positions in FA Premier League soccer. *J. Sports Sci. Med.* **2007**, *6*, 63–70. [PubMed]
13. Rampinini, E.; Impellizzeri, F.M.; Castagna, C.; Coutts, A.J.; Wisløff, U. Technical performance during soccer matches of the Italian serie A league: Effect of fatigue and competitive level. *J. Sci. Med. Sport* **2009**, *12*, 227–233. [CrossRef] [PubMed]
14. Saltin, B. Metabolic fundamentals in exercise. *Med. Sci. Sports Exerc.* **1973**, *5*, 137–146. [CrossRef]
15. Thomas, D.T.; Erdman, K.A.; Burke, L.M. American College of Sports Medicine joint position statement. Nutrition and athletic performance. *Med. Sci. Sports Exerc.* **2016**, *48*, 543–568. [PubMed]
16. Nicholas, C.W.; Tsintzas, K.; Boobis, L.; Williams, C. Carbohydrate-electrolyte ingestion during intermittent high-intensity running. *Med. Sci. Sports Exerc.* **1999**, *31*, 1280–1286. [CrossRef] [PubMed]
17. Cermak, N.; Loon, L. The use of carbohydrates during exercise as an ergogenic aid. *Sports Med.* **2013**, *43*, 1139–1155. [CrossRef] [PubMed]
18. Russell, M.; Kingsley, M. The efficacy of acute nutritional interventions on soccer skill performance. *Sports Med.* **2014**, *44*, 957–970. [CrossRef] [PubMed]
19. Drust, B.; Reilly, T.; Cable, N.T. Physiological responses to laboratory-based soccer-specific intermittent and continuous exercise. *J. Sports Sci.* **2000**, *18*, 885–892. [CrossRef] [PubMed]
20. Nicholas, C.W.; Williams, C.; Lakomy, H.K.; Phillips, G.; Nowitz, A. Influence of ingesting a carbohydrate-electrolyte solution on endurance capacity during intermittent, high-intensity shuttle running. *J. Sports Sci.* **1995**, *13*, 283–290. [CrossRef] [PubMed]
21. Lago-Peñas, C.; Lago-Ballesteros, J.; Dellal, A.; Gómez, M. Game-related statistics that discriminated winning, drawing and losing teams from the spanish soccer league. *J. Sports Sci. Med.* **2010**, *9*, 288–293. [PubMed]
22. Reilly, T. Energetics of high-intensity exercise (soccer) with particular reference to fatigue. *J. Sports Sci.* **1997**, *15*, 257–263. [CrossRef] [PubMed]
23. Hughes, M.; Franks, I. Analysis of passing sequences, shots and goals in soccer. *J. Sports Sci.* **2005**, *23*, 509–514. [CrossRef] [PubMed]
24. Stone, K.J.; Oliver, J.L. The effect of 45 minutes of soccer-specific exercise on the performance of soccer skills. *Int. J. Sports Physiol. Perform.* **2009**, *4*, 163–175. [CrossRef] [PubMed]
25. Harper, L.D.; Briggs, M.A.; McNamee, G.; West, D.J.; Kilduff, L.P.; Stevenson, E.; Russell, M. Physiological and performance effects of carbohydrate gels consumed prior to the extra-time period of prolonged simulated soccer match-play. *J. Sci. Med. Sport* **2016**, *19*, 509–514. [CrossRef] [PubMed]
26. Russell, M.; Benton, D.; Kingsley, M. Influence of carbohydrate supplementation on skill performance during a soccer match simulation. *J. Sci. Med. Sport* **2012**, *15*, 348–354. [CrossRef] [PubMed]
27. Collardeau, M.; Brisswalter, J.; Vercruyssen, M.; Audiffren, M.; Goubault, C. Single and choice reaction time during prolonged exercise in trained subjects: Influence of carbohydrate availability. *Eur. J. Appl. Physiol.* **2001**, *86*, 150–156. [PubMed]
28. Bandelow, S.; Maughan, R.; Shirreffs, S.; Ozgunen, K.; Kurdak, S.; Ersoz, G.; Binnet, M.; Dvorak, J. The effects of exercise, heat, cooling and rehydration strategies on cognitive function in football players. *Scand. J. Med. Sci. Sports* **2010**, *20*, 148–160. [CrossRef] [PubMed]
29. Evans, M.L.; Pernet, A.; Lomas, J.; Jones, J.; Amiel, S.A. Delay in onset of awareness of acute hypoglycemia and of restoration of cognitive performance during recovery. *Diabetes Care* **2000**, *23*, 893–897. [CrossRef] [PubMed]

30. Shirreffs, S.M.; Sawka, M.N.; Stone, M. Water and electrolyte needs for football training and match-play. *J. Sports Sci.* **2006**, *24*, 699–707. [CrossRef] [PubMed]

31. Schönfeld, P.; Reiser, G. Why does brain metabolism not favor burning of fatty acids to provide energy? Reflections on disadvantages of the use of free fatty acids as fuel for brain. *J. Cereb. Blood Flow Metab.* **2013**, *33*, 1493–1499. [CrossRef] [PubMed]

32. Boyle, P.J.; Nagy, R.J.; O'Connor, A.M.; Kempers, S.F.; Yeo, R.A.; Qualls, C. Adaptation in brain glucose uptake following recurrent hypoglycemia. *Proc. Natl. Acad. Sci. USA* **1994**, *91*, 9352–9356. [CrossRef] [PubMed]

33. Kingsley, M.; Penas-Ruiz, C.; Terry, C.; Russell, M. Effects of carbohydrate-hydration strategies on glucose metabolism, sprint performance and hydration during a soccer match simulation in recreational players. *J. Sci. Med. Sport* **2014**, *17*, 239–243. [CrossRef] [PubMed]

34. Holmes, C.S.; Koepke, K.M.; Thompson, R.G.; Gyves, P.W.; Weydert, J.A. Verbal fluency and naming performance in type i diabetes at different blood glucose concentrations. *Diabetes Care* **1984**, *7*, 454–459. [CrossRef] [PubMed]

35. Fanelli, C.; Pampanelli, S.; Epifano, L.; Rambotti, A.M.; Ciofetta, M.; Modarelli, F.; Di Vincenzo, A.; Annibale, B.; Lepore, M.; Lalli, C.; et al. Relative roles of insulin and hypoglycaemia on induction of neuroendocrine responses to, symptoms of, and deterioration of cognitive function in hypoglycaemia in male and female humans. *Diabetologia* **1994**, *37*, 797–807. [CrossRef] [PubMed]

36. Fanelli, C.G.; Epifano, L.; Rambotti, A.M.; Pampanelli, S.; Di Vincenzo, A.; Modarelli, F.; Lepore, M.; Annibale, B.; Ciofetta, M.; Bottini, P.; et al. Meticulous prevention of hypoglycemia normalizes the glycemic thresholds and magnitude of most of neuroendocrine responses to, symptoms of, and cognitive function during hypoglycemia in intensively treated patients with short-term iddm. *Diabetes* **1993**, *42*, 1683–1689. [CrossRef] [PubMed]

37. Ali, A.; Williams, C. Carbohydrate ingestion and soccer skill performance during prolonged intermittent exercise. *J. Sports Sci.* **2009**, *27*, 1499–1508. [CrossRef] [PubMed]

38. De Oliveira, E.P.; Burini, R.C. Food-dependent, exercise-induced gastrointestinal distress. *J. Int. Soc. Sports Nutr.* **2011**, *8*, 12. [CrossRef] [PubMed]

39. Russell, M.; Benton, D.; Kingsley, M. Carbohydrate ingestion before and during soccer match play and blood glucose and lactate concentrations. *J. Athl. Train.* **2014**, *49*, 447–453. [CrossRef] [PubMed]

40. Burke, L.M. Fluid balance during team sports. *J. Sports Sci.* **1997**, *15*, 287–295. [CrossRef] [PubMed]

41. Clarke, N.D.; Drust, B.; MacLaren, D.P.M.; Reilly, T. Strategies for hydration and energy provision during soccer-specific exercise. *Int. J. Sport Nutr. Exerc. Metab.* **2005**, *15*, 625–640. [CrossRef] [PubMed]

42. Clarke, N.D.; Drust, B.; Maclaren, D.P.M.; Reilly, T. Fluid provision and metabolic responses to soccer-specific exercise. *Eur. J. Appl. Physiol.* **2008**, *104*, 1069–1077. [CrossRef] [PubMed]

43. Harper, L.D.; Stevenson, E.J.; Rollo, I.; Russell, M. The influence of a 12% carbohydrate-electrolyte beverage on self-paced soccer-specific exercise performance. *J. Sci. Med. Sport* **2017**, *20*, 1123–1129. [CrossRef] [PubMed]

44. Dougherty, K.A.; Baker, L.B.; Chow, M.; Kenney, W.L. Two percent dehydration impairs and six percent carbohydrate drink improves boys basketball skills. *Med. Sci. Sports Exerc.* **2006**, *38*, 1650–1658. [CrossRef] [PubMed]

45. Carvalho, P.; Oliveira, B.; Barros, R.; Padrão, P.; Moreira, P.; Teixeira, V.H. Impact of fluid restriction and ad libitum water intake or an 8% carbohydrate-electrolyte beverage on skill performance of elite adolescent basketball players. *Int. J. Sport Nutr. Exerc. Metab.* **2011**, *21*, 214–221. [CrossRef] [PubMed]

46. Russell, M.; West, D.J.; Harper, L.D.; Cook, C.J.; Kilduff, L.P. Half-time strategies to enhance second-half performance in team-sports players: A review and recommendations. *Sports Med.* **2016**, *45*, 353–364. [CrossRef] [PubMed]

47. Moseley, L.; Lancaster, G.I.; Jeukendrup, A.E. Effects of timing of pre-exercise ingestion of carbohydrate on subsequent metabolism and cycling performance. *Eur. J. Appl. Physiol.* **2003**, *88*, 453–458. [CrossRef] [PubMed]

48. Astrand, P.; Rodahl, K. *Textbook of Work Physiology: Physiological Bases of Exercise*; McGraw-Hill: New York, NY, USA, 1986.

49. Costill, D.L.; Coyle, E.; Dalsky, G.; Evans, W.; Fink, W.; Hoopes, D. Effects of elevated plasma ffa and insulin on muscle glycogen usage during exercise. *J. Appl. Physiol.* **1977**, *43*, 695–699. [CrossRef] [PubMed]

50. Rose, A.J.; Richter, E.A. Skeletal muscle glucose uptake during exercise: How is it regulated? *Physiology* **2005**, *20*, 260–270. [CrossRef] [PubMed]

51. Jentjens, R.L.; Jeukendrup, A.E. Effects of pre-exercise ingestion of trehalose, galactose and glucose on subsequent metabolism and cycling performance. *Eur. J. Appl. Physiol.* **2003**, *88*, 459–465. [CrossRef] [PubMed]

52. Oosthuyse, T.; Carstens, M.; Millen, A.M.E. Ingesting isomaltulose versus fructose-maltodextrin during prolonged moderate-heavy exercise increases fat oxidation but impairs gastrointestinal comfort and cycling performance. *Int. J. Sport Nutr. Exerc. Metab.* **2015**, *25*, 427–438. [CrossRef] [PubMed]

53. Stevenson, E.J.; Watson, A.; Theis, S.; Holz, A.; Harper, L.D.; Russell, M. A comparison of isomaltulose versus maltodextrin ingestion during soccer-specific exercise. *Eur. J. Appl. Physiol.* **2017**, *117*, 2321–2333. [CrossRef] [PubMed]

54. Coombes, J.S.; Hamilton, K.L. The effectiveness of commercially available sports drinks. *Sports Med.* **2000**, *29*, 181–209. [CrossRef] [PubMed]

55. Sugiura, K.; Kobayashi, K. Effect of carbohydrate ingestion on sprint performance following continuous and intermittent exercise. *Med. Sci. Sports Exerc.* **1998**, *30*, 1624–1630. [CrossRef] [PubMed]

56. Jentjens, R.L.; Cale, C.; Gutch, C.; Jeukendrup, A.E. Effects of pre-exercise ingestion of differing amounts of carbohydrate on subsequent metabolism and cycling performance. *Eur. J. Appl. Physiol.* **2003**, *88*, 444–452. [CrossRef] [PubMed]

57. Short, K.R.; Sheffield-Moore, M.; Costill, D.L. Glycemic and insulinemic responses to multiple preexercise carbohydrate feedings. *Int. J. Sport Nutr.* **1997**, *7*, 128–137. [CrossRef] [PubMed]

58. Towlson, C.; Midgley, A.W.; Lovell, R. Warm-up strategies of professional soccer players: Practitioners' perspectives. *J. Sports Sci.* **2013**, *31*, 1393–1401. [CrossRef] [PubMed]

59. Brouns, F.; Rehrer, N.J.; Saris, W.H.; Beckers, E.; Menheere, P.; ten Hoor, F. Effect of carbohydrate intake during warming-up on the regulation of blood glucose during exercise. *Int. J. Sports Med.* **1989**, *10*, S68–S75. [CrossRef] [PubMed]

60. Shei, R.-J.; Paris, H.L.; Beck, C.P.; Chapman, R.F.; Mickleborough, T.D. Repeated high-intensity cycling performance is unaffected by timing of carbohydrate ingestion. *J. Strength Cond. Res.* **2017**. [CrossRef] [PubMed]

nutrients

MDPI

Article

Amaranthus caudatus Stimulates Insulin Secretion in Goto-Kakizaki Rats, a Model of Diabetes Mellitus Type 2

Silvia Zambrana [1,2], Lena C. E. Lundqvist [3], Virginia Veliz [1], Sergiu-Bogdan Catrina [2,4,5], Eduardo Gonzales [1] and Claes-Göran Östenson [2,*]

[1] Instituto de Investigaciones Farmaco Bioquimicas, Universidad Mayor de San Andres, Avenida Saavedra 2224, La Paz 2314, Bolivia; silvia.zambrana@ki.se (S.Z.); viky9287@gmail.com (V.V.); eduardo.gonzales@gmail.com (E.G.)
[2] Department of Molecular Medicine and Surgery, Karolinska Institutet, Karolinska University Hospital, 171 76 Stockholm, Sweden; Sergiu-Bogdan.Catrina@ki.se
[3] Department of Molecular Sciences, Swedish University of Agricultural Sciences, P.O. Box 7015, 750 07 Uppsala, Sweden; lena.lundqvist@slu.se
[4] Department of Endocrinology, Metabolism and Diabetes, Karolinska University Hospital, 141 86 Stockholm, Sweden
[5] Centrum for Diabetes, Academic Specialist Centrum, 141 86 Stockholm, Sweden
* Correspondence: claes-goran.ostenson@ki.se; Tel.: +4-685-177-6200

Received: 29 November 2017; Accepted: 8 January 2018; Published: 15 January 2018

Abstract: Diabetes Mellitus Type 2 prevalence is increasing worldwide; thus efforts to develop novel therapeutic strategies are required. *Amaranthus caudatus* (AC) is a pseudo-cereal with reported anti-diabetic effects that is usually consumed in food preparations in Bolivia. This study evaluated the anti-diabetic nutraceutical property of an AC hydroethanolic extract that contains mainly sugars and traces of polyphenols and amino acids (as shown by nalysis with liquid chromatography-mass spectrometry (LC-MS) and nuclear magnetic resonance (NMR)), in type 2 diabetic Goto-Kakizaki (GK) rats and healthy Wistar (W) rats. A single oral administration of AC extract (2000 mg/kg body weight) improved glucose tolerance during Oral Glucose Tolerance Tests (OGTT) in both GK rats and in W rats. Long-term treatment (21 days) with AC (1000 mg/kg b.w.) improved the glucose tolerance evaluated by the area under the curve (AUC) of glucose levels during the OGTT, in both GK and W rats. The HbA1c levels were reduced in both GK (19.83%) and W rats (10.7%). This effect was secondary to an increase in serum insulin levels in both GK and W rats and confirmed in pancreatic islets, isolated from treated animals, where the chronic AC exposure increased the insulin production 4.1-fold in GK and 3.7-fold in W rat islets. Furthermore, the effect of AC on in vitro glucose-dependent insulin secretion (16.7 mM glucose) was concentration-dependent up to 50 mg/mL, with 8.5-fold increase in GK and 5.7-fold in W rat islets, and the insulin secretion in perifused GK and W rat islets increased 31 and nine times, respectively. The mechanism of action of AC on insulin secretion was shown to involve calcium, PKA and PKC activation, and G-protein coupled-exocytosis since the AC effect was reduced 38% by nifedipine (L-type channel inhibitor), 77% by H89 (PKA inhibitor), 79% by Calphostine-C (PKC inhibitor) and 20% by pertussis toxin (G-protein suppressor).

Keywords: *Amaranthus caudatus*; nutraceutical; natural product; diabetes mellitus type 2 diabetes; insulin secretion; Goto-Kakizaki rats

1. Introduction

Diabetes Mellitus Type 2 (DMT2) is a metabolic disease characterized by chronically elevated levels of blood glucose, due to impaired insulin secretion and insulin resistance [1,2]. The prevalence of diabetes has been increasing worldwide with the most dramatic increase in low-income countries [3].

According to the International Diabetes Federation (IDF), the number of people with diabetes is predicted to increase to 642 million by 2040 [4]. Moreover, diabetic patients are at higher risk of morbidity and mortality due to the development of diabetes complications [5,6]. Despite the present availability of several anti-diabetic drugs, development of novel therapeutic strategies with low adverse effects and better adherence is needed [7]. Adverse effects, such as hypoglycemias and weight gain, can impair glucose control and reduce treatment compliance.

Natural products are potential sources of novel therapies, which with a good safety profile have become an attractive complement to the regular pharmacological therapies [8]. A number of natural products have been reported to have anti-diabetic effects [9–12], such as α-lipoic acid [9,13], strawberry extracts [14], flavonoids [11]. Thus, the approach of this study is the search for anti-diabetic products using food plants, which may have medicinal properties (nutraceuticals) [15,16]. A nutraceutical is a food or part of a food that has a beneficial pharmaceutical benefit beyond its nutritional value [17].

Based on the use of Bolivian traditional foods, we evaluated the anti-diabetic property of *Amaranthus caudatus* (AC), traditional name kiwicha. AC is a gluten-free pseudocereal [18,19], native of the Bolivian valley region, and its seeds are traditionally consumed in beverages, food preparations such as soups, bread, or as toasted flour (pito). AC is an attractive nutraceutical crop due to its high protein content (rich in lysine), dietary fiber, and bioactive compounds such as tocopherols, phenolic compounds, folate [20,21] squalene, phytates, and vitamins [22]. Among its anti-diabetic effects, AC seed water decoction extract [23] and methanolic extract [24] showed α-amylase inhibitory activity. The methanol extract of AC leaves reduced blood glucose level and improved the lipid profile of rats with streptozotocin-induced diabetes [25].

In the present study, the AC anti-diabetic effect was tested in type 2 diabetic Goto-Kakizaki rats (GK) and in healthy Wistar rats (W). The effect on insulin release was evaluated in pancreatic islets and to explore the in vitro mechanism of action, and inhibitory compounds of insulin secretion pathway were used in batch incubation experiments.

2. Materials and Methods

2.1. Animals

Male healthy Wistar rats (W) and spontaneously type 2 diabetic GK (Goto-Kakizaki) rats (150–300 g) were used in this study. GK rats, originally derived from glucose intolerant W rats, were bred in the animal facilities of the department of Molecular Medicine and Surgery, KI [26], while W rats were purchased from a commercial breeder (Charles River, Sweden). Experiments were done after one week of adaptation in the animal facilities. Animals were kept at 22 °C with alternating 12 h light- dark cycle and free access to food and fresh water. The study was approved by the Laboratory Animal Ethics Committee of the Karolinska Institutet (approval Dnr N50/2014).

2.2. Plant Material

The plant material was collected from local producers in Tomina municipality, Tomina Province, Chuquisaca (latitude 19°25′53.96″ S and longitude 64°15′5.44″ W). One voucher specimen (No. EG-1, Amaranthaceae) was identified and certified by the Herbario Nacional de Bolivia from Universidad Mayor de San Andrés and has been deposited at the Department of Pharmacology at the Instituto de Investigaciones Farmaco Bioquimicas, UMSA, La Paz, Bolivia.

2.3. Plant Extract Preparation

The AC hydro-ethanolic extract was prepared with 200 g of powdered seeds macerated for 48 h with 250 mL of 70% ethanol. The maceration procedure was repeated five times to maximize the extraction yield. Ethanol solvent was evaporated using a rotary evaporator (Heidolph, Schwabach, Germany) and the water fraction was dried under pressure in a freeze dryer (Labconco, Kansas City, MO, USA). Crude extracts obtained had an appearance of a light powder with a yield of 6.5% w/w. For experiments, extracts were dissolved in distillated water and stock solutions were sterilized using 0.2 μm Millipore filter (Sigma-Aldrich, St. Louis, MO, USA).

2.4. LC-MS (Liquid Chromatography-Mass Spectrometry) Analysis

The *AC* extract was dissolved in Milli-Q (Millipore S.A.S., Molsheim, France) water and filtrated through a 0.22 μm filter before performing HPLC-HRMS (High Performance Liquid Chromatography—High Resolution Mass Spectrometry analysis), without any further purification. HPLC analysis was performed using an Agilent 1100 system (Agilent Technologies, Palo Alto, CA, USA) equipped with a Discovery 150 × 4.6 mm reversed phase C18 column (Sigma-Aldrich Supelco, Bellefonte, PA, USA). The mobile phase was composed of water with 0.1% formic acid (A), and acetonitrile (B). A stepwise gradient was used starting with 95% (A): 5% (B), and held there for 5 min, then changed to 80% (B) in 40 min, and finally return to initial conditions 95% (A): 5% (B) in 45 min, with a flow of 0.8 mL/min. For the HRMS detection, a Bruker's MaXis Impact ESI Q-TOF mass spectrometer (Bruker Daltonics GmbH, Bremen, Germany) with sodium formate (positive) as calibrant (positive scanning mode, m/z 50–1500) was used. Ultra-violet (UV) detection was done using an Agilent 1100 series Diode Array Detector (DAD).

The following chemicals were used as standards: Caffeic acid (Sigma-Aldrich, St. Louis, MO, USA); *p*-coumaric acid (Koch-light Laboratories Ltd., Cardiff, UK); ferulic acid (Fluka AG, Buch/SG, Switzerland); vanillic acid (Merck, Kenilworth, NJ, USA); kaempferol (Sigma-Aldrich, St. Louis, MO, USA); myricetin (Sigma-Aldrich, St. Louis, MO, USA); rutin (Sigma-Aldrich, St. Louis, MO, USA); squalene (Fluka AG, Buch/SG, Switzerland); quercetin (Sigma-Aldrich, St. Louis, MO, USA); inositol (Pfanstiehl Chemicals Co., Waukegan, IL, USA); fructose (Pfanstiehl Chemicals Co. Waukegan, IL, USA); glucose (Nutritional Biochemicals Co., Cleveland, OH, USA); maltose (Sigma-Aldrich, St. Louis, MO, USA); raffinose (Kebo, Stockholm, Sweden); isoleucine (Sigma-Aldrich, St. Louis, MO, USA); leucine (Merck, Kenilworth, NJ, USA); phenylalanine (Merck, Kenilworth, NJ, USA); tryptophan (Merck, Kenilworth, NJ, USA); tyrosine (Merck, Kenilworth, NJ, USA).

2.5. NMR Analysis

The *AC* extract was dissolved in Milli-Q water and filtrated through a 0.22 μm filter before being lyophilized, re-dissolved in D2O and submitted to NMR analysis. The NMR data were recorded at 25 °C with a Bruker AVANCE™ III (Bruker BioSpin GmbH, Rheinstetten, Germany) 600 MHz spectrometer equipped with a 5 mm 1H/13C/15N/31P inverse detection QXI probe, with a z-gradient. The ^{13}C and ^{1}H chemical shifts were measured using acetone as an internal standard (δ = 2.225 and 31.05 ppm for proton and carbon respectively). The data were acquired and processed using Bruker software TopSpin 3.1. (Bruker BioSpin GmbH, Rheinstetten, Germany). The ^{1}H–^{13}C HSQC and TOCSY spectra were recorded using standard pulse sequences from the Bruker library. A mixing time of 120 ms was used for the TOCSY experiment.

2.6. Sub-Acute Oral Toxicity

The *AC* extract was evaluated for its potential sub-acute oral toxicity, for 28 days, according to the guidelines set by the Organization for Economic Cooperation and Development (OECD) guideline 407 [27]. Briefly, W rats received the *AC* extract added to the regular chow food in a quantity to reach a daily dose of 1000 mg/kg b.w. During the treatment, changes in skin, fur, eyes, the occurrence of secretions, lacrimation, and piloerection were monitored. At the end point, blood samples were collected to determine hematological and serum biochemical parameters. Body weights did not differ between the groups (Supplementary Materials Figure S1).

2.7. Oral Glucose Tolerance Test (OGTT)

AC hydroethanolic extract (2000, 1000 and 500 mg/kg b.w.) was administrated orally to 10–12 h fasted GK and W rats (n = 6 per group), one hour before the OGTT. The evaluation started with an oral glucose challenge of 2 g/kg b.w. for GK rats and 3 g/kg b.w. for W rats. Blood samples were collected, from the tip of the tail, immediately after the glucose administration (time 0), 30, 60, 90, and 120 min [28]. To measure glycemia, a glucometer Accu-check Aviva (Roche Diagnostic GmbH, Indianapolis, IN, USA) was used. Serum insulin levels were measured at time 0 and 30 min by a radioimmunoassay (RIA) [29]. The placebo group received vehicle (distilled water).

2.8. Long-Term Treatment Evaluation

Long-term oral *AC* treatment with a daily dose during 21 days was evaluated in GK and W rats (*n* = 6 per group). Animals were grouped as follow: group 1: GK rats treated with 1000 mg/kg b.w. of *AC*; group 2: GK rats, treated with vehicle, distilled water; group 3: W rats treated with 1000 mg/kg b.w. of *AC*; group 4: W rats treated with vehicle, distilled water. Body weights (Supplementary Materials Figure S2) and non-fasting glucose levels were measured every third day, and there were no significant differences in body weights between the groups. The OGTT was performed on days 0, 10, and 20; blood samples were collected to measure serum insulin by RIA and glycated hemoglobin (HbA1c) by ELISA (Crystal Chem INC, Elk Grove Village, IL, USA). At the end point, day 21, pancreatic tissue was collected to isolate pancreatic islets to evaluate the insulin secretion.

2.9. Pancreatic Islets Isolation

Pancreatic islets were isolated using collagenase type 1 (Sigma-Aldrich, St. Louis, MO, USA) dissolved in 10 mL of Hank's Balanced Solution (HBSS) (Sigma-Aldrich, St. Louis, MO, USA). Altogether 9 mg collagenase for W and 24 mg collagenase for GK rats, were injected through the bile duct to insufflate the pancreas tissue [30,31]. The tissue was transferred to a test tube that was incubated in a water bath without shaking for 24 min at 37 °C, then collagenase was washed away with HBSS by centrifugation, and digested tissue was filtrated trough a restrainer. Finally, islets were separated from exocrine pancreatic tissue by centrifugation using a mixture of Histopaque 1119 and 1077 (Sigma-Aldrich, St. Louis, MO, USA). Islets were hand-picked using micro pipettes under a stereomicroscope and then cultured overnight at 37 °C, in an atmosphere of 5% CO_2–95% air, in RPMI 1640 medium (SVA, Stockholm, Sweden), supplemented with 30 mg L-glutamine (Sigma-Aldrich, St. Louis, MO, USA), 11 mM glucose (Sigma-Aldrich, St. Louis, MO, USA), antibiotics (100 IU/mL penicillin and 0.1 mg/mL streptomycin) (Invitrogen, CA, USA) and heat-inactivated fetal calf serum (10%) (Sigma-Aldrich, St. Louis, MO, USA).

2.10. Islet Insulin Secretion

Overnight cultured islets were pre-incubated 30–45 min at 37 °C and 3.3 mM glucose in Krebs-Ringer bicarbonate (KRB) buffer (NaCl 118.4 mM, KCl 4.7 mM, $MgSO_4$ 1.2 mM, KH_2PO_4 1.2 mM, $CaCl_2$ 1.9 mM, $NaHCO_3$ 25 mM, 4-(2-hydroxyethyl)-1-piperazineethanesulfonic acid (HEPES) 10 mM and 0.2% bovine serum albumin) (Sigma-Aldrich, St. Louis, MO, USA). Then, batches of three islets of similar size were incubated for 60 min in 300 μL of KRB at 3.3 mM or 16.7 mM glucose, with or without *AC* extract, at 37 °C in a water bath with gentle shaking [28,31]. After incubation, 200 μL of incubation media were collected to new tubes and kept frozen at −20 °C prior to determination of insulin by radioimmunoassay (RIA).

2.11. Islet Perifusion

To explore the effect of *AC* on kinetics of insulin release, batches of 40 or 50 isolated W and GK rat islets were layered between polystyrene beads (Bio-Rad Laboratories, Inc., Hercules, CA, USA) in a perifusion chamber, and KRB buffer was perifused continuously using a peristaltic pump (Ismatec SA, Zurich, Switzerland). To establish the basal insulin secretion rate, islets were perifused with 3.3 mM glucose in KRB for 20 min (−20 to min 0). The KRB buffer content was then changed to 3.3 mM glucose plus *AC* (20 mg/mL), from time 0 to 14 min; to 16.7 mM glucose plus *AC* (20 mg/mL), from time 16 to 30 min; and to 3.3 mM glucose without *AC*, for the last 20 min [28]. Perifusion buffer was collected every 2 min and stored at −20 °C for later insulin determination by RIA. The AUC in presence of *AC* was calculated subtracting the basal value at the beginning of each treatment; for low glucose, time 0 (period 0 to 14 min) and for high glucose, time 16 (period 16 to 30 min) and compared to same periods of untreated islets.

2.12. Mechanisms of Insulin Secretion Induced by AC

To study the mechanism of *AC* effect on insulin secretion, GK and W rat islets were incubated in presence of different compounds that interfere with different steps of the insulin secretion pathway. The effect on the adenosine triphosphate (ATP)-sensitive potassium channels (K-ATP) was evaluated using 0.25 mM diazoxide (DX), an opener of K-ATP channels (Sigma-Aldrich, St. Louis, MO, USA). Islets were incubated in 3.3 or 16.7 mM glucose with *AC* (20 mg/mL) and DX alone or with DX and 50 mM of KCl (to depolarize the β-cells) [32]. The effect on Ca^{2+} channels was evaluated using 10 μM nifedipine (NF) (Sigma-Aldrich, St. Louis, MO, USA), an inhibitor of L-type Ca^{2+} channels. Islets were incubated in 3.3 or 16.7 mM glucose KRB with *AC* (20 mg/mL) plus NF [32]. To evaluate the role of protein kinase A (PKA) and protein kinase C (PKC) on the effect of *AC*, islets were incubated with *AC* (20 mg/mL) with 10 μM H89 (Sigma-Aldrich, St. Louis, MO, USA), a PKA-inhibitor, or with 1.5 μM calphostin-C (Cal-C) (Sigma-Aldrich, St. Louis, MO, USA), a PKC inhibitor, in KRB containing 3.3 mM and 16.7 mM glucose [31,32]. Finally, to explore the role of G-protein-coupled exocytosis, a G-protein suppressor, pertussis toxin (PTx) (Sigma-Aldrich, St. Louis, MO, USA) was used. Islets were pretreated at 37 °C overnight with 100 ng/mL PTx in complete RPMI 1640 culture medium (with additions as given above). After exposure, islets were incubated with 20 mg/mL of *AC* in 3.3 mM or 16.7 mM glucose KRB [32,33]. For all the treatments, 200 μL aliquots of KRB medium were collected and stored at −20 °C for insulin determination by RIA.

2.13. Cytotoxicity

Cellular toxicity was evaluated in batches of W islets exposed to *AC* extract (5–50 mg/mL) in complete RPMI 1640 culture medium (Sigma-Aldrich, St. Louis, MO, USA) during 2 and 24 h at 37 °C. After treatment, cell viability was determined by MTT assay (Sigma-Aldrich, St. Louis, MO, USA) [34].

2.14. Statistical Analysis

Results are presented as mean ± SEM. Statistical differences between groups were analyzed using two-way analysis of variance (ANOVA) for OGTT, serum insulin, glycated hemoglobin, insulin secretion and insulin kinetics evaluations whereas paired Student's t-test was used for AUC analysis. Bonferroni's Post Hoc Test was used for correction of multiple testing. A *p* value of less than 0.05 was considered significant. Data were analyzed using Prism Graph Pad Software (San Diego, CA, USA).

3. Results

3.1. Phytochemical Constituents of AC Extract

The HPLC-HRMS analysis of the *AC* extracts showed more than 59 different types of phytochemicals (Figure 1).

Of the 19 reference standards, only four amino acids could unambiguously be assigned, due to the high amount of sugars and co-eluting peaks. The identified amino acids were iso-leucine, leucine, phenylalanine, and tryptophan. Due to the complexity of the chromatogram no further attempt was made to identify the chemical composition of the phytochemicals constituting the extract. Furthermore, NMR data indicated that the extract consisted primarily of sugars (Figures 2 and 3) and that a minor amount of polyphenols and amino acids were also present. When the extract was dissolved as 20 mg/mL, the glucose concentration was 2 mM.

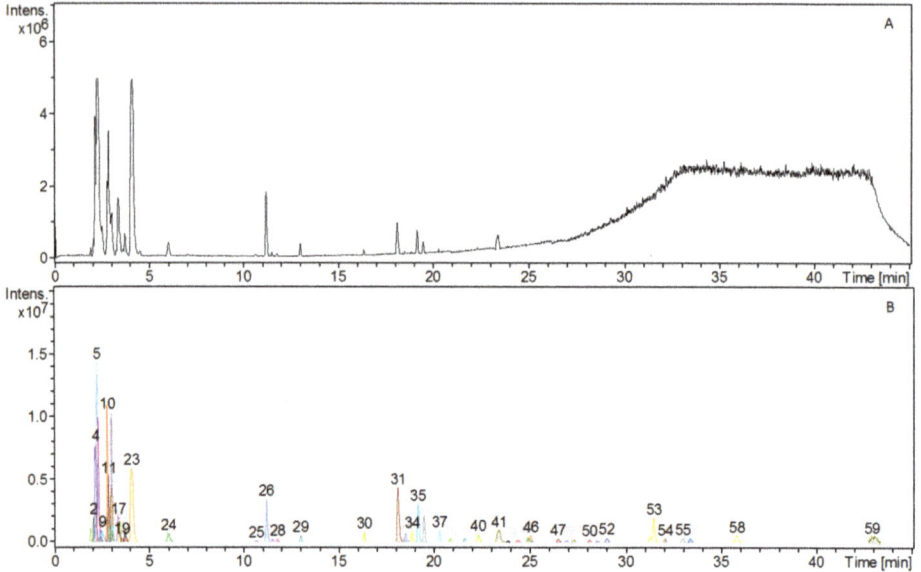

Figure 1. The *Amaranthus caudatus* (*AC*) extract consists of a complex mixture of phytochemicals. A high pressure liquid chromatography (HPLC) chromatogram with (**A**) total ion current (TIC) chromatogram and (**B**) processed chromatogram showing the presence of at least 59 different peaks each corresponding to a compound with a discrete molecular weight. From the retention time and protonated molecular ions [M + H]$^+$ from the MS spectra, some peaks corresponding to amino acids were assigned: Peak 18, iso-leucine MW 131; Peak 21, leucine MW 131; Peak 24, phenylalanine MW 165; and Peak 26, tryptophan MW 204.

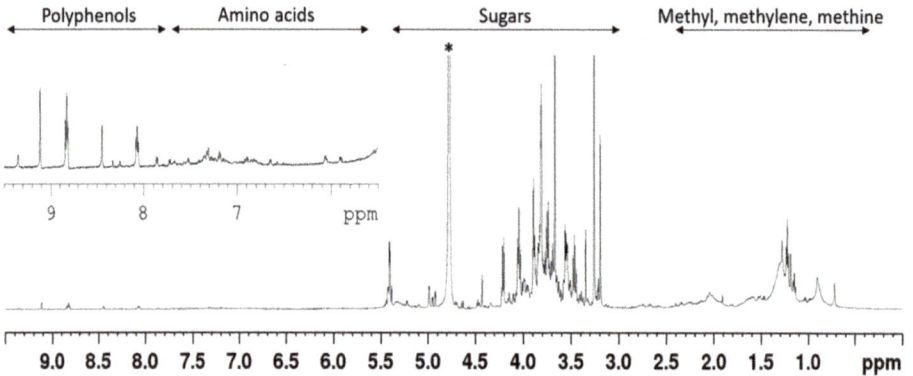

Figure 2. The ^{1}H-NMR (Nuclear Magnetic Resonance) spectrum indicates that the *AC* extract consists primarily of sugars, and of minor amount of polyphenols and amino acids. The extract was analysed by NMR spectrometry in D$_2$O at 25 °C. The ^{1}H-NMR spectrum, indicates the presence of at least four groups of components, with sugars constituting the major fractions of the extract; * denote the solvent (D$_2$O) signal.

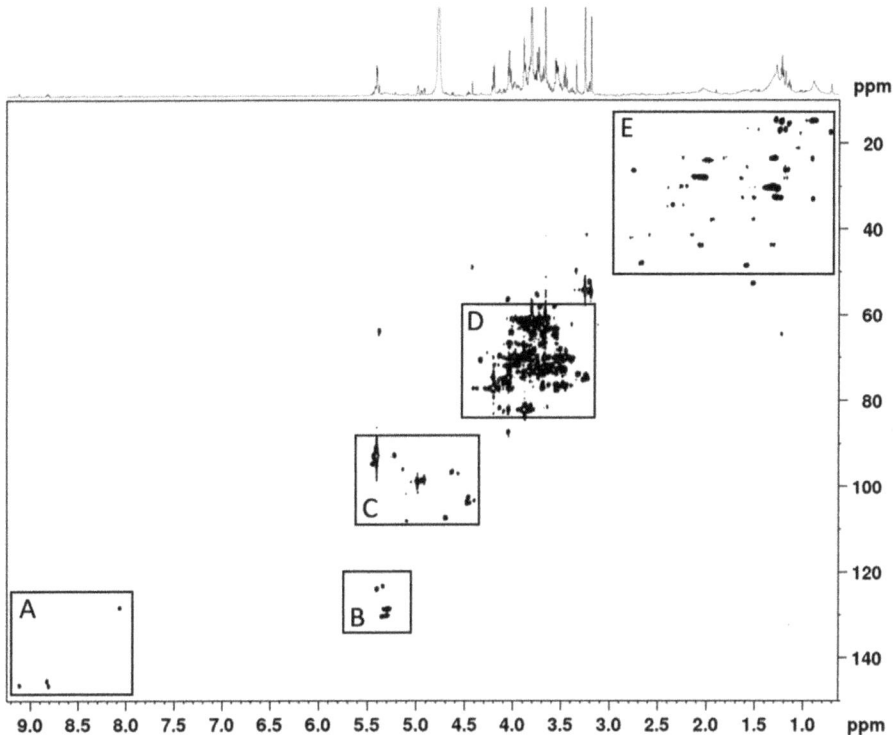

Figure 3. The HSQC (Heteronuclear Single Quantum Coherence) spectrum shows that the *AC* extract consists primarily of sugars, and of minor amount of polyphenols and amino acids. The ^{1}H-^{13}C-HSQC spectrum shows the C_2/H_2 to C_6/H_6 sugar ring signals (region D); the anomeric C_1/H_1 sugars signals (region C); olefinic signals (region B); aromatic/polyphenol signals (region A); and methyl, methylene, methine signals (region E).

3.2. AC Improves Glucose Tolerance in GK and W Rats by Increasing Serum Insulin Levels

AC (2000 mg/kg b.w.) improved the glucose tolerance during the OGTT in GK rats, at 90 min (20.7 ± 0.4 mM) ($p < 0.001$) and at 120 min (16.2 ± 0.3 mM) ($p < 0.01$) after glucose administration as compared to placebo treated GK rats (90 min, 24.9 ± 0.5 mM; 120 min 19.5 ± 0.7 mM) (Figure 4A). *AC* in a lower dose (1000 mg/kg b.w.) improved glucose tolerance only at 90 min (21.4 ± 0.6 mM) ($p < 0.01$). No effect was found with the lowest dose of *AC* tested (500 mg/mL). In W rats, the *AC* effect on glucose tolerance was observed only at the highest dose tested (2000 mg/kg b.w.) starting 30 min after glucose challenge (11.2 ± 0.2 mM) ($p < 0.001$) and during the following time points 60 min (9.4 ± 0.1 mM) ($p < 0.0001$), 90 min (6.1 ± 0.1 mM) ($p < 0.0001$) and 120 min (5.2 ± 0.1 mM) ($p < 0.05$) (Figure 4B).

The *AC* effect during the OGTT was also estimated by the calculation of the AUC of glucose. In GK rats treated with *AC* (2000 mg/kg b.w.) the AUC of glucose was significantly reduced to 1044.6 ± 47.1 mM/120 min compared with the GK rats exposed to placebo (1379.7 ± 132.5 mM/120 min; $p < 0.05$) (Figure 4C). Similarly, in W rats *AC* treatment at the highest dose (2000 mg/kg b.w.) reduced the AUC of glucose (450.5 ± 30.3 mM/120 min) compared to placebo-treated W rats (752.0 ± 27.0 mM/120min; $p < 0.05$) (Figure 4D).

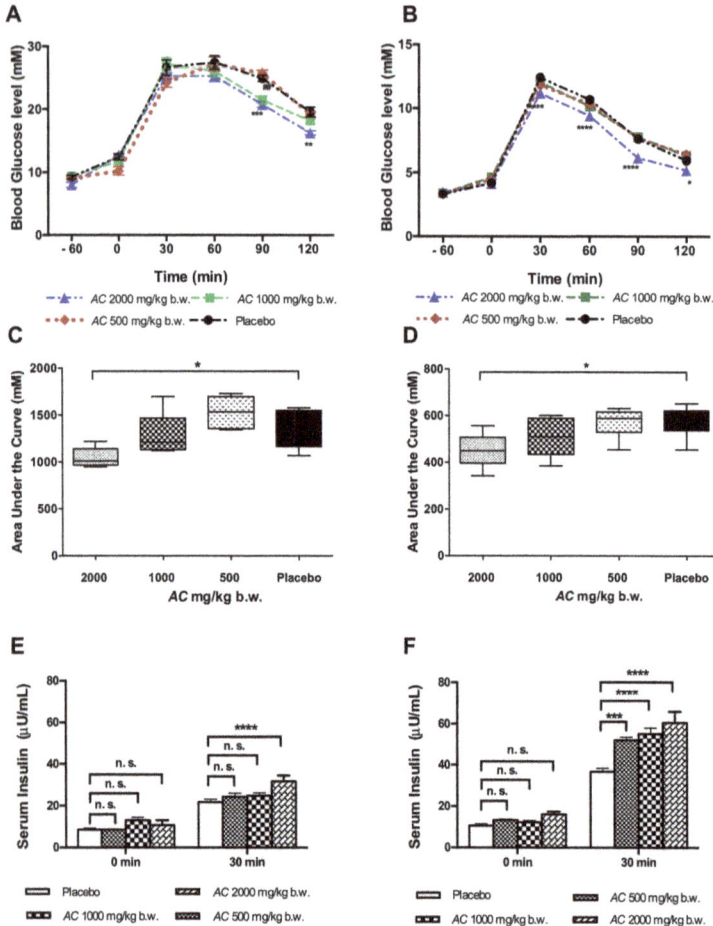

Figure 4. *AC* improves glucose tolerance and increases serum insulin levels. *AC* (500–2000 mg/kg b.w.) effect was evaluated during the OGTT in 12 h-fasted, animals received a single oral administration of *AC* extract one hour before glucose-challenge and glycemia was determined at 0, 30, 60, 90, and 120 min; GK rats (**A**) and W rats (**B**); The AUC of glucose was calculated from time 0 to 120 min in GK rats (**C**) and W rats (**D**); The effect of *AC* on serum insulin during the OGTT in GK rats (**E**) and W rats (**F**) was determined from 0–30 min. Data are presented as means ± SEM ($n = 6$). * $p < 0.05$, ** $p < 0.01$, *** $p < 0.001$, **** $p < 0.0001$ when compared to the placebo group; ## $p < 0.01$.

During the OGTT, *AC* was able to increase serum insulin levels during the first 30 min after glucose challenge. In GK rats, *AC* (2000 mg/kg b.w.) increased insulin levels up to 31.7 ± 2.8 µU/mL compared to placebo (21.7 ± 1.4 µU/mL) ($p < 0.0001$) (Figure 4E). In W rats, *AC* also increased serum insulin levels at all *AC* doses tested 500 mg/kg b.w. (52.0 ± 1.5 µU/mL), 1000 mg/kg b.w. (55.1 ± 2.8 µU/mL) and 2000 mg/kg b.w. (60.6 ± 5.4 µU/mL) compared to placebo (36.8 ± 1.6 µU/mL) ($p < 0.0001$) (Figure 4F).

3.3. Long-Term Oral Treatment with AC Improves Glucose Tolerance and Insulin Secretion

The long-term (21 days) oral treatment with *AC* (1000 mg/kg b.w.) had a non-significant tendency to reduce the non-fasting glucose in GK rats at a few time points, measured every third day during the treatment (Figure 5A). No significant reduction was observed in W rats (Figure 5B). However, when the AUC of non-fasting glucose was calculated, a significant difference was observed between *AC*-treated GK rats (54.4 ± 1.2 mM/19 days) and placebo-treated GK rats (58.2 ± 0.3 mM/19 days) ($p < 0.05$) (Figure 5C).

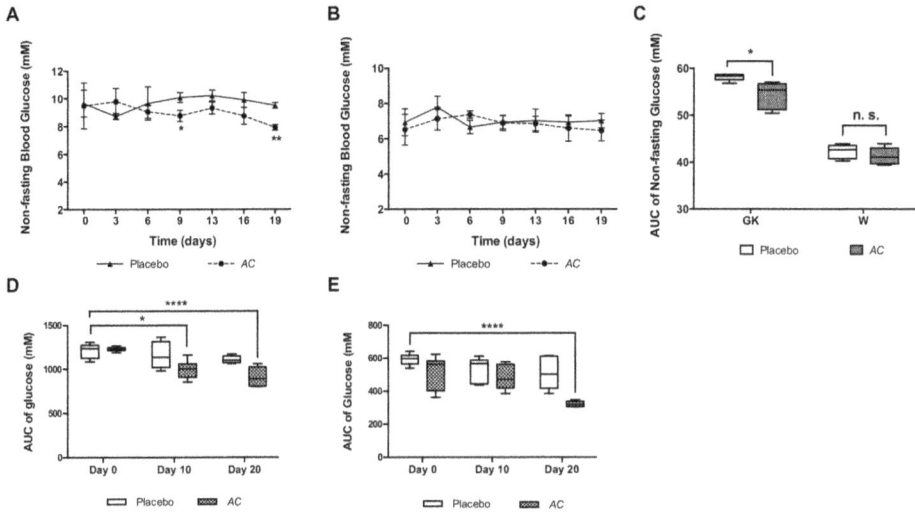

Figure 5. *AC* oral long-term treatment reduces the non-fasting glucose and improves glucose tolerance. The non-fasting glucose was determined every third day of treatment in GK (**A**) and W rats (**B**); the AUC of the non-fasting glucose were calculated from the interval 0–19 days (**C**); The AUC of glucose was calculated for each OGTT performed (day 0, 10 and 20) in GK (**D**) and W rats (**E**). Data are presented as means ± SEM ($n = 6$). * $p < 0.05$, ** $p < 0.01$, **** $p < 0.0001$ when compared to Placebo.

The glucose tolerance was improved by *AC* treatment reflected by the AUC of the serum glucose during OGTT in GK rats at day 10 (996.5 ± 42.5 mM/120 min) ($p < 0.05$) and day 20 (910.0 ± 47.3 mM/120 min) ($p < 0.0001$) compared to placebo GK rats at day zero (1209.0 ± 39.0 mM/120 min) (Figure 5D). In W rats *AC* reduced the AUC only at day 20 (321.3 ± 6.8 mM/120 min) compared to placebo W rats at day zero (593.3 ± 14.6 mM/120 min) ($p < 0.0001$) (Figure 5E).

The effect of *AC* on glucose tolerance was also detected by analyzing differences between each time point along the OGTT. In GK rats, at day 10 and 20 compared to day zero of treatment, during the intervals of 60–120 min of the OGTT ($p < 0.01$–$p < 0.0001$) and during the interval 90–120 min of the OGTT ($p < 0.05$–$p < 0.0001$), when compared to placebo-treated GK rats (Supplementary Materials Table S1). Similarly, in W rats where *AC* improved glucose tolerance at day 10 (90 and 120 min of the OGTT) and day 20 (30–120 min of the OGTT), compared to day zero of treatment ($p < 0.05$–$p < 0.0001$) and at day 20 (30–120 min of the OGTT) ($p < 0.05$–$p < 0.0001$), when compared to placebo-treated W rats (Supplementary Materials Table S1).

The plasma HbA1c levels were reduced by 19.8% in GK rats treated with *AC* at day 20 (6.3 ± 0.35%) compared to placebo (7.8 ± 0.46%) ($p < 0.05$) and by 27.6% compared to day zero (8.7 ± 0.50%) ($p < 0.01$) (Figure 6A). In W rats HbA1c levels were reduced by 2.2% (2.4 ± 0.01) at day 10 and by 10.7% at day 20 (2.3 ± 0.05%) ($p < 0.0001$) and compared to placebo W rats (2.5 ± 0.02, day 10 and 2.6 ± 0,02, day 20) (Figure 6B).

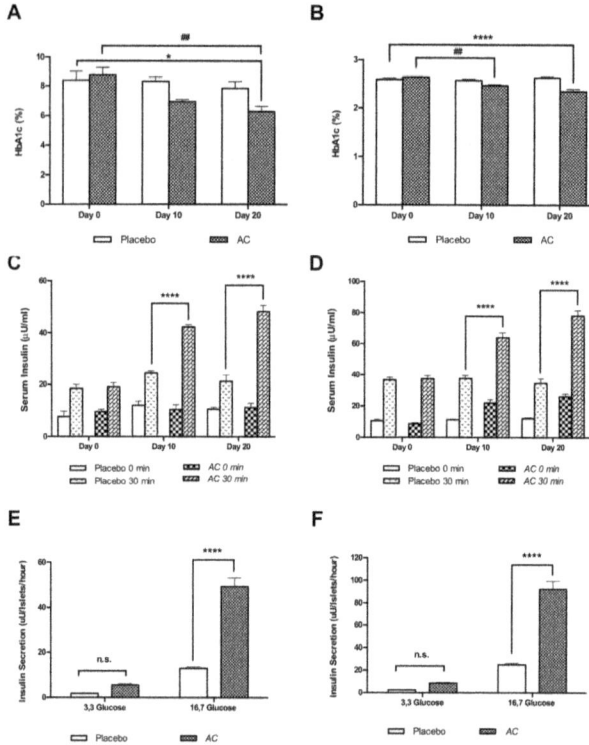

Figure 6. *AC* oral long-term treatment reduces the plasma HbA1c, increases serum insulin, and improved insulin secretion in pancreatic islets of treated animals. The plasma HbA1c were measured during each OGTT in GK rats (**A**) and in W rats (**B**); Serun Insulin were measured during each OGTT in GK rats (**C**) and in W rats (**D**) Data are presented as means ± SEM (*n* = 6). Pancreatic islets were isolated at the end point of treatment (day 21) from GK rats (**E**) and W rats (**F**) were cultured at low (3.3 mM) and high (16.7 mM) glucose. Data are presented as means ± SEM (*n* = 4). Insulin concentration was measured by RIA. * $p < 0.05$, **** $p < 0.0001$ when compared to placebo group; ## $p < 0.01$, when compared to values from the same group.

Serum insulin levels in GK rats treated with *AC* were increased during the OGTT 1.7-fold at day 10 ($p < 0.0001$) and 2.3-fold at day 20 ($p < 0.0001$) (Figure 6C), and in W rats 1.6-fold at day 10 ($p < 0.0001$) and 2.2-fold at day 20 ($p < 0.0001$) (Figure 6D). Furthermore, *AC* augmented insulin secretion in pancreatic islets isolated from treated animals. Thus, insulin secretion increased 4.1-fold at high glucose (16.7 mM) in islets isolated from *AC* treated GK rats, compared to the secretion in islets isolated from placebo treated GK rats ($p < 0.0001$) (Figure 6E). Similarly, *AC* treatment in W rats increased islet insulin secretion at 16.7 mM glucose 3.7-fold, compared to the secretion in islets from placebo treated W rats ($p < 0.0001$) (Figure 6F).

The sub-acute toxicity studies did not show significant differences between *AC*-treated W rats and placebo group of the hematological indicators i.e., red and white blood cells number, hematocrit and hemoglobin, and serum biochemical parameters i.e., triglycerides, cholesterol, glucose, creatinine, alkaline phosphatase, aspartate aminotransferase, and alanine transferase (Supplementary Materials Table S2).

3.4. AC Stimulates In Vitro Insulin Secretion in a Concentration-Dependent Manner

In GK rat islets, insulin secretion was stimulated by 10, 20 and 50 mg/mL *AC* at 3.3 mM glucose, and by 5, 10, 20 and 50 mg/mL *AC* at 16.7 mM glucose in a concentration-dependent manner, with a maximal 7.9-fold increase at low glucose (3.3 mM) ($p < 0.0001$) and 8.5-fold increase at high glucose (16.7 mM), ($p < 0.0001$) compared to untreated islets (Figure 7A). In W rats islets, the insulin secretion was increased in all concentrations tested up to 50 mg/mL, 5.1-fold (3.3 mM) and 5.7-fold (16.7 mM) compared to untreated islets (Figure 7B). Significant differences were found in lower concentrations until 10 mg/mL (3.3 mM) and 5 mg/mL (16.7 mM) in both GK and W rat islets. No cytotoxic effect was observed in tested concentrations (Supplementary Materials Figure S3).

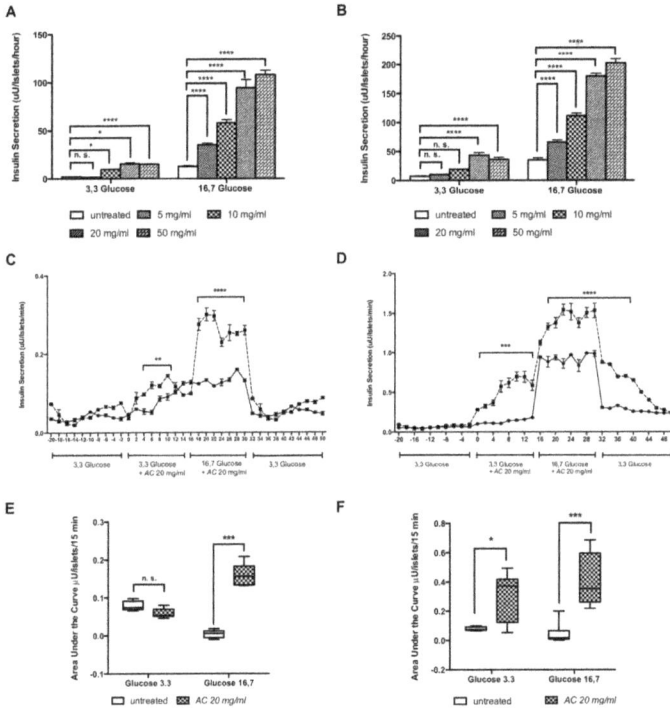

Figure 7. *AC* stimulates the in vitro insulin secretion in pancreatic islets. Insulin secretion was evaluated in GK rat islets (**A**) and W rat islets (**B**) cultured at low (3.3 mM) and high (16.7 mM) glucose in presence of *AC* (5–200 mg/mL). Data are presented as means ± SEM ($n = 8$). Batches of (50) GK (**C**) and Wistar (**D**) rats islets were perifused with 3.3 mM glucose, from time 0 to 14 min, and with 16.7 mM glucose, from time 16 to 30 min, in presence ---■--- or absence ---●--- of *AC* (20 mg/mL). The AUC of the insulin secretion from the intervals at 3.3 mM and 16.7 mM of glucose in presence or absence of *AC* was calculated for GK (**E**) and W (**F**). rats islets. Data are presented as means ± SEM ($n = 4$). * $p < 0.05$, ** $p < 0.01$, *** $p < 0.001$, **** $p < 0.0001$ when compared to untreated islets.

3.5. The AC Effect on Kinetics of Insulin Secretion

To monitor the *AC* effect on the kinetics of insulin secretion, isolated islets were perifused with KRB buffer containing *AC* extract (20 mg/mL). Significant differences were found during the period of 4 to 12 min ($p < 0.01$), when GK rat islets were perfused with *AC* at 3.3 mM glucose and during 16 to 32 min ($p < 0.0001$), when the islets were perfused with *AC* at 16.7 mM glucose, compared to the respective time point from untreated GK rat islets (Figure 7C). Similar pattern was observed in W rat perifused islets during the time period of 0 to 14 min (3.3 mM) ($p < 0.001$) and 18 to 40 min (16.7 mM) ($p < 0.0001$) (Figure 7D).

The effect on insulin secretion was also observed when the AUC of insulin secretion was calculated. In GK islet perifused with 16.7 mM of glucose (14–30 min), AC increased insulin secretion 37.2-fold ($p < 0.001$) (Figure 7E) and in W rats with 3.9-fold at 3.3 mM of glucose (0–14 min) ($p < 0.05$) and with 9.2-fold at 16.7 mM (16–30 min) ($p < 0.001$) (Figure 7F).

3.6. AC Stimulates Insulin Secretion through the Activation of PKC and PKA Systems and Partially by L-Type Calcium Channels and G Protein-Coupled Exocytosis

Diazoxide (DX), a selective ATP-sensitive K^+-channel opener, did not reduce the effect of AC on insulin secretion. Not even combined exposure to DX and high concentrations of KCl to induce a transient membrane depolarization altered the AC effect, in GK or W rat islets (Supplementary Materials Figure S4).

Nifedipine (NF), an inhibitor of L-type voltage-dependent Ca^{2+} channels, reduced the insulin secretion with 38% in GK rat islets exposed to 16.7 mM glucose ($p < 0.001$) (Figure 8A), and by 44% respectively 45% in W rat islets exposed to 3.3 mM glucose respectively to 16.7 mM glucose ($p < 0.001$) (Figure 8B).

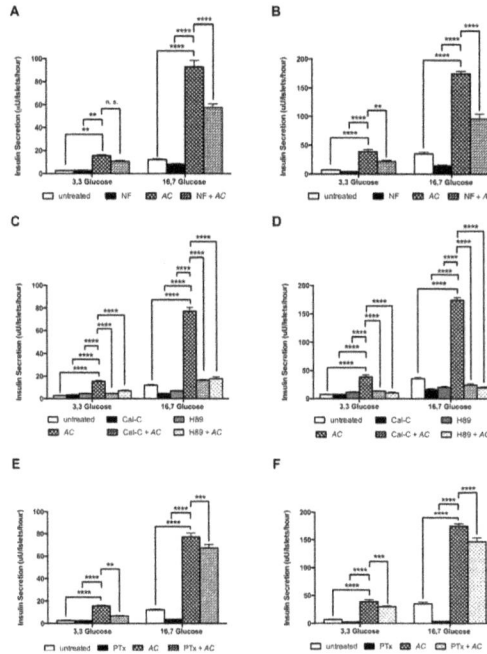

Figure 8. AC stimulates insulin secretion through PKC and PKA systems, L-type calcium channels and G protein-coupled exocytosis. The insulin secretion was evaluated in islets cultured at low (3.3 mM) and high (16.7 mM) glucose in presence of AC (20 mg/mL), and the different inhibitors. 10 μM NF in GK (**A**) and W rats islets (**B**); 1.5 μM Cal-C or 10 μM H89 in GK (**C**) and W rats islets (**D**); and 100 ng/mL PTx in GK (**E**) and W rats islets (**F**). Insulin concentration was measured by radioimmunoassay (RIA). Data are presented as means ± SEM ($n = 8$). ** $p < 0.01$, *** $p < 0.001$, **** $p < 0.0001$ when compared to islets treated with AC alone.

To assess whether the AC effect is dependent on PKA and PKC activation, we investigated the influence of H89, an inhibitor of cyclic adenosine monophosphate (cAMP)-dependent PKA and of

Cal-C, a PKC inhibitor on the *AC* modulation of insulin secretion. In GK rat islets, H89 reduced insulin secretion by 55% and by 77% at 3,3 mM and 16.7 mM glucose, respectively ($p < 0.0001$) and Cal-C reducted by 71% (3.3 mM) and by 79% (16.7 mM) ($p < 0.0001$) (Figure 8C). In W rat islets, H89 reduced insulin by 75% and by 89% at 3.3 mM and 16.7 mM glucose, respectively ($p < 0.0001$) and Cal-C reduction was 66% (3.3 mM) and 87% (16.7 mM) ($p < 0.0001$) (Figure 8D). Finally, to explore whether the *AC* effect was mediated by exocytotic G-proteins, pertussis toxin (PTx) an inhibitor of G-proteins via ADP-ribosylation was used. We found that PTx inhibited the *AC* effect in GK rat islets by 57% at 3.3 mM ($p < 0.01$) and by 20% ($p < 0.001$) at 16.7 mM glucose (Figure 8E). Similar effects were found in W rat islets, with reductions of 23% ($p < 0.0001$) and 16% ($p < 0.001$) at 3.3 mM and 16.7 mM of glucose, respectively (Figure 8F).

4. Discussion

One of the main features of DMT2 is the hyperglycemia that contributes to diabetes complications. Therefore, the restoration of glucose homeostasis is a main therapeutic target [7,35]. We have demonstrated that a crude extract of *AC* seeds improved glucose tolerance in the type 2 diabetic GK rats. *AC* stimulates insulin secretion, an effect mediated by PKC and PKA systems and partially by L-type calcium channel and insulin G protein-coupled exocytosis in the β-cells.

The LC-MS analysis showed that the *AC* extract consists of complex components including sugars, and minor amounts of polyphenols and amino acids. Similar composition has been described previously for *AC* and other species from the *Amaranthus* family [22]. Interestingly, polyphenols have been demonstrated to exert anti-diabetic effects [12]. Noteworthy, the glucose concentration in *AC* extract at 20 mg/mL concentration (as used in most in vitro experiments) was only 2 mM, i.e., too low to significantly enhance insulin secretion from islets exposed to 5, 10 or 20 mg/mL of the extract.

The in vivo experiments showed that a single oral administration of *AC* contributed to control the hyperglycemia in GK rats, by reducing the non-fasting glycaemia and improving the glucose tolerance through increase of the serum insulin levels. *AC* also reduced the postprandial hyperglycemia in healthy W rats by increasing serum insulin levels, however, without inducing a hypoglycemic state.

In addition, *AC* reduced the percentage of HbA1c in GK and W rats, indicating that *AC* treatment improved long-term glycemic control. Treated animals showed high serum insulin levels suggesting that the main target for the beneficial effect of *AC* is an increase of the insulin release. Indeed, the *AC* long-term treatment improved glucose-stimulated insulin secretion of pancreatic islets isolated from treated animals. The sub-acute toxicity experiments did not show any toxic effect after 28 days; thus, the treatment conditions used appear to be safe. However, further studies to test β-cell function and the impact of *AC* treatment on diabetes complications in animal models are needed.

Others have found that a methanol extract of *AC* leaves decreases blood glucose levels in rats with streptozotocin (STZ)-induced type 1 diabetes and in normal rats after seven days of treatment [25], however by an effect suggested to be exerted through inhibition of α-amylase [23,24]. Similar effects were reported for other species from the same *Amaranthus* genus, *Amaranthus dubius* [36], *Amaranthus Tricolour* L. [37], *Amaranthus viridis* [38,39] and *Amaranthus spinosus* Linn [40]. It is noteworthy that our investigation is the first to describe an *AC* effect on insulin secretion.

The in vitro evaluations showed that *AC* increased the insulin secretion in batch-incubated islets, an effect that was concentration-dependent and present already at rather low concentrations of the *AC* extract (lower than 20 mg/mL), for both GK and W rat islets incubated at high glucose levels. This may imply that the stimulation of insulin secretion can be achieved without or with a small risk of having hypoglycemic or toxic effects. It is also of interest that the *AC* effect on insulin secretion was comparable with the effect at the high glucose concentration of the β-cell secretagogue drug glibenclamide (Supplementary Materials Figure S5). Moreover, the *AC* effect was observed when assessing the kinetics of insulin secretion in islet perfusions. Thus, in the presence of *AC* basal insulin secretion at low glucose conditions was partially enhanced and was further stimulated by *AC* at high glucose concentrations. When *AC* was removed from the perfusion buffer, the insulin secretion returned gradually to basal levels, supporting the view that the *AC* effect is reversible and does not cause insulin leakage by any toxic effect on the β-cells. A similar effect has been described for other plant extracts and isolated compounds from natural resources [28]. In this context it is noteworthy that the *AC* extract may exert stimulatory effects on β-cell mass or islet insulin content. However, it seems unlikely that such effects are involved in its effects on insulin secretion, since the action is immediate as shown in isolated islets. Furthermore, we do not think effects on food intake are implicated, since *AC* extract treatment up to 28 days did not impact on body weight development.

When blood glucose levels are elevated after a meal, glucose is transported into the pancreatic β-cell by glucose transporter 2 (GLUT2) and metabolized via glycolysis and the Krebs cycle to produce ATP [41,42]. This results in increased cytosolic ATP/adenosine diphosphate (ADP) ratios, that leads to closure of potassium-sensitive (K-ATP) channels and further depolarization of the β-cell membrane. This in turn activates the voltage-dependent calcium (L-type Ca^{2+}) channels with calcium entry into the β-cell, producing an increased cytosolic calcium concentration that stimulates insulin secretion [1,41].

To study the mechanism behind the *AC* effect on glucose-induced insulin secretion, we first evaluated the role of K-ATP channels. As expected, GK and W rat islets treated with diazoxide, which maintains K-ATP channels open [32], showed a reduction of insulin secretion. *AC* did not change the effect of diazoxide in either GK or W rat islets, suggesting that *AC* does not modulate the closure of K-ATP channels. Moreover, the *AC* effect was not altered at 16.7 glucose in GK and W rat islets depolarized by a high concentration of KCl in the presence of diazoxide. This finding suggests that the *AC* effect reached the near-maximal threshold and is not primarily dependent on membrane depolarization.

To explore the mediation of L-type Ca^{2+}-channels for the effect of *AC*, we used nifedipine, an L-type Ca^{2+} channel blocker, that blocks Ca^{2+}-influx into the β-cell and thereby inhibits insulin secretion [31,33]. In the presence of nifedipine, the *AC* effect at 16.7 mM glucose on GK and W rat islets was partially reduced compared to islets incubated with *AC* alone, suggesting that the *AC* effect is partially dependent on activation of L-type Ca^{2+}-channels.

In addition, other intracellular signals mediated by second messengers, cAMP and diacylglycerol (DAG), can induce insulin release through phosphorylation by PKA and PKC [43]. When applying the PKA inhibitor, H89, and the PKC inhibitor, calphostin-C, the *AC* effect was significantly suppressed, in both GK and W rat islets, indicating that the *AC* effect on β-cells involves the activation of both the PKA and PKC systems. Finally, guanine nucleotide-binding proteins (G-proteins) are involved in signaling pathways of insulin secretion in pancreatic β-cells. Among the different G-proteins, Ge are involved in exocytosis [31,41]. In GK and W rat islets pre-treated with pertussis toxin, an inhibitor of G proteins via ADP-ribosylation, the stimulatory effect on insulin secretion by *AC* was partially reduced, indicating that the *AC* effect partially involves G protein-coupled insulin exocytosis.

Based on our findings, PKC and PKA activation appears to be important for the *AC* effect. Both enzymes phosphorylate proteins required for the initial steps of the exocytotic process [43]. One mechanism of kinase activation is mediated by Ca^{2+}-activation of the receptor-coupled enzyme phospholipase C (PLC). PLC hydrolyzes the plasma membrane phospholipid phosphatidylinositol bisphosphate (PIP2) into the second messengers, diacylglycerol (DAG) and inositol trisphosphate (IP3). DAG activates PKC, and IP3 liberates Ca^{2+} from the endoplasmic reticulum [42,43]. Since an increase in cytosolic Ca^{2+} takes part in the activation of kinases, the partial inhibition of the *AC* effect by nifedipine could be explained by the lack of calcium entrance. Alternatively, the activation of adenylate cyclase by Gs protein is required for PKA activation [31,32] and PLC activation is also mediated by Gq protein [41]. Therefore, the inhibition of the *AC* effect by PTx might be explained by blocking of the synthesis of cAMP and PLC activation [41]. PKA and PKC promote insulin secretion by increasing the total number of vesicles available for insulin release and which are highly sensitive to Ca^{2+}, the so called highly calcium-sensitive pool (HCSP) [43]. Thus, we hypothesize that the *AC* effect may be explained by the recruitment of more insulin granules into the HCSP through the activation of PKA and PKC.

5. Conclusions

We demonstrated that *AC* stimulates insulin secretion in GK and W rats by modulating steps of the glucose-dependent insulin secretion, e.g., by protein kinases A and C, effects that are dependent on intracellular increase in calcium and that partially involve the G protein-coupled exocytosis membrane proteins. The improved glucose tolerance in diabetic GK rats indicates that *AC* might be a candidate for use as a nutraceutical therapy in type 2 diabetes in man.

Supplementary Materials: The following are available online at http://www.mdpi.com/2072-6643/10/1/94/s1, Table S1, Effect of AC long-term treatment on the OGTT test performed at day 0, day 10 and day 20 of treatment in GK and W rats, Table S2, Effect on hematological and biochemical parameters of male Wistar rats after 28 days of *AC* treatment, Figure S1, Effect on body weights of Wistar rats after 28 days of *AC* treatment, Figure S2, Effect of *AC* long-term treatment (20 days) on body weights, Figure S3, Cytotoxic effect of *AC* on W pancreatic islets, Figure S4, The *AC* effect is not mediated by ATP-dependent potassium channels, Figure S5, AC effect on insulin secretion was comparable with the effect of glibenclamide.

Acknowledgments: Special thanks to Elisabeth Noren-Krog for excellent technical support. This research was funded by the Swedish International Development Cooperation Agency, SIDA.

Author Contributions: S.Z. and V.V. conducted all the biological experiments. L.C.E.L. performed extract characterization. E.G., S.-B.C. and C.-G.O. designed the experiments. S.Z., L.C.E.L., E.G., S.-B.C. and C.-G.O. wrote the paper.

Conflicts of Interest: The authors declare no conflict of interest.

References

1. Ostenson, C.G. The pathophysiology of type 2 diabetes mellitus: An overview. *Acta Physiol. Scand.* **2001**, *171*, 241–247. [CrossRef] [PubMed]
2. Kahn, S.E.; Cooper, M.E.; Del Prato, S. Pathophysiology and treatment of type 2 diabetes: Perspectives on the past, present, and future. *Lancet* **2014**, *383*, 1068–1083. [CrossRef]
3. World Health Organization. *Global Report on Diabetes*; World Health Organization: Geneva, Switzerland, 2016.
4. Ogurtsova, K.; da Rocha Fernandes, J.D.; Huang, Y.; Linnenkamp, U.; Guariguata, L.; Cho, N.H.; Cavan, D.; Shaw, J.E.; Makaroff, L.E. IDF diabetes atlas: Global estimates for the prevalence of diabetes for 2015 and 2040. *Diabetes Res. Clin. Pract.* **2007**, *128*, 40–50. [CrossRef] [PubMed]
5. Brownlee, M. The pathobiology of diabetic complications: A unifying mechanism. *Diabetes* **2005**, *54*, 1615–1625. [CrossRef] [PubMed]
6. Dahlstrom, E.; Sandholm, N. Progress in defining the genetic basis of diabetic complications. *Curr. Diabetes Rep.* **2017**, *17*, 80. [CrossRef] [PubMed]
7. Thrasher, J. Pharmacologic management of type 2 diabetes mellitus: Available therapies. *Am. J. Cardiol.* **2017**, *120*, S4–S16. [CrossRef] [PubMed]
8. Rios, J.L.; Francini, F.; Schinella, G.R. Natural products for the treatment of type 2 diabetes mellitus. *Planta Med.* **2015**, *81*, 975–994. [CrossRef] [PubMed]
9. Kim, C.S.; Sohn, E.J.; Kim, Y.S.; Jung, D.H.; Jang, D.S.; Lee, Y.M.; Kim, D.H.; Kim, J.S. Effects of kiom-79 on hyperglycemia and diabetic nephropathy in type 2 diabetic Goto-Kakizaki rats. *J. Ethnopharmacol.* **2007**, *111*, 240–247. [CrossRef] [PubMed]
10. Magrone, T.; Perez de Heredia, F.; Jirillo, E.; Morabito, G.; Marcos, A.; Serafini, M. Functional foods and nutraceuticals as therapeutic tools for the treatment of diet-related diseases. *Can. J. Physiol. Pharmacol.* **2013**, *91*, 387–396. [CrossRef] [PubMed]
11. Zhang, B.-W.; Sang, Y.-B.; Sun, W.-L.; Yu, H.-S.; Ma, B.-P.; Xiu, Z.-L.; Dong, Y.-S. Combination of flavonoids from *Oroxylum Indicum* seed extracts and acarbose improves the inhibition of postprandial blood glucose: In vivo and in vitro study. *Biomed. Pharmacother.* **2017**, *91*, 890–898. [CrossRef] [PubMed]
12. Ota, A.; Ulrih, N.P. An overview of herbal products and secondary metabolites used for management of type two diabetes. *Front. Pharmacol.* **2017**, *8*, 436. [CrossRef] [PubMed]
13. Rosa, P.-G.; Jesús, S.-G. A critical review of bioactive food components, and of their functional mechanisms, biological effects and health outcomes. *Curr. Pharm. Des.* **2017**, *23*, 1–11.

14. Mandave, P.; Khadke, S.; Karandikar, M.; Pandit, V.; Ranjekar, P.; Kuvalekar, A.; Mantri, N. Antidiabetic, lipid normalizing, and nephroprotective actions of the strawberry: A potent supplementary fruit. *Int. J. Mol. Sci.* **2017**, *18*, 124. [CrossRef] [PubMed]

15. Cicero, A.F.; Tartagni, E.; Ertek, S. Nutraceuticals for metabolic syndrome management: From laboratory to benchside. *Curr. Vasc. Pharmacol.* **2013**, *12*, 565–571. [CrossRef]

16. Das, L.; Bhaumik, E.; Raychaudhuri, U.; Chakraborty, R. Role of nutraceuticals in human health. *J. Food Sci. Technol.* **2012**, *49*, 173–183. [CrossRef] [PubMed]

17. Santini, A.; Tenore, G.C.; Novellino, E. Nutraceuticals: A paradigm of proactive medicine. *Eur. J. Pharm. Sci.* **2017**, *96*, 53–61. [CrossRef] [PubMed]

18. Wijngaard, H.H.; Arendt, E.K. Buckwheat. *Cereal Chem. J.* **2006**, *83*, 391–401. [CrossRef]

19. Alvarez-Jubete, L.; Arendt, E.K.; Gallagher, E. Nutritive value and chemical composition of pseudocereals as gluten-free ingredients. *Int. J. Food Sci. Nutr.* **2009**, *60* (Suppl. 4), 240–257. [CrossRef] [PubMed]

20. Ramos Diaz, J.M.; Kirjoranta, S.; Tenitz, S.; Penttilä, P.A.; Serimaa, R.; Lampi, A.-M.; Jouppila, K. Use of amaranth, quinoa and kañiwa in extruded corn-based snacks. *J. Cereal Sci.* **2013**, *58*, 59–67. [CrossRef]

21. Bruni, R.; Medici, A.; Guerrini, A.; Scalia, S.; Poli, F.; Muzzoli, M.; Sacchetti, G. Wild *Amaranthus Caudatus* seed oil, a nutraceutical resource from ecuadorian flora. *J. Agri. Food Chem.* **2001**, *49*, 5455–5460. [CrossRef]

22. Venskutonis, P.R.; Kraujalis, P. Nutritional components of *Amaranth* seeds and vegetables: A review on composition, properties, and uses. *Compr. Rev. Food Sci. Food Saf.* **2013**, *12*, 381–412. [CrossRef]

23. Conforti, F.; Statti, G.; Loizzo, M.R.; Sacchetti, G.; Poli, F.; Menichini, F. In vitro antioxidant effect and inhibition of α-amylase of two varieties of *Amaranthus Caudatus* seeds. *Biol. Pharm. Bull.* **2005**, *28*, 1098–1102. [CrossRef] [PubMed]

24. Kumar, A.; Lakshman, K.; Jayaveera, K.N.; Sheshadri Shekar, D.; Narayan Swamy, V.B.; Khan, S.; Velumurga, C. In vitro α-amylase inhibition and antioxidant activities of methanolic extract of *Amaranthus Caudatus* Linn. *Oman Med. J.* **2011**, *26*, 166–170. [CrossRef] [PubMed]

25. Girija, K.; Lakshman, K.; Udaya, C.; Sabhya, S.G.; Divya, T. Anti-diabetic and anti-cholesterolemic activity of methanol extracts of three species of *Amaranthus. Asian Pac. J. Trop. Biomed.* **2011**, *1*, 133–138. [CrossRef]

26. Ostenson, C.G.; Khan, A.; Abdel-Halim, S.M.; Guenifi, A.; Suzuki, K.; Goto, Y.; Efendic, S. Abnormal insulin secretion and glucose metabolism in pancreatic islets from the spontaneously diabetic GK rat. *Diabetologia* **1993**, *36*, 3–8. [CrossRef] [PubMed]

27. OECD. *Test No. 407: Repeated Dose 28-Day Oral Toxicity Study in Rodents*; OECD Publishing: Paris, France, 2008.

28. Lokman, F.E.; Gu, H.F.; Wan Mohamud, W.N.; Yusoff, M.M.; Chia, K.L.; Ostenson, C.G. Antidiabetic effect of oral borapetol B compound, isolated from the plant tinospora crispa, by stimulating insulin release. *Evid. Based Complement. Altern. Med.* **2013**, *2013*, 727602. [CrossRef] [PubMed]

29. Herbert, V.; Lau, K.S.; Gottlieb, C.W.; Bleicher, S.J. Coated charcoal immunoassay of insulin. *J. Clin. Endocrinol. Metab.* **1965**, *25*, 1375–1384. [CrossRef] [PubMed]

30. Hoa, N.K.; Phan, D.V.; Thuan, N.D.; Ostenson, C.G. Insulin secretion is stimulated by ethanol extract of *Anemarrhena asphodeloides* in isolated islet of healthy Wistar and diabetic Goto-Kakizaki rats. *Experimental Clin. Endocrinol. Diabetes* **2004**, *112*, 520–525. [CrossRef] [PubMed]

31. Hoa, N.K.; Norberg, A.; Sillard, R.; Van Phan, D.; Thuan, N.D.; Dzung, D.T.; Jornvall, H.; Ostenson, C.G. The possible mechanisms by which phanoside stimulates insulin secretion from rat islets. *J. Endocrinol.* **2007**, *192*, 389–394. [CrossRef] [PubMed]

32. Lokman, E.F.; Gu, H.F.; Wan Mohamud, W.N.; Östenson, C.-G. Evaluation of antidiabetic effects of the traditional medicinal plant *Gynostemma pentaphyllum* and the possible mechanisms of insulin release. *Evid. Based Complement. Altern. Med.* **2015**, *2015*, 120572. [CrossRef] [PubMed]

33. Pelletier, J.; Domingues, N.; Castro, M.M.; Ostenson, C.G. In vitro effects of bis(1,2-dimethyl-3-hydroxy-4-pyridinonato)oxidovanadium(iv), or vo(dmpp)2, on insulin secretion in pancreatic islets of type 2 diabetic Goto-Kakizaki rats. *J. Inorg. Biochem.* **2016**, *154*, 29–34. [CrossRef] [PubMed]

34. Mosmann, T. Rapid colorimetric assay for cellular growth and survival: Application to proliferation and cytotoxicity assays. *J. Immunol. Methods* **1983**, *65*, 55–63. [CrossRef]

35. Chang, C.L.T.; Lin, Y.; Bartolome, A.P.; Chen, Y.-C.; Chiu, S.-C.; Yang, W.-C. Herbal therapies for type 2 diabetes mellitus: Chemistry, biology, and potential application of selected plants and compounds. *Evid. Based Complement. Altern. Med.* **2013**, *2013*, 378657. [CrossRef] [PubMed]

36. Montero-Quintero, K.C.; Moreno-Rojas, R.; Molina, E.A.; Colina-Barriga, M.S.; Sanchez-Urdaneta, A.B. Effect of consumption of bread with amaranth (*Amaranthus Dubius* Mart. ex Thell.) on glycemic response and biochemical parameters in Sprague dawley rats. *Nutr. Hosp.* **2014**, *31*, 313–320. [PubMed]

37. Rahmatullah, M.; Hosain, M.; Rahman, S.; Rahman, S.; Akter, M.; Rahman, F.; Rehana, F.; Munmun, M.; Kalpana, M.A. Antihyperglycaemic and antinociceptive activity evaluation of methanolic extract of whole plant of *Amaranthus Tricolour* L. (Amaranthaceae). *Afr. J. Tradit. Complement. Altern. Med.* **2013**, *10*, 408–411. [PubMed]

38. Ashok Kumar, B.S.; Lakshman, K.; Jayaveea, K.N.; Sheshadri Shekar, D.; Saleemulla, K.; Thippeswamy, B.S.; Veerapur, V.P. Antidiabetic, antihyperlipidemic and antioxidant activities of methanolic extract of *Amaranthus Viridis* Linn in alloxan induced diabetic rats. *Exp. Toxicol. Pathol.* **2012**, *64*, 75–79. [CrossRef] [PubMed]

39. Krishnamurthy, G.; Lakshman, K.; Pruthvi, N.; Chandrika, P.U. Antihyperglycemic and hypolipidemic activity of methanolic extract of *Amaranthus Viridis* leaves in experimental diabetes. *Indian J. Pharmacol.* **2011**, *43*, 450–454. [PubMed]

40. Sangameswaran, B.; Jayakar, B. Anti-diabetic, anti-hyperlipidemic and spermatogenic effects of *Amaranthus Spinosus* Linn. on streptozotocin-induced diabetic rats. *J. Nat. Med.* **2008**, *62*, 79–82. [CrossRef] [PubMed]

41. Ahren, B. Islet G protein-coupled receptors as potential targets for treatment of type 2 diabetes. *Nat. Rev. Drug Discov.* **2009**, *8*, 369–385. [CrossRef] [PubMed]

42. Barker, C.J.; Berggren, P.O. New horizons in cellular regulation by inositol polyphosphates: Insights from the pancreatic β-cell. *Pharmacol. Rev.* **2013**, *65*, 641–669. [CrossRef] [PubMed]

43. Wan, Q.-F.; Dong, Y.; Yang, H.; Lou, X.; Ding, J.; Xu, T. Protein kinase activation increases insulin secretion by sensitizing the secretory machinery to Ca^{2+}. *J. Gen. Physiol.* **2004**, *124*, 653–662. [CrossRef] [PubMed]

nutrients

MDPI

Review

Carbohydrate Counting in Children and Adolescents with Type 1 Diabetes

Giorgia Tascini, Maria Giulia Berioli, Laura Cerquiglini, Elisa Santi, Giulia Mancini, Francesco Rogari, Giada Toni and Susanna Esposito *

Pediatric Clinic, Department of Surgical and Biomedical Sciences, Università degli Studi di Perugia, 06132 Perugia, Italy; giorgia.tascini@gmail.com (G.T.); mgiuliaberioli@gmail.com (M.G.B.); laura.cerquiglini@ospedale.perugia.it (L.C.); elisa.santi1988@gmail.com (E.S.); giu87manci@gmail.com (G.M.); rogari.francesco@virgilio.it (F.R.); toni.giada@gmail.com (G.T.)
* Correspondence: susanna.esposito@unimi.it; Tel.: +39-075-578-4417; Fax: +39-075-578-4415

Received: 26 December 2017; Accepted: 16 January 2018; Published: 22 January 2018

Abstract: Carbohydrate counting (CC) is a meal-planning tool for patients with type 1 diabetes (T1D) treated with a basal bolus insulin regimen by means of multiple daily injections or continuous subcutaneous insulin infusion. It is based on an awareness of foods that contain carbohydrates and their effect on blood glucose. The bolus insulin dose needed is obtained from the total amount of carbohydrates consumed at each meal and the insulin-to-carbohydrate ratio. Evidence suggests that CC may have positive effects on metabolic control and on reducing glycosylated haemoglobin concentration (HbA_{1c}). Moreover, CC might reduce the frequency of hypoglycaemia. In addition, with CC the flexibility of meals and snacks allows children and teenagers to manage their T1D more effectively within their own lifestyles. CC and the bolus calculator can have possible beneficial effects in improving post-meal glucose, with a higher percentage of values within the target. Moreover, CC might be integrated with the counting of fat and protein to more accurately calculate the insulin bolus. In conclusion, in children and adolescents with T1D, CC may have a positive effect on metabolic control, might reduce hypoglycaemia events, improves quality of life, and seems to do so without influencing body mass index; however, more high-quality clinical trials are needed to confirm this positive impact.

Keywords: carbohydrate counting; glycosylated haemoglobin; hypoglycaemia; insulin; type 1 diabetes

1. Introduction

Type 1 diabetes (T1D), formerly called insulin-dependent diabetes or juvenile diabetes, is an autoimmune disorder characterised by a severe deficiency or absence of endogenous insulin, which results in chronic hyperglycaemia. The achievement of optimal glucose control is facilitated by intensive insulin treatment in patients with T1D [1,2]. The current recommendations are based on the basal bolus paradigm, with subcutaneous long-acting insulin or continuous subcutaneous infusion of rapid-acting insulin to cover basal requirements and with rapid-acting insulin to prevent or correct glucose excursions. Frequent self-measurements of capillary blood glucose allow for insulin dose adjustments, with the aim of achieving good metabolic control [3]. Furthermore, continuous glucose monitoring (CGM) provides information about the glucose level every 5 min and allows patients to view their glycaemic data either on their insulin pump or on a separate receiver [4].

Despite advances in medical treatment and technology, nutritional therapy continues to be a cornerstone of diabetes care [5]. Nutritional recommendations for a healthy lifestyle for the general population are also appropriate for children and adolescents with T1D, with the only difference compared to healthy peers being the need for insulin therapy. The International Society for Paediatric and Adolescent Diabetes (ISPAD) Consensus Guidelines of 2014 stressed the importance of and special issues applying to nutritional management for children with diabetes [6]. The Guidelines recommended that a proper diet should allow for optimal growth, maintain an ideal weight, and prevent acute and chronic complications. The approximate energy intake and essential nutrients should be distributed as follows: carbohydrates 50–55%; fat 30–35%; and protein 10–15%.

Carbohydrates are the primary macronutrient that has an effect on the postprandial glycaemic response; their dietary intake should not be limited to ensure proper growth in children and adolescents with T1D. Adjusting the insulin dose to carbohydrate intake could produce potential improvements in glycaemic control and quality of life. Carbohydrate counting (CC) is a meal-planning tool for patients with T1D, based on an awareness of foods that contain carbohydrates and their effect on blood glucose. The bolus insulin dose needed is obtained from the total amount of carbohydrates consumed at each meal and the insulin-to-carbohydrate ratio (ICR) [7]. The current guidelines recommend the algorithms for prandial insulin be based on the carbohydrate amount taken in during a meal. They recognise the additional benefits to glycaemic control due to the use of the glycaemic index (GI) and that fats and proteins in the diet may influence postprandial glycaemia [6,8,9]. The aim of this review is to analyse the impact of CC in children and adolescents with T1D.

2. Educational Program on Carbohydrate Counting (CC)

With the discovery of insulin, researchers recognised that the total amount of dietary carbohydrates should be used to determine the need for insulin at meals. In the 1980s, the conventional insulin therapy imposed very rigid eating patterns with restricted and controlled carbohydrates portions. Subsequently, in the 1990s, the Diabetes Control and Complication Trial (DCTT) was a turning point for the treatment of diabetes. From this study, it was found that intensive insulin therapy, using multiple daily injection (MDI) or insulin pumps (CSII), improved glycaemic control and reduced complications related to the disease. In the DCTT, CC, one of the methods of meal planning, was considered effective in achieving good glycaemic control and allowing more flexibility in food choices [10,11].

An experienced team, including the role of the dietitian, is essential for the patient with T1D and his or her family so that they can then partake in the educational program and learn the method of CC as well as healthy-eating principles [11]. Children and those who care for them can learn to estimate the amount of carbohydrates with reasonable accuracy; however, continuing education therapy is needed. Carbohydrates have the most significant impact on raising postprandial blood glucose. A careful counting of carbohydrates will lead to the correct calculation of the required insulin dose, which in turn will lead to normalising postprandial glycaemia [12]. The American Academy of Nutrition and Dietetics have identified three levels of CC [8]:

- Level 1, Patients should understand that carbohydrates raise blood glucose, and they must be encouraged to consume a consistent amount of carbohydrates per meal. Simple methods to quantify the carbohydrates must be provided: gram increments of carbohydrates, 10–12 g carbohydrate portions, and 15 g carbohydrate exchanges. In particular, this level is helpful for patients in therapy with twice-daily insulin doses and who need consistent carbohydrate intake [13];
- Level 2, Pattern management. This represents an intermediate step, in which patients learn to evaluate changes in blood glucose compared to carbohydrates consumed, which are modified by insulin and physical activity. Therefore, patients supported by paediatric teams can make changes to their insulin. However, ever more frequently, paediatric teams use other methods, such as carbohydrate intake or the insulin-to-carb ratio (ICR);
- Level 3, Advanced CC. Patients in MDI or CSII learn ICR and how to use it.

3. Insulin-to-Carb Ratio (ICR), Insulin Sensitivity Factor (ISF), and Glycaemic Index (GI)

The insulin-to-carb ratio (ICR) is individualised and depends on one's sensitivity to insulin, which is how many grams of carbohydrates 1 unit of insulin covers. ICR allows children to obtain their insulin needs at mealtimes on the basis of the carbohydrates that will be consumed at that time, their blood glucose level, and their anticipated physical activity [6,9]. For the identification of the carbohydrates-to-insulin ratio, the empirical method of the 500 rule or 300–450 rule, for very young children who need less than 10 units of insulin a day, is often used [14–17]. It consists of dividing 500, 300, or 450 by the total daily insulin dose (TDD). ICR is not constant during the day; it tends to be higher in the morning, lower for lunch, and higher in the evening [18]. ICR varies more in children than in adults, which is linked to the more frequent changes occurring in the daily activity of the child [9].

The insulin sensitivity factor (ISF) is a correction algorithm of pre-prandial glycaemia. In particular, ISF indicates how many mmol/L (or mg/dL) 1 unit of insulin lowers blood glucose by; it is obtained by dividing 1800 (rapid analogue) or 1500 (regular insulin) by the TDD [19]. ISF also needs to be individualised for each patient. Regarding infants and toddlers, ISF is generally higher, approximately 100–150 mg/dL [19]. A recent study showed that the bolus insulin for meals was more than the one calculated by the 500 rule, in particular at breakfast and in pre-pubescent children. Moreover, for corrections, insulin need was slightly less than that calculated by the 100 rule (100 divided by total daily insulin dose) [19].

The GI expresses the glycaemic response after eating a known amount of carbohydrates contained in food, but in relation to the same amount of carbohydrates contained in white bread. The glycaemic area measured 120 min after food intake is expressed as a percentage of the standard. Using the GI and eating low-GI food may produce modest benefits in the control of postprandial hyperglycaemia. Moreover, to consider both the quality and the amount of carbohydrates, the concept of glycaemic load was introduced [20]. In patients with T1D, GI should be used with a method of carbohydrate quantification [21].

3.1. Glycaemic Control

In previous studies conducted in adult subjects, glycaemic variations were assessed after the ingestion of different amounts and qualities of carbohydrates: complex carbohydrates in a mixed meal and simple sugar dextrose [22]. Blood sugar was higher, but for a shorter time, after the dextrose load than after the mixed meals. The total amount released by the artificial pancreas to return glycaemia to basal levels did not differ when exposed to different types of carbohydrates, but a difference did occur in the release kinetics. Furthermore, there was a linear relation between carbohydrate intake and insulin need.

A systematic review and meta-analysis was recently performed to evaluate the effectiveness of CC on glycaemic control in people with T1D [22]. It included 10 studies, published from 2000 to 2014, involving 773 participants. Four studies enrolled children and adolescents [23–26], and six studies included adults [27–32]. Overall, CC, compared with other diabetes diet methods or usual diabetes diets, improved HbA$_{1c}$ values, in line with a previous systematic review and a meta-analysis [28,33,34]. A non-significant improvement of HbA$_{1c}$ was found in the subgroup analysis restricted to trials that compared CC with other T1D diet methods [14].

One systematic review and meta-analysis aimed to summarise all available evidence from both observational and randomised controlled studies on the effects of advanced CC on glycaemic control, psychosocial measures, body weight, and severe hypoglycaemic events in patients of all age groups with T1D [26]. Six randomised controlled trials and 21 before/after studies were included; four uncontrolled trials (Table 1) [35–38] and one randomised controlled trial (Table 2) [39] were conducted in children and adolescents with a mean age ± standard deviation of 12 ± 4 years. In only one study, in which 28 adolescents were included, the improvement in HbA$_{1c}$ was not demonstrated after advanced CC. These 28 patients switched from a prescriptive to a flexible meal plan and insulin. According to the authors, this result could be related to inaccurate CC, incorrect calculation of insulin dose, or inappropriate insulin injection technique.

Table 1. Characteristics of uncontrolled studies with carbohydrate counting (CC) used in children and adolescents with type 1 diabetes (T1D).

Authors and Year	Design	Follow-Up (Months)	Sample Size	Insulin Regimen	Insulin Type (Short-Acting/Long-Acting)	Age (Years)	T1D Duration (Years)	HbA$_{1c}$ at Baseline (mmol/mol)	HbA$_{1c}$ at Baseline (%)	Psychosocial Measures	Weight (kg)/BMI (kg/m^2)/BMI SDS at Baseline	Hypoglycaemia Reporting
					Uncontrolled Trials							
Abaci et al., 2009 [35]	Before/after	12	9	MDI	Analogue/NPH	15	4	78	9.3	Not reported	—/—/1.1	Yes
Alemzadeh et al., 2003 [36]	Before/after	12	44	MDI	Analogue/NPH or analogue	11	5	78	9.3	Not reported	—/19.3/—	Yes
Alemzadeh et al., 2005 [37]	Before/after	12	35	MDI	Analogue/analogue	5	3	73	8.8	Not reported	—/17.1/—	Yes
Hayes et al., 2012 [38]	Before/after	9	28	MDI	Human or analogue/analogue	14 (median)	4 (median)	61 (median)	7.7 (median)	DQOL-Y	—/—/0.99	Not reported

BMI: body mass index; BMI SDS: body mass index standard deviation score; DQOL-Y: Diabetes Quality of Life for Youth scale; HbA1c: glycosylated haemoglobin concentration; MDI: multiple dose injections; NPH: neutral protamine Hagedorn.

Table 2. Characteristics of controlled, randomised study with carbohydrate counting (CC) used in children and adolescents with type 1 diabetes (T1D).

Author and Year	Country	Population	No. of Patients	Intervention	Control	HbA$_{1c}$ (%) (M ± SD) Intervention/Control	Hypoglycaemia (M ± SD)	Insulin Dose (U/kg) (M ± SD)	BMI (M ± SD)	Follow Up
Enander et al. [39]	Sweden	Children and young people	45; 26/30; 14/15	Dietary education in carbohydrate counting	Dietary education in the traditional methodology (the plate exchange method)	7.43 ± 0.83 to 7.69 ± 1.00 7.70 ± 1.00 to 8.00 ± 1.00	-	0.78 ± 0.24 to 0.80 ± 0.19 0.81 ± 0.22 to 0.83 ± 0.22	-	12 months

BMI: body mass index; HbA1c: glycosylated haemoglobin concentration.

Bell et al. only included one study of children in their systematic review [28]. The results favoured the alternative approach, i.e., a low glycaemic index over CC. However, the improvement of the glycaemic control found in adults needed to be interpreted with care, as more studies had a concurrent intensification of insulin therapy in the intervention group.

Compliance and the ability to accurately estimate the amount of carbohydrates consumed by children and adolescents with T1D is another aspect that should be considered. The greater ability of parents to count carbohydrates led to lower values of HbA_{1c} in their children and adolescents [40–44]. At the same time, there are wide variations in this ability, with studies where participants may estimate carbohydrates to within 10–15 g or within 15–20% of the actual amount, while in others, only half could accurately evaluate the carbohydrate amount or had large differences in calculations [28,40–45].

An important aspect that should not be overlooked is the effect of time on changes in HbA_{1c}. The effects of diabetes self-management training programmes were short-lived. Thus, as recommended in many countries, regular reinforcement of advanced CC is critical in patient management [41,46,47]. In the review by Schmidt et al., the study with the longest period of evaluation was also the study with the smallest improvement in HbA_{1c} [34]. The accuracy of counting carbohydrates is associated with an improvement in glycaemic control; at the same time, intense glucose monitoring is an independent variable related to the reduction in HbA_{1c} [43,48].

Considering the available data, in several studies the CC method has been compared to a control arm consisting of usual or standard care rather than alternative methods for calculating bolus insulin. In addition, observational studies generally conclude showing no improvement in glycaemic control. Furthermore, results in the paediatric and adolescent population seem better than those observed in adults, probably also for the psychological aspects associated with CC. All these considerations highlight the need of further well-designed, double-blind large studies that confirm the role of CC in patients with T1D of different age groups.

3.2. Hypoglycaemia

The calculation of carbohydrate amount for the adjustment of insulin dose is useful for better glycaemic control, and, for the practical CC, the small quantification errors are permissive. A 10 g variation in the carbohydrate quantity did not induce differences in blood glucose. Furthermore, patients may have episodes of hypo- and hyperglycaemia for an overestimation or underestimation of 20 g in a meal of 60 g carbohydrates [35,37,49]. Adolescents may be less accurate in the calculation and have a decay of glycaemic control. Hypoglycaemia is not reported or defined uniformly [40]. Uncontrolled studies showed a significant reduction in severe hypoglycaemia in children with T1D using CC [29,30]. Other data suggested that CC did not significantly decrease hypoglycaemia events [14]. In a before/after study in nine children, an insignificant increase in non-severe hypoglycaemia was observed [28]. CC might reduce the frequency of hypoglycaemia; probably, the reduction in HbA_{1c} concentration is a result of stabilised glycaemic control rather than just an overall lowering of blood glucose [15].

3.3. Growth, Weight, and Psychological Effects

Intensive and appropriate insulin therapy and, consequently, better metabolic control have enabled an improvement in the growth prognosis of paediatric patients with T1D [30,50]. Modern T1D care prevents abnormalities of the GH-IGF-I axis, which means that children with T1D can achieve a final height similar to their healthy peers [51].

Some clinicians believe that flexibility in the diet, which can be permitted by CC, may lead to the development of obesity [8,9]. The DAFNE study demonstrated that a freedom diet did not increase cardiovascular risk factors, including obesity [26]. Paediatric studies have shown that therapy MDI associated with the calculation of carbohydrates does not cause obesity in children with T1D. Some studies documented a decrease in body mass index (BMI), and some studies reported an increase; nevertheless, only minor variations in weight after the use of CC were found [30,52–54]. The increased weight documented in children and adolescents with T1D was attributed to changes in growth occurring during puberty progression [29]. In another study, Marigliano et al. evaluated the impact of CC and nutritional education on changes in dietary habits, body fat composition, and fat body distribution in diabetic children in therapy with CSII [55]. The results revealed that CC and a proper nutrition education did not have a negative effect on dietary habits, body composition, or body fat distribution. Moreover, a reduction of intake of fat and protein, with an increase in carbohydrates intake, was observed in a subgroup of subjects showing a significant improvement in metabolic control.

Females at a young age with T1D face a higher risk of eating behaviour disorders and, consequently, are more likely to have impaired metabolic control and diabetic complications [56]. Furthermore, a cross-sectional case-controlled study has shown that there was an increased frequency of food disturbances in pre-teen and early-teenage girls with T1D [57]. Rigid dietary patterns that do not conform to the habits and needs of patients might have an effect on the psychological development of children with T1D. An Australian 9-month before/after study that evaluated diabetes-related quality of life reported significant improvements in patient-rated life satisfaction after the introduction of CC. Often, the lack of flexibility in the management of meals and snacks is one of the worst aspects of living with diabetes, and adolescents especially ignore dietary advice. Thus, CC is an important tool allowing children and teenagers to manage their T1D more effectively within their own lifestyles [1,31]. This psychological effect of CC could also be useful during the honeymoon period and the transition from adolescence to adulthood.

In addition, it is extremely important to highlight that for the children and adolescents with T1D, the psychological aspects, such as the quality of life, adherence, and motivation, are key points of good glycaemic control. The reports of the Dose Adjustment for Normal Eating (DAFNE) study emphasised the positive effect of CC on the quality of life [58]. So far, many kinds of structural education programs for T1D have been reported [59,60]. Most of these programs included CC and showed the improvement of patients' adherence, as well as motivation. Therefore, CC has a positive effect on patients' adherence and motivation too. In addition, it has been observed that CC also has a positive effect on the parents and caregivers, who are the key persons for the treatment of infants and young children with T1D [59,60].

3.4. Bolus Calculator

The bolus calculator is a useful tool both for patients with T1D who require MDI and patients with an insulin pump, which allows the calculation of the prandial insulin dose based on the amount of carbohydrates, the pre-prandial blood glucose value, and active insulin [61]. Its use is associated with improved metabolic control and greater satisfaction in treating adults, children, and adolescents [62]. Few studies found the bolus calculator to be effective in children using CSII. The bolus calculator and CC may have beneficial effects in reducing HbA_{1c} by improving post-meal glucose, with a higher percentage of values within the target range [63–65]. Moreover, an insulin pump offers the possibility to tailor insulin delivery to the meal composition through the use of different bolus options: normal, dual wave, or square wave. Evidence suggests that children who apply at least one D-W/S-W bolus a day are better at achieving the recommended HbA_{1c} level [66]. In an 18-month observational study, Rabbone et al. investigated the effect of CC, with or without an automated bolus calculator, in children treated with MDI. This study showed that bolus calculator use led to greater improvements in HbA_{1c} [67].

3.5. Fat and Protein Counting

Learning about CC is important because it also increases knowledge about other macronutrients [57]. Low carbohydrate diets have been popularised by social media for those who wish to lose weight [68]. However, the promotion of a low carbohydrate diet in lay media is in contrast to published paediatric diabetes guidelines that endorse a balanced diet from a variety of foods for optimal growth and development in children with T1D [69]. De Bock et al. described a series of six cases where the adoption of a low carbohydrate diet in children impacted growth and cardiovascular risk factors with potential long-term sequelae [69]. Meals enriched in fat and protein result in a prolonged blood glucose rise by 3–4 h after food ingestion, and frequently, a relative insulin resistance is observed [70]. The American Diabetes Association recommends an appropriate glycaemic impact education regarding protein and fat. Recent studies and the use of continuous glucose monitoring have shown that post-prandial glucose excursion may occur because of fat, protein, and GI [71,72]. CGM and increasing focus on the daily lives of patients highlight the limitations of the traditional CC method, used to calculate the meal-time insulin dose. From the new insights about the effect of dietary macronutrients on post-prandial glucose control, it appears that, to improve the calculation of the insulin bolus, CC should be integrated with the counts of fats and proteins. On the other hand, it has been demonstrated that bolus calculation for high fat feeding prevents late rises in postprandial triglycerides and tumour necrosis factor alpha, thus improving cardiovascular risk profile [73]. Therefore, the development of suitable and usable algorithms is necessary, without forgetting the importance of educational therapy for a successful translation of fat/protein counting in real life [74].

4. Conclusions

Current guidelines recommend that the algorithms for prandial insulin dose be based on the carbohydrate amount taken with a meal. CC and intensive insulin therapy enable patients to have greater freedom in the management of the meals or snacks and activities of daily living, without forgetting the principles of healthy eating. A paediatric interdisciplinary T1D care team is fundamental. It has to provide education, monitoring, and support to the children and their caregivers, bearing in mind crucial aspects of children and adolescents with T1D: physical and psychological development and T1D complications.

CC may improve metabolic control and reduce HbA$_{1c}$ concentration. Moreover, with developments in medical technology and with the new insights into the effect of dietary macronutrients on post-prandial glucose control, it appears that, to improve the calculation of the insulin bolus, the CC should be integrated with the counts of fats and proteins. Research into suitable and usable algorithms is necessary. Overall, CC may have a positive effect on metabolic control, reduce hypoglycaemia events, improve quality of life, and not influence BMI. However, more high-quality clinical trials are needed to confirm this positive impact.

Acknowledgments: This review, including the costs to publish in open access, was supported by a grant from the World Association of Infectious Diseases (WAidid).

Author Contributions: Giorgia Tascini wrote the first draft of the manuscript; Maria Giulia Berioli and Laura Cerquiglini revised the text; Elisa Santi, Giulia Mancini, Francesco Rogari and Giada Toni gave a support in the literature review; Susanna Esposito critically revised the text and made substantial scientific contributions. All the authors approved the final version of the manuscript.

Conflicts of Interest: The authors declare no conflicts of interest.

References

1. Switzer, S.M.; Moser, E.G.; Rockler, B.E.; Garg, S.K. Intensive insulin therapy in patients with type 1 diabetes mellitus. *Endocrinol. Metab. Clin. N. Am.* **2012**, *41*, 89–104. [CrossRef] [PubMed]
2. Tamborlane, W.V.; Sikes, K.A. Insulin therapy in children and adolescents. *Endocrinol. Metab. Clin. N. Am.* **2012**, *41*, 145–160. [CrossRef] [PubMed]
3. Malik, F.S.; Taplin, C.E. Insulin therapy in children and adolescents with type 1 diabetes. *Pediatr. Drugs* **2014**, *16*, 141–150. [CrossRef] [PubMed]
4. Golicki, D.T.; Golicka, D.; Groele, L.; Pankowska, E. Continuous glucose monitoring system in children with type 1 diabetes mellitus: A systematic review and meta-analysis. *Diabetologia* **2008**, *51*, 233–240. [CrossRef] [PubMed]

5. Smart, C.E.; Aslander-van de Vliet, E.; Waldron, S. ISPAD Clinical Practice Consensus Guidelines 2009 Compendium: Nutritional management in children and adolescents with diabetes. *Pediatr. Diabetes* **2009**, *10*, 100–117. [CrossRef] [PubMed]
6. Smart, C.E.; Annan, F.; Bruno, L.P.C.; Higgins, L.A.; Acerini, C.L. ISPAD Clinical Practice Consensus Guidelines 2014 Compendium: Nutritional management in children and adolescents with diabetes. *Pediatr. Diabetes* **2014**, *15*, 135–153. [CrossRef] [PubMed]
7. Gillespie, S.J.; Kulkarni, K.D.; Daly, A.E. Using carbohydrate counting in diabetes clinical practice. *J. Am. Diet. Assoc.* **1998**, *98*, 897–905. [CrossRef]
8. Kawamura, T. The importance of carbohydrate counting in the treatment of children with diabetes. *Pediatr. Diabetes* **2007**, *8*, 57–62. [CrossRef] [PubMed]
9. Sheard, N.F.; Clark, N.G.; Brand-Miller, J.C.; Franz, M.J.; Pi-Sunyer, F.X.; Mayer-Davis, E.; Kulkarni, K.; Geil, P. Dietary carbohydrate (amount and type) in the prevention and management of diabetes: A statement by the American diabetes association. *Diabetes Care* **2004**, *27*, 2266–2271. [CrossRef] [PubMed]
10. THE DCCT Research Group. Nutrition interventions for intensive therapy in the Diabetes Control and Complications Trial. *J. Am. Diet. Assoc.* **1993**, *93*, 768–772.
11. Rabbone, I.; Canova, A.; Tuli, G.; Gioia, E.; Sicignano, S.; Cerutti, F. The calculation of carbohydrates in type 1 diabetes in children. *G. Ital. Diabetol. Metab.* **2011**, *31*, 150–154. (In Italian)
12. Deeb, A.; Al Hajeri, A.; Alhmoudi, I.; Nagelkerke, N. Accurate carbohydrate counting is an important determinant of postprandial glycemia in children and adolescents with type 1 diabetes on insulin pump therapy. *J. Diabetes Sci. Technol.* **2017**, *11*, 753–758. [CrossRef] [PubMed]
13. Wolever, T.M.; Hamad, S.; Chiasson, J.L.; Josse, R.G.; Leiter, L.A.; Rodger, N.W.; Ross, S.A.; Ryan, E.A. Day-to-day consistency in amount and source of carbohydrate associated with improved blood glucose control in type 1 diabetes. *J. Am. Coll. Nutr.* **1999**, *18*, 242–247. [CrossRef] [PubMed]
14. Slama, G.; Klein, J.C.; Delage, A.; Ardila, E.; Lemaignen, H.; Papoz, L.; Tchobroutsky, G. Correlation between the nature and amount of carbohy- drate in meal intake and insulin delivery by the artificial pancreas in 24 insulin-dependent diabetics. *Diabetes* **1981**, *30*, 101–105. [CrossRef] [PubMed]
15. Halfon, P.; Belkhadir, J.; Slama, G. Correlation between amount of carbohydrate in mixed meals and insulin delivery by artificial pancreas in seven IDDM subjects. *Diabetes Care* **1989**, *12*, 427–429. [CrossRef] [PubMed]
16. Walsh, J.; Roberts, R.; Bayle, T.; Varma, C.B. *Using Insulin, Everything You Need for Success with Insulin*; Torrey Pines Press: San Diego, CA, USA, 2003.
17. Walsh, J.; Roberts, R. *Pumping Insulin*, 4th ed.; Torrey Pines Press: San Diego, CA, USA, 2000.
18. Danne, T. Current practice of insulin pump therapy in children and adolescents-the Hannover recipe. *Pediatr. Diabetes* **2006**, *7*, 25–31. [CrossRef] [PubMed]
19. Hanas, R.; Adolfsson, P. Bolus calculator settings in well-controlled prepubertal children using insulin pumps are characterized by low insulin to carbohydrate ratios and short duration of insulin action time. *J. Diabetes Sci. Technol.* **2017**, *11*, 247–252. [CrossRef] [PubMed]
20. Barclay, A.W.; Petocz, P.; McMillan-Price, J.; Flood, V.M.; Prvan, T.; Mitchell, P.; Brand-Miller, J.C. Glycemic index, glycemic load, and chronic disease risk—A meta-analysis of observational studies. *Am. J. Clin. Nutr.* **2008**, *87*, 627–637. [PubMed]
21. Craig, M.E.; Twigg, S.M.; Donaghue, K.C. National Evidence-Based Clinical Care Guidelines for Type 1 Diabetes in Children, Adolescents and Adults. Available online: https://www.google.com.hk/url?sa= t&rct=j&q=&esrc=s&source=web&cd=1&ved=0ahUKEwjkvJndgOHYAhVItpQKHSCHAj0QFgglMAA& url=https%3A%2F%2Fdiabetessociety.com.au%2Fdownloads%2FType1guidelines(7Feb11).pdf&usg= AOvVaw1sAlGP3LfMJDEqvaxMzrbE (accessed on 19 December 2017).
22. Gupta, L.; Khandelwal, D.; Kalra, S. Applied carbohydrate counting. *J. Pak. Med. Assoc.* **2017**, *67*, 1456–1457. [PubMed]
23. Gilbertson, H.R.; Brand-Miller, J.C.; Thorburn, A.W.; Evans, S.; Chondros, P.; Werther, G.A. The effect of flexible low glycemic index dietary advice versus measured carbohydrate exchange diets on glycemic control in children with type 1 diabetes. *Diabetes Care* **2001**, *24*, 1137–1143. [CrossRef] [PubMed]
24. Goksen, D.; Altinok, Y.A.; Ozen, S.; Demir, G.; Darcan, S. Effects of carbohydrate counting method on metabolic control in children with type 1 diabetes mellitus. *J. Clin. Res. Pediatr. Endocrinol.* **2014**, *1*, 74–78. [CrossRef] [PubMed]

25. De Albuquerque, I.Z. Carbohydrate counting, nutritional status and metabolic profile of adolescents with type 1 diabetes mellitus. *Sci. Med.* **2014**, *24*, 21–24.

26. Enander, R.; Gundevall, C.; Strömgren, A.; Chaplin, J.; Hanas, R. Carbohydrate counting with a bolus calculator improves post-prandial blood glucose levels in children and adolescents with type 1 diabetes using insulin pumps. *Pediatr. Diabetes* **2012**, *13*, 545–551. [CrossRef] [PubMed]

27. Kalergis, M.; Pacaud, D.; Strychar, I.; Meltzer, S.; Jones, P.J.; Yale, J.F. Optimizing insulin delivery: Assessment of three strategies in intensive diabetes management. *Diabetes Obes. Metab.* **2000**, *2*, 299–305. [CrossRef] [PubMed]

28. Bell, K.J.; Barclay, A.W.; Petocz, P.; Colagiuri, S.; Brand-Miller, J.C. Efficacy of carbohydrate counting in type 1 diabetes: A systematic review and meta-analysis. *Lancet Diabetes Endocrinol.* **2014**, *2*, 133–140. [CrossRef]

29. Trento, M.; Trinetta, A.; Kucich, C.; Grassi, G.; Passera, P.; Gennari, S.; Paganin, V.; Tedesco, S.; Charrier, L.; Cavallo, F.; et al. Carbohydrate counting improves coping ability and metabolic control in patients with type 1 diabetes managed by group care. *J. Endocrinol. Investig.* **2011**, *34*, 101–105. [CrossRef] [PubMed]

30. Schmidt, S.; Meldgaard, M.; Serifovski, N.; Storm, C.; Christensen, T.M.; Gade-Rasmussen, B.; Nørgaard, K. Use of an automated bolus calculator in MDI-treated type 1 diabetes: The BolusCal Study, a randomized controlled pilot study. *Diabetes Care* **2012**, *35*, 984–990. [CrossRef] [PubMed]

31. Scavone, G.; Manto, A.; Pitocco, D.; Gagliardi, L.; Caputo, S.; Mancini, L.; Zaccardi, F.; Ghirlanda, G. Effect of carbohydrate counting and medical nutritional therapy on glycemic control in Type 1 diabetic subjects: A pilot study. *Diabet. Med.* **2010**, *27*, 477–479. [CrossRef] [PubMed]

32. DAFNE Study Group. Training in flexible, intensive insulin management to enable dietary freedom in people with type 1 diabetes: Dose adjustment for normal eating (DAFNE) randomised controlled trial. *BMJ* **2002**, *325*, 746.

33. Fu, S.; Li, L.; Deng, S.; Zan, L.; Liu, Z. Effectiveness of advanced carbohydrate counting in type 1 diabetes mellitus: A systematic review and meta-analysis. *Sci. Rep.* **2016**, *6*, 37067. [CrossRef] [PubMed]

34. Schmidt, S.; Schelde, B.; Nørgaard, K. Effects of advanced carbohydrate counting in patients with type 1 diabetes: A systematic review. *Diabet. Med.* **2014**, *31*, 886–896. [CrossRef] [PubMed]

35. Abaci, A.; Atas, A.; Unuvar, T.; Bober, E.; Buyukgebiz, A. The effect of carbohydrate counting on metabolic control in patients with type 1 diabetes mellitus. *Gulhane Med. J.* **2009**, *51*, 1–5.

36. Alemzadeh, R.; Palma-Sisto, P.; Parton, E.; Totka, J.; Kirby, M. Beneficial effects of flexible insulin therapy in children and adolescents with type 1 diabetes mellitus. *Acta Diabetol.* **2003**, *40*, 137–142. [CrossRef] [PubMed]

37. Alemzadeh, R.; Berhe, T.; Wyatt, D.T. Flexible insulin therapy with glargine insulin improved glycemic control and reduced severe hypoglycemia among preschool-aged children with type 1 diabetes mellitus. *Pediatrics* **2005**, *115*, 1320–1324. [CrossRef] [PubMed]

38. Hayes, R.L.; Garnett, S.P.; Clarke, S.L.; Harkin, N.M.; Chan, A.K.; Ambler, G.R. A flexible diet using an insulin to carbohydrate ratio for adolescents with type 1 diabetes—A pilot study. *Clin. Nutr.* **2012**, *31*, 705–709. [CrossRef] [PubMed]

39. Gupta, L.; Khandelwal, D.; Kalra, S. Carbohydrate counting-1: South Asian framework. *J. Pak. Med. Assoc.* **2017**, *67*, 1296–1298. [PubMed]

40. Bishop, F.; Maahs, D.; Spiegel, G. The carbohydrate counting in adolescents with type 1 diabetes (CCAT) study. *Diabetes Spectr.* **2009**, *22*, 56–62. [CrossRef]

41. Smart, C.E.; Ross, K.; Edge, J.A.; King, B.R.; McElduff, P.; Collins, C.E. Can children with Type 1 diabetes and their caregivers estimate the carbohydrate content of meals and snacks? *Diabet. Med.* **2010**, *27*, 348–353. [CrossRef] [PubMed]

42. Rabasa-Lhoret, R.; Garon, J.; Langelier, H.; Poisson, D.; Chiasson, J.L. Effects of meal carbohydrate content on insulin requirements in type 1 diabetic patients treated intensively with the basal-bolus (ultralente-regular) insulin regimen. *Diabetes Care* **1999**, *22*, 667–673. [CrossRef] [PubMed]

43. Mehta, S.N.; Quinn, N.; Volkening, L.K.; Laffel, L.M. Impact of carbo- hydrate counting on glycemic control in children with type 1 diabetes. *Diabetes Care* **2009**, *32*, 1014–1016. [CrossRef] [PubMed]

44. Nebel, I.T.; Bluher, M.; Starcke, U.; Muller, U.A.; Haak, T.; Paschke, R. Evaluation of a computer based interactive diabetes education program designed to train the estimation of the energy or carbohydrate contents of foods. *Patient Educ. Couns.* **2002**, *46*, 55–59. [CrossRef]

45. Ahola, A.J.; Makimattila, S.; Saraheimo, M.; Mikkilä, V.; Forsblom, C.; Freese, R.; Groop, P.H. FinnDIANE Study Group. Many patients with type 1 diabetes estimate their prandial insulin need inappropriately. *J. Diabetes* **2010**, *2*, 194–202. [CrossRef] [PubMed]
46. Funnell, M.M.; Brown, T.L.; Childs, B.P.; Haas, L.B.; Hosey, G.M.; Jensen, B.; Maryniuk, M.; Peyrot, M.; Piette, J.D.; Reader, D.; et al. National standards for diabetes self-management education. *Diabetes Care* **2012**, *35*, S101–S108. [CrossRef] [PubMed]
47. Norris, S.L.; Lau, J.; Smith, S.J.; Schmid, C.H.; Engelgau, M.M. Self-management education for adults with type 2 diabetes: A meta-analysis of the effect on glycemic control. *Diabetes Care* **2002**, *25*, 1159–1171. [CrossRef] [PubMed]
48. Plank, J.; Kohler, G.; Rakovac, I.; Semlitsch, B.M.; Horvath, K.; Bock, G.; Kraly, B.; Pieber, T.R. Long-term evaluation of a structured outpatient education programme for intensified insulin therapy in patients with type 1 diabetes: A 12-year follow-up. *Diabetologia* **2004**, *47*, 1370–1375. [CrossRef] [PubMed]
49. Smart, C.E.; Ross, K.; Edge, J.A.; Collins, C.E.; Colyvas, K.; King, B.R. Children and adolescents on intensive insulin therapy maintain postprandial glycaemic control without precise carbohydrate counting. *Diabet. Med.* **2009**, *26*, 279–285. [CrossRef] [PubMed]
50. Salardi, S.; Tonioli, S.; Tassoni, R.; Tellarini, M.; Mazzanti, I.; Caccianti, E. Growth and growth factors in diabetes mellitus. *Arch. Dis. Child.* **1987**, *62*, 57–62. [CrossRef] [PubMed]
51. Chiarelli, F.; Giannini, C.; Mohn, A. Growth, growth factors and diabetes. *Eur. J. Endocrinol.* **2004**, *151*, U109–U117. [CrossRef] [PubMed]
52. Dias, V.M.; Pandini, J.A.; Nunes, R.R.; Sperandei, S.L.; Portella, E.S.; Cobas, R.A.; Gomes, M.B. Effect of the carbohydrate counting method on glycemic control in patients with type 1 diabetes. *Diabetol. Metab. Syndr.* **2010**, *2*, 54. [CrossRef] [PubMed]
53. Pitocco, D.; Rizzi, A.; Scavone, G.; Tanese, L.; Zaccardi, F.; Manto, A.; Ghirlanda, G. Fields of application of continuous subcutaneous insulin infusion in the treatment of diabetes and implications in the use of rapid-acting insulin analogues. *Minerva Endocrinol.* **2013**, *38*, 321–328. [PubMed]
54. Laurenzi, A.; Bolla, A.M.; Panigoni, G.; Doria, V.; Uccellatore, A.; Peretti, E.; Saibene, A.; Galimberti, G.; Bosi, E.; Scavini, M. Effects of carbohydrate counting on glucose control and quality of life over 24 weeks in adult patients with type 1 diabetes on continuous subcutaneous insulin infusion: A randomized, prospective clinical trial (GIOCAR). *Diabetes Care* **2011**, *34*, 823–827. [CrossRef] [PubMed]
55. Marigliano, M.; Morandi, A.; Maschio, M.; Sabbion, A.; Contreas, G.; Tomasselli, F.; Tommasi, M.; Maffeis, C. Nutritional education and carbohydrate counting in children with type 1 diabetes treated with continuous subcutaneous insulin infusion: The effects on dietary habits, body composition and glycometabolic control. *Acta Diabetol.* **2013**, *50*, 959–964. [CrossRef] [PubMed]
56. Rydall, A.C.; Rodin, G.M.; Olmsted, M.P.; Devenyi, R.G.; Daneman, D. Disordered eating behavior and microvascu-lar complications in young women with insulin-dependent diabetes mellitus. *N. Engl. J. Med.* **1997**, *336*, 1905–1906. [CrossRef] [PubMed]
57. Colton, P.; Olmsted, M.; Daneman, D.; Rydall, A.; Rodin, G. Disturbed eating behavior and eating disorders in preteen and early teenage girls with type 1 diabetes: A case-controlled study. *Diabetes Care* **2004**, *27*, 1654–1659. [CrossRef] [PubMed]
58. Cooke, D.; O'Hara, M.C.; Beinart, N.; Heller, S.; La Marca, R.; Byrne, M.; Mansell, P.; Dinneen, S.F.; Clark, M.; Bond, R.; et al. NIHR DAFNE Study Group. Linguistic and psychometric validation of the Diabetes-Specific Quality-of-Life Scale in U.K. English for adults with type 1 diabetes *Diabetes Care* **2013**, *36*, 1117–1125. [CrossRef] [PubMed]
59. Kime, N.H.; Waldron, S.; Webster, E.; Lange, K.; Zinken, K.; Danne, T.; Aschemeier, B.; Sumnik, Z.; Cinek, O.; Raposo, J.F.; et al. Pediatric diabetes training for healthcare professionals in Europe: Time for change. *Pediatr. Diabetes* **2017**. [CrossRef] [PubMed]
60. Coates, V.E.; Horigan, G.; Davies, M.; Davies, M.T. Exploring why young people with Type 1 diabetes decline structured education with a view to overcoming barriers. *Diabet. Med.* **2017**, *34*, 1092–1099. [CrossRef] [PubMed]
61. Bruttomesso, D.; Cipponeri, E. The binomial cho counting and home glycemic self-control: Is it applied? *Il Giornale di AMD* **2013**, *16*, 144–149. (In Italian)
62. Colin, I.M.; Paris, I. Glucose meters with built-in automated bolus calculator: Gadget or real value for insulin-treated diabetic patients? *Diabetes Ther.* **2013**, *4*, 1–11. [CrossRef] [PubMed]

63. Vallejo Mora, M.D.R.; Carreira, M.; Anarte, M.T.; Linares, F.; Olveira, G.; González Romero, S. Bolus calculator reduces hypoglycemia in the short term and fear of hypoglycemia in the long term in subjects with type 1 diabetes (CBMDI Study). *Diabetes Technol. Ther.* **2017**, *19*, 402–409. [CrossRef] [PubMed]

64. Pankowska, E.; Blazik, M. Bolus calculator with nutrition database software, a new concept of prandial insulin programming for pump users. *J. Diabetes Sci. Technol.* **2010**, *4*, 571–576. [CrossRef] [PubMed]

65. Shashaj, B.; Busetto, E.; Sulli, N. Benefits of a bolus calculator in pre- and postprandial glycaemic control and meal flexibility of paediatric patients using continuous subcutaneous insulin infusion (CSII). *Diabet. Med.* **2008**, *25*, 1036–1042. [CrossRef] [PubMed]

66. Pańkowska, E.; Szypowska, A.; Lipka, M.; Szpotańska, M.; Błazik, M.; Groele, L. Application of novel dual wave meal bolus and its impact on glycated hemoglobin A1c level in children with type 1 diabetes. *Pediatr. Diabetes* **2009**, *10*, 298–303. [CrossRef] [PubMed]

67. Rabbone, I.; Scaramuzza, A.E.; Ignaccolo, M.G.; Tinti, D.; Sicignano, S.; Redaelli, F.; De Angelis, L.; Bosetti, A.; Zuccotti, G.V.; Cerutti, F. Carbohydrate counting with an automated bolus calculator helps to improve glycaemic control in children with type 1 diabetes using multiple daily injection therapy: An 18-month observational study. *Diabetes Res. Clin. Pract.* **2014**, *103*, 388–394. [CrossRef] [PubMed]

68. Krebs, J.D.; Parry Strong, A.; Cresswell, P.; Reynolds, A.N.; Hanna, A.; Haeusler, S. A randomised trial of the feasibility of a low carbohydrate diet vs standard carbohydrate counting in adults with type 1 diabetes taking body weight into account. *Asia Pac. J. Clin. Nutr.* **2016**, *25*, 78–84. [PubMed]

69. De Bock, M.; Lobley, K.; Anderson, D.; Davis, E.; Donaghue, K.; Pappas, M.; Siafarikas, A.; Cho, Y.H.; Jones, T.; Smart, C. Endocrine and metabolic consequences due to restrictive carbohydrate diets in children with type 1 diabetes: An illustrative case series. *Pediatr. Diabetes* **2017**. [CrossRef] [PubMed]

70. Ahern, J.A.; Gatcomb, P.M.; Held, N.A.; Petit, W.A., Jr.; Tamborlane, W.V. Exaggerated hyperglycemia after a pizza meal in well-controlled diabetes. *Diabetes Care* **1993**, *16*, 578–580. [CrossRef] [PubMed]

71. Herman, W.H. Approaches to glycemic treatment Sec. 7: In standards of medical care in diabetes. *Diabetes Care* **2015**, *38*, S41–S48.

72. Bell, K.J.; Smart, C.E.; Steil, G.M.; Brand-Miller, J.C.; King, B.; Wolpert, H.A.; Kirstine, J. Impact of fat, protein, and glycemic index on postprandial glucose control in type 1 diabetes: Implications for intensive diabetes management in the continuous glucose monitoring era. *Diabetes Care* **2015**, *38*, 1008–1015. [CrossRef] [PubMed]

73. Campbell, M.D.; Walker, M.; Ajjan, R.A.; Birch, K.M.; Gonzalez, J.T.; West, D.J. An additional bolus of rapid-acting insulin to normalise postprandial cardiovascular risk factors following a high-carbohydrate high-fat meal in patients with type 1 diabetes: A randomised controlled trial. *Diabetes Vasc. Dis. Res.* **2017**, *14*, 336–344. [CrossRef] [PubMed]

74. Kordonouri, O.; Hartmann, R.; Remus, K.; Bläsig, S.; Sadeghian, E.; Danne, T. Benefit of supplementary fat plus protein counting as compared with conventional carbohydrate counting for insulin bolus calculation in children with pump therapy. *Pediatr. Diabetes* **2012**, *13*, 540–544. [CrossRef] [PubMed]

nutrients

MDPI

Article

Type-4 Resistant Starch in Substitution for Available Carbohydrate Reduces Postprandial Glycemic Response and Hunger in Acute, Randomized, Double-Blind, Controlled Study

Maria L. Stewart [1,*], **Meredith L. Wilcox** [2], **Marjorie Bell** [2], **Mary A. Buggia** [2] and **Kevin C. Maki** [2]

1   Global Nutrition R & D, Ingredion Incorporated, 10 Finderne Ave, Bridgewater, NJ 08807, USA
2   Midwest Biomedical Research, Center for Metabolic and Cardiovascular Health, 489 Taft Ave Suite 202, Glen Ellyn, IL 60137, USA; mwilcox@mbclinicalresearch.com (M.L.W.); mbell@mbclinicalresearch.com (M.B.); mbuggia@mbclinicalresearch.com (M.A.B.); kmaki@mbclinicalresearch.com (K.C.M.)
*   Correspondence: maria.stewart@ingredion.com; Tel.: +1-908-685-5470

Received: 28 December 2017; Accepted: 22 January 2018; Published: 26 January 2018

Abstract: Resistant starch (RS) is a type of dietary fiber that has been acknowledged for multiple physiological benefits. Resistant starch type 4 (RS4) is a subcategory of RS that has been more intensively studied as new types of RS4 emerge in the food supply. The primary aim of this randomized, double-blind, controlled study was to characterize the postprandial glucose response in healthy adults after consuming a high fiber scone containing a novel RS4 or a low fiber control scone without RS4. Secondary aims included assessment of postprandial insulin response, postprandial satiety, and gastrointestinal tolerance. The fiber scone significantly reduced postprandial glucose and insulin incremental areas under the curves (43–45% reduction, 35–40% reduction, respectively) and postprandial glucose and insulin maximum concentrations (8–10% and 22% reduction, respectively). The fiber scone significantly reduced hunger and desire to eat during the 180 min following consumption and yielded no gastrointestinal side effects compared with the control scone. The results from this study demonstrate that a ready-to-eat baked-good, such as a scone, can be formulated with RS4 replacing refined wheat flour to yield statistically significant and clinically meaningful reductions in blood glucose and insulin excursions. This is the first study to report increased satiety after short-term RS4 intake, which warrants further investigation in long-term feeding studies.

Keywords: resistant starch type 4; dietary fiber; post-prandial; blood glucose; insulin; capillary glucose; venous glucose; glycemic response; satiety; gastrointestinal tolerance

1. Introduction

Dietary fiber encompasses a wide range of non-digestible carbohydrates with multiple physiological benefits, and it is noted as a short-fall nutrient in Western diets [1,2]. One such physiological benefit is improved blood glucose control. Postprandial blood glucose management has been well-documented among viscous fibers, such as oat beta-glucan, due to attenuated glucose absorption in the small intestine [3]. Decreased postprandial blood glucose is also observed when fibers, such as resistant starch (RS), replace available carbohydrate in food formulations [4]. Postprandial blood glucose control has long been recognized as a predictor of diabetes development. More recently, poor postprandial blood glucose control correlated with the presence of coronary heart disease [5], thus demonstrating the value of improved postprandial blood glucose control.

As noted previously, RS can reduce postprandial blood glucose, particularly when replacing refined wheat flour in product formulations [4,6]. The majority of clinical research on RS evaluated the effects of resistant starch type-2, which is a granular, native starch, resistant to digestion, and resistant starch type-3, which is a retrograded starch that resists digestion. Fewer clinical studies have been conducted on resistant starch type-4 (RS4, chemically modified starch that resists digestion by intestinal enzymes). The category of RS4 is diverse, with a range of starch bases and chemical modifications existing in the food supply. These attributes can affect functionality in a food product, and digestibility and fermentability after consumption [7]. Due to these differences, it is critical to evaluate food applications and physiological effects of each, specific type of RS4. Within the category of RS4, the most studied type is phosphated distarch phosphate [8–13]. Fewer studies have been conducted on distarch phosphate [14–16], hydroxypropyl distarch phosphate [17,18] and only one study has been conducted on RS4 that is acid hydrolyzed and heat treated, to date [19].

Among the aforementioned studies, the most frequent, primary outcome was postprandial blood glucose. Additional outcomes of the acute studies included satiety, energy expenditure, and substrate utilization. Long-term studies of RS4 examined its effects on blood glucose, blood lipids, and gut microbiota. The primary aim of the present study was to characterize the postprandial blood glucose response in healthy adults to a novel RS4 (acid hydrolyzed and heat treated maize-based RS) in a ready-to-eat baked good. Secondary aims were to evaluate postprandial insulin response, satiety, and gastrointestinal tolerance. We hypothesized that replacement of digestible carbohydrate from refined wheat flour with RS4 would reduce postprandial blood glucose.

2. Materials and Methods

2.1. Study Design and Study Visits

This double-blind, randomized, controlled study was conducted in accordance with the Declaration of Helsinki (2000). The protocol was reviewed and received ethical approval before initiation of the trial (Aspire IRB, Santee, CA, USA; protocol number MB-1702). All subjects provided informed, written consent before any study procedures were conducted. The subjects attended 3 study visits (visit 1, day 7, screening; visit 2 day 0, treatment 1; visit 3, day 7, treatment 2). Female subjects of child bearing ability attended study visits during the follicular phase of their menstrual cycles. On visit 2, subjects were randomly assigned to one of two treatment sequences. Prior to visit 2, subjects completed a 24-h diet record. Subjects arrived to the study center fasted on the morning of visits 2 and 3. Baseline satiety was assessed 30 min prior to study product consumption, and baseline blood samples were taken 15 min prior to study product consumption. Subjects consumed the study scone and 240 mL of water within 10 min. Subjects were offered an additional 178 mL of water after they consumed the study scone. Blood measurements and satiety assessment were conducted during the 180 min following the onset of scone consumption. Prior to visit 3, the subjects were instructed to replicate their diet, based on the 24-h diet record from visit 2. Further details are provided in Section 2.4.

2.2. Study Subjects

Subjects were recruited from the Boca Raton area in Florida, USA. Subjects were enrolled in the study after meeting inclusion (age 18–74 years, body mass index 18.5–29.99 kg/m^2, general good health, women of child bearing potential had to be willing to commit to a medically approved form of contraception for the duration of the study) and exclusion criteria (fasting capillary glucose ≥ 5.55 mmol/L, major trauma or surgical event within 3 months of screening, current or recent history of drug or alcohol abuse, ≥ 4.5 kg weight change in past 2 months prior to screening, uncontrolled hypertension, recent use of antibiotics or signs or symptoms of active infection, extreme dietary habits, recent consumption of foods fortified with and/or containing supplements containing probiotics or who had used medications known to influence carbohydrate metabolism, gastrointestinal motility, satiety, appetite, taste, sense of smell, or weight, women only who are currently pregnant or lactating). Subjects agreed to certain study restrictions such as having no plans to change smoking habits or other nicotine use during the study, abstaining from use of tobacco products during study visits, willingness to maintain body weight and habitual diet throughout trial, attempting to replicate the pre-visit 2 diet during the 24 h pre-testing period before visit 3, abstaining from alcohol consumption, and abstaining from vigorous physical activity for 24 h prior and during test visits.

2.3. Study Foods

The fiber scone contained VERSAFIBE™ 2470 resistant starch (Ingredion Incorporated, Bridgewater, NJ, USA), which was the primary source of fiber in the scone. VERSAFIBE™ 2470 resistant starch is a resistant starch type 4 with 70% dietary fiber (AOAC 2009.01). VERSAFIBE™ 2470 resistant starch is produced from food grade high-amylose maize starch. Digestibility of the high-amylose maize starch is decreased through acid hydrolysis and heat treatment, which results in increased RS4 and total dietary fiber (TDF) in the finished product, VERSAFIBE 2470 resistant starch. Because the ingredient is produced from high-amylose maize starch, the nondigestible carbohydrate (fiber) in VERSAFIBE 2470 resistant starch is RS4. There are no nonstarch polysaccharides in the ingredient.

The fiber scone and control scone were matched for weight, fat, protein, sugar and total carbohydrate (Table 1). The portion size was selected based on typical scones in the United States marketplace. Nutrient composition of the scones was analyzed by Medallion Labs (Minneapolis, MN, USA) using standard methods of analysis (fats AOAC 996.06, protein AOAC 992.15, sugar AOAC 977.20, and dietary fiber AOAC 2009.01). Total carbohydrate was determined through calculation by difference. Energy content was calculated using Atwater factors (9 kcal/g fat, 4 kcal/d protein, 4 kcal/g available carbohydrate). Control scone and fiber scone formulations are available in Supplementary Table S1. The fiber and control scones were identical in appearance. The scones were packaged in identical opaque envelopes with a numeric code for identification. Neither the study subjects nor the investigators knew the identity of the scones. The subjects rated the scones on palatability after consumption.

Table 1. Nutrient composition of study scones.

Per Serving, As-Eaten	Control Scone	Fiber Scone
Weight (g)	83.9	84.1
Calories (kcal)	328	270
Fat (g)	16.0	14.4
Saturated fat (g)	5.0	4.7
Protein (g)	7.1	6.1
Total Carbohydrates (g)	42.8	46.4
Available Carbohydrates (g)	38.8	28.9
Dietary Fiber (g) *	4.0	17.5
Sugars (g)	14.8	14.9

* VERSAFIBE™ 2470 resistant starch provided 16.5 g dietary fiber in the Fiber Scone.

2.4. Measurements

2.4.1. Capillary Glucose

The TRUEtrack™ Blood Glucose Monitoring System and the TRUE METRIX® Blood Glucose Meter (Trivida Health Inc., Fort Lauderdale, FL, USA) were used for determination of capillary glucose at screening (visit 1, day 7) and at test visits 2 and 3 (days 0 and 7). At screening, a single fasting glucose test was performed, and at test visits, the capillary glucose was measured at $t = -15 \pm 2$ min, 15, 30, 45, 60, 90, 120 and 180 min ± 2 min, where $t = 0$ was the start of study product consumption.

The test-retest % CV for tests on different days was similar for the fasting capillary (9.0%) and venous (10.4%) samples.

2.4.2. Venous Glucose and Insulin

Venous samples were drawn from an indwelling catheter that was placed at least 10 min prior to the first sample collected. The catheter was flushed regularly with normal saline to maintain patency. At visits 2 and 3 (days 0 and 7), venous samples for measurement of plasma glucose and insulin concentrations were collected at $t = -15$ min $\pm$ 2 min and at $t = 15, 30, 45, 60, 90, 120$ and 180 min $\pm$ 2 min, where $t = 0$ min was the start of study product consumption. The Cleveland Heart Lab (Cleveland, OH, USA) conducted the plasma glucose and insulin analyses. Glucose was measured using a hexokinase/glucose-6-phosphate dehydrogenase enzymatic assay on an automated assay instrument (Roche Cobas Mira Plus Chemistry System, Roche Diagnostic Systems, Indianapolis, IN, USA), and insulin was measured with an electrochemiluminescence immunoassay (Cobas e 411 Immunoassay Analyzer, Roche Diagnostics, Indianapolis, IN, USA) [20].

2.4.3. Satiety Visual Analog Scales (VAS)

At visits 2 and 3 (days 0 and 7), satiety VAS ratings were assessed for fullness, hunger, desire to eat, and prospective consumption at $t = -30, 30, 60, 90, 120, 150,$ and 180 min $\pm$ 2 min, where $t = 0$ min was the start of study product consumption. The questions were "How hungry do you feel?" with anchors at 0 and 100 mm of "Not hungry at all" and "As hungry as I've ever felt"; "How full do you feel?" with anchors of "Not full at all" and "As full as I've ever felt"; "How strong is your desire to eat?" with anchors of "Not at all strong" and "As strong as I've ever felt"; "How much food do you think you can eat?" with anchors of "Nothing at all" and "A large amount."

2.4.4. Gastrointestinal (GI) Tolerability Questionnaire

A GI Tolerability Questionnaire [21] was administered at visits 2 and 3 (days 0 and 7) to assess the presence and severity of selected GI symptoms including nausea, GI rumblings, abdominal pain, bloating, flatulence, and diarrhea during the 0–180 min time period, where $t = 0$ min was the start of study product consumption. GI symptoms were scored as follows: 0 = none, 1 = no more than usual, 2 = somewhat more than usual, and 3 = much more than usual.

2.4.5. Palatability

At the end of visits 2 and 3 (days 0 and 7), subjects completed a study product palatability questionnaire [22,23] in which they rated the study products on a scale from 1 (dislike extremely) to 10 (like extremely) on appearance, texture, flavor, and overall acceptance of the scones; on a scale from 1 (not at all healthy) to 10 (extremely healthy) on how healthy they thought the scone was; and on a scale from 1 (very little) to 10 (very nutritious) on how nutritious they thought the scone was. Subjects also responded to statements of "I recommend the scone to family and friends" and "I tolerated the muffin well, with no complaints" by marking rating of strongly disagree (-2), disagree (-1), no opinion (0), agree (1), or strongly agree (2).

2.5. Data Analysis

2.5.1. Outcome Variables

The primary outcome variable was the difference between treatment conditions in the incremental area under the curve (iAUC) for venous glucose from 0 to 120 min. The secondary outcome variables were the differences between treatments in the following parameters: venous glucose iAUC from 0 to 180 min; capillary glucose iAUC from 0 to 120 min and from 0 to 180 min; insulin iAUC from 0 to 120 min and from 0 to 180 min; maximal concentrations (Cmax) for glucose (venous and capillary) and insulin (venous); and the hunger, fullness, desire to eat and prospective consumption net incremental area under the curve (niAUC) from 0 to 120 min. For the iAUC calculation of the venous and capillary glucose and venous insulin measures, the pre-consumption measurement at $t = -15$ min was counted as $t = 0$. For the niAUC calculation of the satiety VAS measures, the pre-consumption measurement at $t = -30$ min was counted as $t = 0$. Areas under the curve were calculated using the trapezoidal rule [24].

2.5.2. Sample Size

An evaluable sample of 29 subjects was expected to provide 80% power to detect a difference of 10% in the primary outcome variable of venous glucose iAUC$_{0-120 \text{ min}}$ between treatment conditions. This calculation was based on a paired t-test with an expected standard deviation of 18.5% based on previous studies conducted by the investigators. No adjustment to the alpha level was planned for comparisons between treatments for secondary outcome variables. A total of 36 subjects were randomized allowing for anticipated subject attrition.

2.5.3. Statistical Analyses

All tests of significance were assessed at alpha = 0.05, 2-sided. Statistical analyses were conducted using SAS for Windows (version 9.3, Cary, NC, USA). Missing data were not imputed and only observed data were included in the statistical models. Descriptive statistics (number of subjects, mean, standard error of the mean (SEM), standard deviation, median, interquartile limits, minimum and maximum for continuous variables; counts and percentages for categorical variables) were calculated by study product. Baseline comparability of sequence groups was assessed by analysis of variance (ANOVA) with treatment sequence as a fixed effect for continuous variables. For categorical variables, baseline comparability was assessed by chi-square test.

For analyses of the continuous outcome variables and continuous palatability scores, differences in responses between study products were assessed using SAS Proc Mixed repeated measures ANOVA. Initial repeated measures ANOVA models contained terms for treatment, sequence, and period with subject as a random effect. Models were reduced in a stepwise manner until only significant ($p < 0.05$) terms or treatment remained in the model. Assumptions of normality of residuals were investigated for each response measurement. In cases when the normality assumption was rejected at the 1% level with the Shapiro–Wilk test, an analysis using ranks was performed. Differences between study products in the frequency of scores of 2 or greater (somewhat more or much more than usual) on the GI tolerability questionnaire were assessed using McNemar's test.

Sensitivity analyses were performed to assess evidence of any condition by sequence interaction. Because no evidence was present for heterogeneity of response by sequence, the data from the two treatment sequences were pooled for analyses by study product.

3. Results

3.1. Study Subjects

Forty-three persons were screened for the study, but of those, six did not meet the inclusion/exclusion criteria and one met the screening criteria but was later screen failed because of scheduling conflicts with visits 2 and 3. A total of 36 subjects were randomized to a treatment sequence. One subject was terminated from the study before its completion because of inadequate compliance with study instructions. There were 32 subjects included in the glucose and insulin analyses because of missing samples from at least one of the test visits; 35 subjects were included in the satiety VAS, GI tolerability, and palatability analyses. Subject flow through the study is shown in Figure 1.

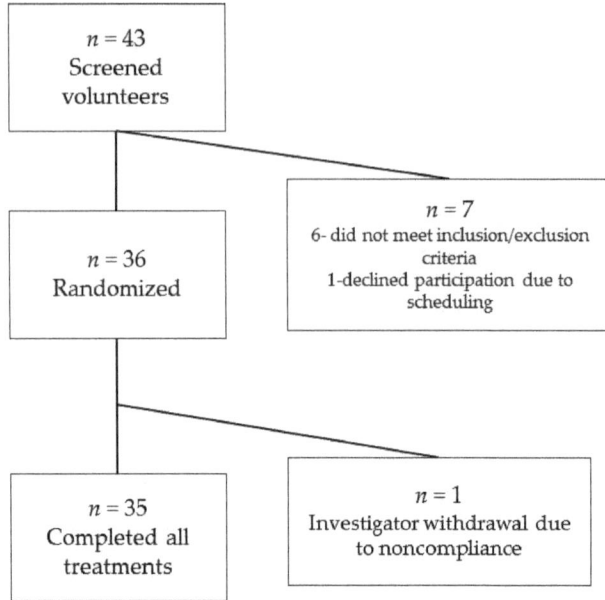

Figure 1. Subject flow through study.

Study subject demographics are reported in Table 2. The subjects were primarily female, with a mean age of 46.2 years and a mean body mass index of 26.1 kg/m². Each subject was given a copy of his or her diet record for the 24 h prior to test 1 and asked to replicate intake for the same period prior to test 2. The prior 24-h diet record returned on the day of test 2 was reviewed and any material deviations were noted. None of the subjects had deviations sufficiently large to warrant exclusion from the per protocol analysis in the judgment of the study director.

Table 2. Subject demographics.

Variable	All Participants ($n = 35$) [1]
Age (y)	46.2 ± 2.2
Sex (male/female)	12/23
Race (white/nonwhite)	22/13
Ethnicity (non-Hispanic/Hispanic)	26/9
Weight (kg)	73.9 ± 2.1
Body mass index (kg/m²)	26.1 ± 0.5
Fasting capillary glucose (mmol/L)	5.04 ± 0.09

[1] Values for continuous variables are mean $\pm$ standard error of the mean; values for categorical values are frequencies.

3.2. Blood Glucose and Insulin

The time-course graph of venous and capillary glucose and venous insulin concentrations is presented in Figure 2, and iAUC and Cmax are shown in Table 3. Venous and capillary glucose iAUC$_{0-120\,min}$ and iAUC$_{0-180\,min}$ mean values were all 43–45% lower after consumption of the fiber scone compared to control scone ($p < 0.05$). Venous insulin iAUC$_{0-120\,min}$ and iAUC$_{0-180\,min}$ mean values were 35% and 40% lower, respectively, after consumption of the fiber scone compared to control scone ($p < 0.05$). Mean venous and capillary glucose and venous insulin Cmax levels were also significantly lower after consumption of the fiber scone compared to control scone (8%, 10% and 22% lower, respectively; $p < 0.05$).

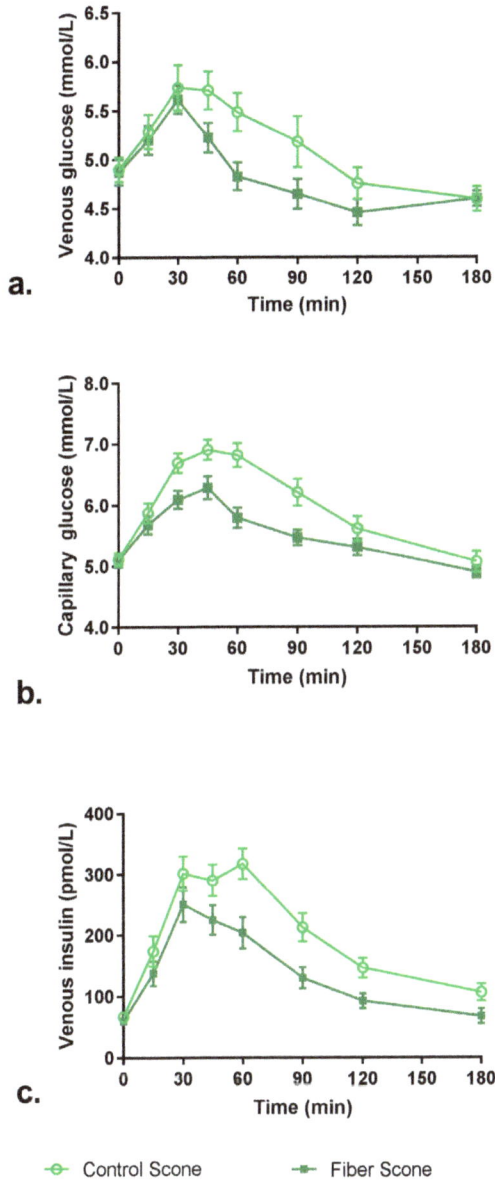

Figure 2. Mean post-prandial glucose and insulin concentrations over 180 min: (**a**) venous glucose; (**b**) capillary glucose (**c**) venous insulin. Error bars represent the standard error of the mean.

Table 3. Post-prandial glucose and insulin iAUC and Cmax ($n = 32$) [1,2].

Parameter	Control Scone	Fiber Scone	*p*-Value [3]
Venous blood glucose			
iAUC$_{0-120 \text{ min}}$ (min × mmol/L)	69.65 ± 9.05	38.41 ± 5.77	0.0014
iAUC$_{0-180 \text{ min}}$ (min × mmol/L) [4]	84.75 ± 11.43	48.29 ± 9.55	0.0039
Cmax (mmol/L) [4]	6.44 ± 0.18	5.88 ± 0.12	0.0039
Capillary blood glucose			
iAUC$_{0-120 \text{ min}}$ (min × mmol/L) [4]	139.14 ± 14.60	79.75 ± 8.10	0.0004
iAUC$_{0-180 \text{ min}}$ (min × mmol/L) [4]	171.88 ± 19.65	94.63 ± 10.66	0.0003
Cmax (mmol/L) [4]	7.49 ± 0.19	6.72 ± 0.16	0.0002
Venous insulin			
iAUC$_{0-120 \text{ min}}$ (min × pmol/L)	19,229 ± 1865	12,592 ± 1686	0.0005
iAUC$_{0-180 \text{ min}}$ (min × pmol/L)	23,850 ± 2138	14,192 ± 1901	<0.0001
Cmax (pmol/L) [4]	392 ± 28	305 ± 26	0.0008

[1] iAUC = incremental area under the curve, Cmax = maximum concentration. [2] Values are mean ± standard error of the mean. [3] *p*-values derived from repeated measure analysis of variance (ANOVA) with subject included as a random effect. [4] Values were ranked prior to ANOVA.

3.3. Satiety VAS Scores

Results for niAUC satiety VAS scores are shown in Table 4. Hunger and desire to eat niAUC$_{0-180 \text{ min}}$ mean values were significantly lower after consumption of the fiber scone compared to control scone ($p < 0.05$). Fullness and prospective consumption niAUC$_{0-180 \text{ min}}$ mean values did not differ significantly between study products.

Table 4. Satiety visual analog scores ($n = 35$) [1,2].

Parameter	Control Scone	Fiber Scone	*p*-Value [3]
Hunger niAUC$_{0-180 \text{ min}}$ (mm × min)	−1173 ± 896	−2840 ± 780	0.0316
Fullness niAUC$_{0-180 \text{ min}}$ (mm × min)	3926 ± 753	4028 ± 898	0.8858
Desire to eat niAUC$_{0-180 \text{ min}}$ (mm × min)	−1184 ± 918	−3046 ± 773	0.0135
Prospective consumption niAUC$_{0-180 \text{ min}}$ (mm × min)	−2098 ± 799	−2847 ± 672	0.1619

[1] mm = millimeters, min = minutes, niAUC = net incremental area under the curve; [2] Values are mean ± standard error of the mean; [3] *p*-values derived from repeated measures analysis of variance with subject included as a random effect.

3.4. GI Tolerability Questionnaire

Results for the analyses of the frequency of scores of 2 or greater (somewhat more than usual and much more than usual) on the components of the GI tolerability questionnaire are shown in Table 5. There were no significant differences between study products in the number of subjects with ratings of somewhat more than usual or much more than usual for nausea, bloating, GI rumblings, flatulence, abdominal pain, or diarrhea.

Table 5. Gastrointestinal Tolerability: Frequency of scores ≥ 2 ($n = 35$) [1,2,3].

Parameter	Control Scone	Fiber Scone	*p*-Value [4]
Nausea	0 (0) [3]	0 (0)	1.000
Bloating	1 (2.9)	0 (0)	1.000
GI Rumblings [4]	4 (11.4)	1 (2.9)	0.2482
Flatulence	1 (2.9)	0 (0.0)	1.000
Abdominal Pain	0 (0.0)	0 (0.0)	1.000
Diarrhea	0 (0.0)	0 (0.0)	1.000

[1] GI = gastrointestinal; [2] Scoring system: 0 = none, 1 = no more than usual, 2 = somewhat more than usual, 3 = much more than usual; [3] Values are n (%); [4] *p*-value generated from McNemar's test for frequencies of values ≥ 2.

3.5. Study Product Palatability

Results for the analyses of study product palatability are shown in Table 6. There was no significant difference between study products for any of the characteristics evaluated on the palatability questionnaire including appearance, texture, flavor, acceptance, healthiness or nutritiousness of the fiber scone and control scones. There were also no significant differences between study products for subjects' agreement with the statements "I recommend the scone to family and friends" and "I tolerated the scone well, with no complaints".

Table 6. Palatability Evaluation (*n* = 35) [5].

Parameter	Control Scone	Fiber Scone	*p*-Value [6]
Appearance [1]	7.74 ± 0.32	7.63 ± 0.32	0.4875
Texture [1]	6.97 ± 0.34	6.40 ± 0.36	0.0743
Flavor [1]	7.11 ± 0.04	6.69 ± 0.43	0.1185 [7]
Acceptance [1]	7.17 ± 0.36	6.77 ± 0.43	0.2271 [7]
Healthy [2]	5.91 ± 0.46	5.71 ± 0.44	0.4205
Nutritious [3]	5.57 ± 0.44	5.23 ± 0.44	0.2092 [7]
Recommend [4]	0.17 ± 0.20	0.09 ± 0.21	0.5712
Tolerated [4]	0.89 ± 0.20	0.97 ± 0.16	0.8601 [7]

[1] Scoring system: 1 (dislike extremely) to 10 (like extremely) on the appearance, texture, flavor, and overall acceptance of the scones; [2] Scoring system: 1 (not at all healthy) to 10 (extremely healthy); [3] Scoring system: 1 (very little) to 10 (very nutritious). [4] "I recommend the scone to family and friends" and "I tolerated the scone well, with no complaints" by marking rating of −2 (strongly disagree), −1 (disagree), 0 (no opinion), +1 (agree), or +2 (strongly agree); [5] Values are mean ± standard error of the mean; [6] *p*-value derived from repeated measures analysis of variance (ANOVA) that included subject as a random effect; [7] Individual values were ranked prior to ANOVA.

4. Discussion

The present study demonstrates that replacing a portion of refined wheat flour in a food with RS4 can reduce postprandial glucose and insulin concentrations and increase selected measures of satiety. The blood glucose and insulin findings align with results from a previous study administering the same RS4 in a muffin top to healthy adults [19]. In the present study, the percent reductions in $iAUC_{0-120\,min}$ glucose (venous and capillary) and $iAUC_{0-120\,min}$ insulin were greater than in the previously published study, which may be attributed to the higher dietary fiber content of the RS4-containing scone in the present study. Both study foods replaced a portion of refined wheat flour with VERSAFIBE 2470 resistant starch. This replacement resulted in an increase in TDF content and a decrease in available carbohydrate content which contributed to the reduced glucose and insulin responses. The absolute blood glucose values typically differ when measured intravenously or through capillary sampling, with capillary sampling typically displaying lower variation [25]. It is expected that the venous values will be a bit lower than the capillary values because capillary blood will have a value intermediate between that of venous and arterial blood. The present data from individual time points align with this previous observation.

The intended application of the novel RS4 in the present study is to replace digestible starch (e.g., wheat flour) in food products. Accordingly, the study was designed to provide data on the glycemic and insulinemic responses to food products prepared with non-resistant starch (control) or RS4 resistant starch. Therefore, the trial was designed in a manner that matched total carbohydrate content and portion size rather than available carbohydrate content. This deliberate choice in study design provides information relevant to the intended application of the ingredient studied (RS4). Glycemic response, which was evaluated in the present study, is often confused with glycemic index, which requires that the control food and test food be matched for 50 g available carbohydrate. These two concepts have been discussed in detail elsewhere [26].

Other types of RS4 have also been effective in lowering blood glucose compared to a control treatment. Four acute studies of RS4 containing distarch phosphate or phosphated distarch phosphate demonstrated reduction in postprandial blood glucose when administered in a beverage [8,14], a bar [9], and a cookie [16]. An acute study of maize hydroxypropyl distarch phosphate RS4 (38 g) in pancakes administered to healthy adults reported a significant reduction in postprandial glucose and insulin compared to control (0–180 min) [17]. A similarly designed study using the same RS4 (40 g) in pancakes reported no difference in blood glucose or insulin values over 180 min [18]. It should be noted that the RS4-containing pancake in both studies provided 0 g fiber. The inconsistency in blood glucose response to different RS4s emphasizes the need for evaluating health effects from consuming specific types of RS4, especially as new RS4s are developed.

Three studies examined the effect of a different type of RS4 on acute satiety response [11,14,18], and each study examined a different type of RS4. Satiety ratings in response to wheat phosphated distarch phosphate in a bar (10 g) [11] and potato distarch phosphate in a beverage (30 g) [14], and maize hydroxypropyl distarch phosphate in pancakes (40 g) [18], did not differ from those in control treatments. In contrast, the present study demonstrated that acid-hydrolyzed and heat-treated RS4 reduced hunger and desire to eat during the 180 min following consumption. The mechanism of action for this short-term effect is unclear and warrants further research. Additionally, long-term studies of RS4 on energy intake and weight management are needed to understand the full range of responses to consumption of this fiber.

Effects of long-term intake of wheat-based RS4 (phosphated distarch phosphate) were examined in two studies and published in three articles [10,12,13]. Martinez et al. (2010) compared the effects of high amylose maize resistant starch (RS2) and wheat phosphated distarch phosphate (RS4) on the gut microbiota [10]. While both types of RS impacted the gut microbiota, their effects were distinct, which reinforces the hypothesis that different types of RS may yield different health effects. A study in the Hutterite community, which has a high prevalence of metabolic syndrome, demonstrated that inclusion of RS4 did not affect measures of glycemic health, but improved blood lipids, waist circumference, and body composition [12]. The amount of RS4 or fiber consumed was not reported, which limits the ability to draw conclusions on long-term intake. An analysis of a subsample from the aforementioned study reported changes in gut microbiota composition and increased fecal short-chain fatty acid concentrations, suggesting that the gut microbiota may be a mediator for the changes in metabolic biomarkers [13]. This paper reported mean nutrient intake at baseline and after the two intervention periods. However, this was a subsample of the previous study and it is unclear whether the dietary data applies to the larger group in the previously published paper.

The glycemic response findings in the present study are supported by results from many previous studies of RS4. This study demonstrates that a reduction in post-prandial blood glucose and insulin can be achieved using an RS4 that is created through a novel chemical modification (acid hydrolyzed and heat-treated, instead of phosphate cross-linked). This is the first study to report significantly increased satiety (hunger, desire to eat) after consuming RS4. There are a few limitations of this study: (1) the study was conducted in healthy individuals, so we cannot draw conclusions related to chronic disease risk or management of existing conditions such as diabetes mellitus; and (2) the study foods (low fiber scone and high fiber scone) may not be culturally relevant to all countries and, thus, additional research is needed to evaluate glycemic and satiety response to culturally appropriate foods. Future research should examine the long-term effects of different RS4s to better understand the range of health effects from these ingredients.

5. Conclusions

In conclusion, this study demonstrated that a baked good, fortified with RS4 (acid hydrolyzed and heat treated maize starch) that replaced available carbohydrate from wheat flour, reduced postprandial plasma glucose response (venous and insulin) and correspondingly reduced plasma insulin. The study foods were administered in a practical portion size, emulating what consumers might find in the marketplace. This is the first study to report changes in two measures of satiety, hunger and desire to eat, after consuming RS4. The high fiber scone was well-tolerated and did not change gastrointestinal symptoms. Resistant starch type 4 has the potential for helping to fill the fiber gap in the Western diets while improving glycemic health.

Supplementary Materials: The following are available online at www.mdpi.com/2072-6643/10/2/129/s1, Table S1: Scone Formulations.

Acknowledgments: This study was funded by Ingredion Incorporated.

Author Contributions: M.L.S. conceived and designed the study and contributed to writing the paper; K.C.M. was the study director and contributed to study design. M.B. completed the statistical analysis. M.A.B. was the principal investigator and supervising physician, she contributed to study design, data collection and data interpretation. M.L.W. contributed to writing the paper. All authors have reviewed and approved the paper.

Conflicts of Interest: M.L.S. is an employee of Ingredion Incorporated.

References

1. Dahl, W.J.; Stewart, M.L. Position of the academy of nutrition and dietetics: Health implications of dietary fiber. *J. Acad. Nutr. Diet.* **2015**, *115*, 1861–1870. [CrossRef] [PubMed]
2. Stephen, A.M.; Champ, M.M.; Cloran, S.J.; Fleith, M.; van Lieshout, L.; Mejborn, H.; Burley, V.J. Dietary fibre in Europe: Current state of knowledge on definitions, sources, recommendations, intakes and relationships to health. *Nutr. Res. Rev.* **2017**, *30*, 1–42. [CrossRef] [PubMed]
3. Tosh, S.M. Review of human studies investigating the post-prandial blood-glucose lowering ability of oat and barley food products. *Eur. J. Clin. Nutr.* **2013**, *67*, 310–317. [CrossRef] [PubMed]
4. Robertson, M.D. Dietary-resistant starch and glucose metabolism. *Curr. Opin. Clin. Nutr. Metab. Care* **2012**, *15*, 362–367. [CrossRef] [PubMed]
5. Jiang, J.; Zhao, L.; Lin, L.; Gui, M.; Aleteng, Q.; Wu, B.; Wang, S.; Pan, B.; Ling, Y.; Gao, X. Postprandial blood glucose outweighs fasting blood glucose and hba1c in screening coronary heart disease. *Sci. Rep.* **2017**, *7*, 14212. [CrossRef] [PubMed]
6. Lockyer, S.; Nugent, A.P. Health effects of resistant starch. *Nutr. Bull.* **2017**, *42*, 10–42. [CrossRef]
7. Maningat, C.C.; Seib, P.A. RS4-type resistant starch: Chemistry, functionality and health benefits. In *Resistant Starch: Sources, Applications and Health Benefits*; Shi, Y.C., Maningat, C.C., Eds.; John Wiley & Sons: Chichester, UK, 2013; pp. 43–77.
8. Haub, M.D.; Hubach, K.L.; Al-Tamimi, E.K.; Ornelas, S.; Seib, P.A. Different types of resistant starch elicit different glucose reponses in humans. *J. Nutr. Metab.* **2010**, *2010*, 230501. [CrossRef] [PubMed]
9. Al-Tamimi, E.K.; Seib, P.A.; Snyder, B.S.; Haub, M.D. Consumption of cross-linked resistant starch ($RS4_{XL}$) on glucose and insulin responses in humans. *J. Nutr. Metab.* **2010**, *2010*, 651063. [CrossRef] [PubMed]
10. Martínez, I.; Kim, J.H.; Duffy, P.R.; Schlegel, V.L.; Walter, J. Resistant starches types 2 and 4 have differential effects on the composition of the fecal microbiota in human subjects. *PLoS ONE* **2010**, *5*, e15046. [CrossRef] [PubMed]
11. Karalus, M.; Clark, M.; Greaves, K.A.; Thomas, W.; Vickers, Z.; Kuyama, M.; Slavin, J. Fermentable fibers do not affect satiety or food intake by women who do not practice restrained eating. *J. Acad. Nutr. Diet.* **2012**, *112*, 1356–1362. [CrossRef] [PubMed]
12. Nichenametla, S.N.; Weidauer, L.A.; Wey, H.E.; Beare, T.M.; Specker, B.L.; Dey, M. Resistant starch type 4-enriched diet lowered blood cholesterols and improved body composition in a double blind controlled cross-over intervention. *Mol. Nutr. Food Res.* **2014**, *58*, 1365–1369. [CrossRef] [PubMed]
13. Upadhyaya, B.; McCormack, L.; Fardin-Kia, A.R.; Juenemann, R.; Nichenametla, S.; Clapper, J.; Specker, B.; Dey, M. Impact of dietary resistant starch type 4 on human gut microbiota and immunometabolic functions. *Sci. Rep.* **2016**, *6*, 28797. [CrossRef] [PubMed]
14. Haub, M.D.; Louk, J.A.; Lopez, T.C. Novel resistant potato starches on glycemia and satiety in humans. *J. Nutr. Metab.* **2012**, *2012*, 478043. [CrossRef] [PubMed]
15. Dahl, W.J.; Ford, A.L.; Ukhanova, M.; Radford, A.; Christman, M.C.; Waugh, S.; Mai, V. Resistant potato starches (type 4 RS) exhibit varying effects on laxation with and without phylum level changes in microbiota: A randomised trial in young adults. *J. Funct. Foods* **2016**, *23*, 1–11. [CrossRef]
16. Stewart, M.L.; Zimmer, J.P. A high fiber cookie made with resistant starch type 4 reduces post-prandial glucose and insulin responses in healthy adults. *Nutrients* **2017**, *9*, 237. [CrossRef] [PubMed]

17. Shimotoyodome, A.; Suzuki, J.; Kameo, Y.; Hase, T. Dietary supplementation with hydroxypropyl-distarch phosphate from waxy maize starch increases resting energy expenditure by lowering the postprandial glucose-dependent insulinotropic polypeptide response in human subjects. *Br. J. Nutr.* **2011**, *106*, 96–104. [CrossRef] [PubMed]

18. Gentile, C.L.; Ward, E.; Holst, J.J.; Astrup, A.; Ormsbee, M.J.; Connelly, S.; Arciero, P.J. Resistant starch and protein intake enhances fat oxidation and feelings of fullness in lean and overweight/obese women. *Nutr. J.* **2015**, *14*, 113. [CrossRef] [PubMed]

19. Stewart, M.L.; Zimmer, J.P. Post-prandial glucose and insulin response to high-fiber muffin top containing resistant starch type 4 in healthy adults: A double-blind, randomized, controlled trial. *Nutrition* **2018**, in press.

20. Cassidy, J.P.; Luzio, S.D.; Marino, M.T.; Baughman, R.A. Quantification of human serum insulin concentrations in clinical pharmacokinetic or bioequivalence studies: What defines the "best method"? *Clin. Chem. Lab. Med.* **2012**, *50*, 663–666. [CrossRef] [PubMed]

21. Maki, K.C.; Rains, T.M.; Kelley, K.M.; Cook, C.M.; Schild, A.L.; Gietl, E. Fibermalt is well tolerated in healthy men and women at intakes up to 60 g/d: A randomized, double-blind, crossover trial. *Int. J. Food Sci. Nutr.* **2013**, *64*, 274–281. [CrossRef] [PubMed]

22. Mialon, V.S.; Clark, M.R.; Leppard, P.I.; Cox, D.N. The effect of dietary fibre information on consumer responses to breads and "English" muffins: A cross-cultural study. *Food Qual. Preference* **2002**, *13*, 1–12. [CrossRef]

23. Baixauli, R.; Salvador, A.; Martínez-Cervera, S.; Fiszman, S.M. Distinctive sensory features introduced by resistant starch in baked products. *LWT Food Sci. Technol.* **2008**, *41*, 1927–1933. [CrossRef]

24. Brouns, F.; Bjorck, I.; Frayn, K.N.; Gibbs, A.L.; Lang, V.; Slama, G.; Wolever, T.M. Glycaemic index methodology. *Nutr. Res. Rev.* **2005**, *18*, 145–171. [CrossRef] [PubMed]

25. Wolever, T.M.; Vorster, H.H.; Bjorck, I.; Brand-Miller, J.; Brighenti, F.; Mann, J.I.; Ramdath, D.D.; Granfeldt, Y.; Holt, S.; Perry, T.L.; et al. Determination of the glycaemic index of foods: Interlaboratory study. *Eur. J. Clin. Nutr.* **2003**, *57*, 475–482. [CrossRef] [PubMed]

26. Augustin, L.S.; Kendall, C.W.; Jenkins, D.J.; Willett, W.C.; Astrup, A.; Barclay, A.W.; Bjorck, I.; Brand-Miller, J.C.; Brighenti, F.; Buyken, A.E.; et al. Glycemic index, glycemic load and glycemic response: An International Scientific Consensus Summit from the International Carbohydrate Quality Consortium (ICQC). *Nutr. Metab. Cardiovasc. Dis.* **2015**, *25*, 795–815. [CrossRef] [PubMed]

nutrients

MDPI

Article

The Influence of Pre-Exercise Glucose versus Fructose Ingestion on Subsequent Postprandial Lipemia

Tsung-Jen Yang [1], Chih-Hui Chiu [2], Mei-Hui Tseng [3], Cheng-Kang Chang [3] and Ching-Lin Wu [4],*

[1] Department of Physical Education, National Taiwan Normal University, Taipei 106, Taiwan;
    andy32437@yahoo.com.tw
[2] Graduate Program in Department of Exercise Health Science, National Taiwan University of Sport,
    Taichung 404, Taiwan; loveshalom@hotmail.com
[3] Sport Science Research Center, National Taiwan University of Sport, Taichung 404, Taiwan;
    tsengmh2009@gmail.com (M.-H.T.); wspahn@seed.net.tw (C.-K.C.)
[4] Graduate Institute of Sports and Health Management, National Chung Hsing University,
    Taichung 402, Taiwan
*   Correspondence: psclw@dragon.nchu.edu.com; Tel.: +886-4-22840845 (ext. 602)

Received: 29 November 2017; Accepted: 25 January 2018; Published: 29 January 2018

Abstract: Ingestion of low glycemic index (LGI) carbohydrate (CHO) before exercise induced less insulin response and higher fat oxidation than that of high GI (HGI) CHO during subsequent exercise. However, the effect on the subsequent postprandial lipid profile is still unclear. Therefore, the aim of this study was to investigate ingestion of CHO drinks with different GI using fructose and glucose before endurance exercise on the subsequent postprandial lipid profile. Eight healthy active males completed two experimental trials in randomized double-blind cross-over design. All participants ingested 500 mL CHO (75 g) solution either fructose (F) or glucose (G) before running on the treadmill at 60% VO_2max for 1 h. Participants were asked to take an oral fat tolerance test (OFTT) immediately after the exercise. Blood samples were obtained for plasma and serum analysis. The F trial was significantly lower than the G trial in TG total area under the curve (AUC; 9.97 ± 3.64 vs. 10.91 ± 3.56 mmol × 6 h/L; $p = 0.033$) and incremental AUC (6.57 ± 2.46 vs. 7.14 ± 2.64 mmol/L × 6 h, $p = 0.004$). The current data suggested that a pre-exercise fructose drink showed a lower postprandial lipemia than a glucose drink after the subsequent high-fat meal.

Keywords: glycemic index; triacylglycerol; high-density lipoprotein; fat oxidation; oral fat tolerance test

1. Introduction

An increase in postprandial plasma triacylglycerol (TG) concentrations was suggested to cause damage on vascular subcutaneous cells and vascular walls [1]. The postprandial lipemia phenomena may last for 6 to 8 h [2], suggesting that a high postprandial TG concentration in the circulation is likely lasting an entire day after breakfast is ingested. A number of studies indicated that an increase in postprandial TG level correlates positively with the mortality rate and risk of cardiovascular disease (CVD) [3–5]. In order to reduce high postprandial TG concentrations, several studies proposed that endurance exercise was effective for lowering postprandial TG concentrations [6–8].

A high-carbohydrate (CHO) diet increases the storage and use of muscle glycogen, and improves exercise performance [9,10]. However, a high CHO diet might cause a rise in very-low-density lipoprotein (VLDL) concentration [11–13] and a reduction in the level of high-density lipoprotein cholesterol (HDL-C) [14], which were considered to increase the risk of CVD. Previous study showed exercise intervention might elevate the level of HDL-C, increase lipoprotein lipase (LPL) activity [15], increase the transportation of blood lipids into the muscle cells for storage or use, and lower postprandial TG concentrations [11,16]. Katsanos and colleagues [17] showed that following moderate-intensity endurance exercise there was significantly lower insulin concentration and TG area under the curve (AUC) over 6 h after ingestion of a high-fat meal when compared to the no exercise trial. Therefore, insulin action may play one of the key factors in influencing postprandial TG levels [18,19].

A rise in insulin concentration by following CHO ingestion results in an increase in the rate of CHO oxidation, and the rate of fat oxidation inhibited [18]. Previous studies examined ingesting CHO meals with a low glycemic index (LGI) and a high glycemic index (HGI) before exercise on substrate utilization during exercise [20–22]. The results suggested that ingestion of the LGI meal induced lower insulin response and showed a significantly higher rate of fat oxidation than that of the HGI meal during subsequent exercise [20,22]. Ingesting CHO with a distinct GI stimulates insulin response, leading to changes in the fat oxidation rate during exercise, which possibly exerts varying degrees of influence on postprandial lipid metabolism when the body is recovering from the exercise. Kaviani and colleagues [23] reported that ingestion of a post-exercise LGI meal induced lower postprandial TG concentrations when compared to that of the HGI meal. However, how the pre-exercise CHO influenced the subsequent postprandial lipid profile is still unclear.

An extant study verified that exercise intervention effectively lowered the increased level of blood lipids due to CHO ingestion [24,25]. We hypothesize that ingestion of an LGI drink prior the exercise may induce higher fat oxidation than ingestion of an HGI drink during exercise and subsequently induces a higher plasma TG clearance rate during the postprandial period after a high fat meal. To date, no studies have been conducted on postprandial lipid profiles immediately after exercise with ingestion of pre-exercise GI CHO. Therefore, the purpose of the present study was to determine the effect of fructose versus glucose pre-exercise drinks and exercise intervention on the subsequent postprandial lipid profile.

2. Materials and Methods

Eight healthy active males voluntary participated in the present study (age 23.1 ± 0.7 years, weight 68.9 ± 2.0 kg, and maximal oxygen consumption (VO_2max) 47.7 ± 1.6 mL/kg/min). This study was conducted in the Sports Science Research Center of National Taiwan University of Sport, with the approval of the Human Subject Committee of National Taiwan University of Sport (NTCPE-95-01). Participants were given their written informed consent after complete understood the study design and possible risks. All participants completed the health history questionnaire before undertaking the experiments.

2.1. Experimental Design

A randomized double-blind cross-over experimental design was adopted and the trial order was counter-balanced for this study. All of the participants underwent two experimental trials separated at least seven days. The participants were asked to ingest CHO-containing either a fructose or glucose drink 30 min before running on the treadmill for 1 h at 60% VO_2max. Immediately after the exercise the participants were asked to take an oral fat tolerance test (OFTT), which asked participants to ingest a high-fat meal to observe postprandial lipemia for 6 h [26]. The participants were asked to record their diet for three days before the first trial and were required to repeat the same diet before the next trial. A standardized lunch and dinner was served for the participants on the day before the main trial. The participants were asked to avoid any heavy physical activity or exercise three days before the main trials. In addition, they were asked to refrain from smoking and ingesting alcohol- and caffeine-containing beverages before the experiment.

2.2. Preliminary Measurements

Two preliminary tests were conducted: running economy (RE) test and VO_2max measurement.

RE test: The RE protocol was a four-stage test running on a treadmill (Medtrack ST65, Quinton, Seattle, Washington, USA) at an initial speed of 7.0–8.0 km/h and increase at 1.0–1.5 km/h every 4 min. The oxygen uptake (VO_2) was measured 1 min before the end of each stage using a gas analyzer (Vmax Series 29C, Sensor Medics, Yorba Linda, CA, USA). The four VO_2 measurements were substituted in a linear regression equation to calculate the relationship between VO_2 and running speed [22].

VO_2max test: The speed of the treadmill was set at a constant pace. The initial slope of the treadmill was set at 3.5% and increased at 2.5% every 3 min. Participants were encouraged to complete every stage of the exercise until volitional fatigue. The VO_2max criteria were a plateau in VO_2, heart rate coming close to the age-predicted maximal heart rate, and respiratory exchange ratio (RER) ≥ 1.15.

2.3. Test Drink, Oral Fat Tolerance Test, and Lunch and Dinner before Experiment Day

Participants ingested different GI carbohydrate drinks that we provided in the present study: (a) a high-GI drink (GI = 100; HGI); and (b) a low-GI drink (GI = 40; LGI). Carbohydrate source: HGI was 75 g glucose (Wako Pure Chemical Industries, Ltd., Osaka, Japan), or LGI was 75 g fructose (Shimakyu's Pure Chemicals, Osaka, Japan) in 500 mL water.

An oral fat tolerance test meal included white bread, whipping cream, nuts, butter, and cereal. The meal provided as based on the body weight of the participants and contained 1.2 g/kg fat, 1.1 g/kg CHO, 0.33 g/kg protein, and 16.5 kcal/kg [26]. All of the foods were purchased from the same supermarket, and the foods were of the same brand. The calorie value of the foods, as well as CHO, protein, and fat contents, were calculated according to the nutritional label by the manufacturer.

The standardized lunch and dinner before the main trial were described in the previous study [26] and were purchased from the same convenience store. The lunch provided 840.0 ± 57.0 kcal, with 50.7 ± 0.3% energy from carbohydrate (106.5 ± 7.4 g), 31.5 ± 0.5% from fat (29.4 ± 1.8 g), and 17.8 ± 0.5% from protein (37.5 ± 3.2 g). The standard dinner offered 692 kcal, with 50% energy from carbohydrate (86.5 g), 32% from fat (24.6 g), and 18% from protein (31.1 g). The calculated GI value was 68.8.

2.4. Protocol

Participants were given the same lunch and dinner one day before the start of the experiment. They were instructed to report to the laboratory after overnight fasting for 12 h. After their height and weight were measured, they were asked to ingest a fructose (F) or glucose (G) drink (500 mL). Following 30 min of a resting period, the participants ran at 60% VO_2max for 1 h on the treadmill. After the exercise was completed, fasting blood specimens were drawn at 0 h from the antecubital vein by a catheter. Subsequently, the participants were asked to ingest an OFTT meal within 20 min, and their blood specimens were collected at 0.5, 1, 2, 3, 4, 5, and 6 h after the meal. The participants were required to sit quietly in the laboratory to avoid any physical activities during the 6 h postprandial period. The environmental temperature was maintained at 22 °C–25 °C and a humidity of 50–60%.

2.5. Blood Sample Collection and Analysis

A catheter (Venflon 20G, Ohmeda, Sweden) was connected to the three-way stopcock (Connecta Ltd., Helsingborg, Sweden) with a 10-cm long tube for collecting 10 mL blood samples at each time point. The 10-cm long blood tube was regularly washed with sterile sodium chloride solution (0.9% *w/v*) to prevent blood coagulation in the tube. A non-heparinized tube was used to collect 2 mL of blood sample, and it was allowed to stand for 1 h to wait for the blood to coagulate. Another tube containing ethylenediaminetetraacetic acid (EDTA) was used to collect 8 mL of blood sample. The collected sample was then centrifuged (Eppendorf 5810, Hamburg, Germany) at 4 °C at 2000 rpm for 20 min. The extracted serum and plasma samples were stored at −70 °C in a freezer before analysis. Plasma glucose (Shino, Tokyo, Japan), TG (Wako, Osaka, Japan), non-esterified fatty acid (NEFA; Wako, Neuss, Germany), glycerol (Randox, Co., Antrim, UK), and HDL-C (Kyowa, Osaka, Japan) were measured using an automated biochemical analyzer (Hitachi 7020, Tokyo, Japan). Electrochemiluminescence (Elecsys 2010, Roche Diagnostics, Basel, Switzerland) immunoassay was used to analyze the serum insulin concentrations (Roche Diagnostics, Mannheim, Germany). The intra-assay coefficients of the variation of the blood sample measurement were: TG: CV(%) = 4.9; insulin: CV(%) = 2.8; NEFA: CV(%) = 4.51; glycerol: CV(%) = 6.42; glucose: CV(%) = 6.9; and HDL-C: CV(%) = 4.9.

2.6. Statistical Analyses

All collected data was presented as mean $\pm$ SD. Changes in blood samples were analyzed by a two-way ANOVA with repeated measures. The Bonferroni post hoc test for comparison with two groups for each time point was used when the ANOVA showed a significant interaction effect (condition $\times$ time). The blood biochemistry concentrations over the time AUCs were analyzed using a paired T-test. The differences between F and G trials were evaluated by Cohen's effect size (ES). The analysis was performed with SPSS 23.0. A p-value less than 0.05 was considered statistically significant.

3. Results

3.1. Plasma Triacylglycerol

Plasma TG IAUC (Figure 1a) was significantly lower in the F than in the G trial (6.57 $\pm$ 2.46 vs. 7.14 $\pm$ 2.64 mmol/L $\times$ 6 h, $p = 0.004$); plasma TG total AUC (Figure 1b) was significantly lower in the F compared with the G trial (9.97 $\pm$ 3.64 vs. 10.91 $\pm$ 3.56 mmol/L $\times$ 6 h, $p = 0.033$). Plasma TG concentration over 6 h (Figure 1c) showed no significant difference between trials by time interaction (condition $\times$ time, $p = 0.628$; condition, $p = 0.342$; time, $p < 0.001$).

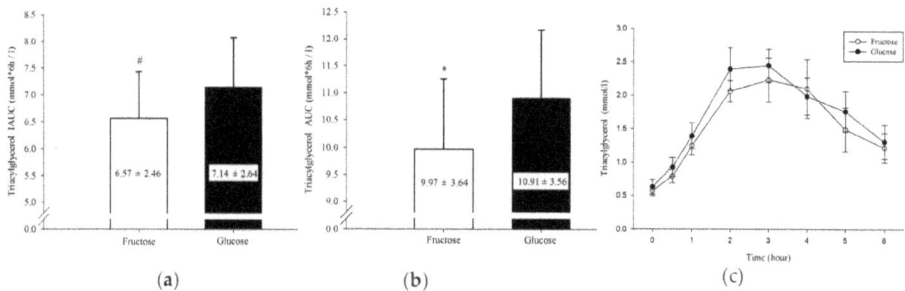

Figure 1. Triacylglycerol (TG) incremental area under the curve (**a**) and TG area under the curve (**b**) in 6 h and postprandial TG concentration over 6 h (**c**). [#] F was significantly lower than G ($p = 0.004$). [*] F was significantly lower than G ($p = 0.033$).

3.2. Serum Insulin and Plasma Glucose

Serum insulin concentration and plasma glucose concentration over the 6 h are displayed in Figure 2. There was no significant difference in serum insulin (condition $\times$ time, $p = 0.557$; condition, $p = 0.893$; Figure 2a). Plasma glucose concentration showed no significant difference between trials (condition $\times$ time, $p = 0.191$; condition, $p = 0.763$; Figure 2b). There were no significant differences between trials in serum insulin AUC and glucose AUC ($p = 0.717$; $p = 0.951$; Table 1).

3.3. Plasma Non-Esterified Fatty Acids (NEFA), Glycerol

Plasma NEFA concentration was significantly higher before OFTT (after exercise) in the F trial than in the G trial (0.39 $\pm$ 0.1 vs. 0.24 $\pm$ 0.06 mmol/L; $p = 0.011$; Table 2). Plasma NEFA concentration over 6 h showed no significant difference between trials (condition $\times$ time, $p = 0.052$; condition, $p = 0.563$; time, $p < 0.001$; Figure 3a). There was no difference between trials in plasma NEFA AUC (F: 3.05 ± 0.45; G: 3.06 ± 0.54, $p = 0.962$; Table 1).

Plasma glycerol concentration was significantly higher before OFTT (after exercise) in the F trial than in the G trial (168.75 ± 38.86 vs. 131.75 ± 45.30 µmol/L; $p = 0.015$; Table 2). Plasma glycerol concentration over 6 h showed no significant difference between trials (condition $\times$ time, $p = 0.141$; condition, $p = 0.064$; time, $p < 0.001$; Figure 3b). There was no difference between trials in plasma glycerol AUC (F: 395.84 ± 69.55; G: 363.19 ± 64.67, $p = 0.192$; Table 1).

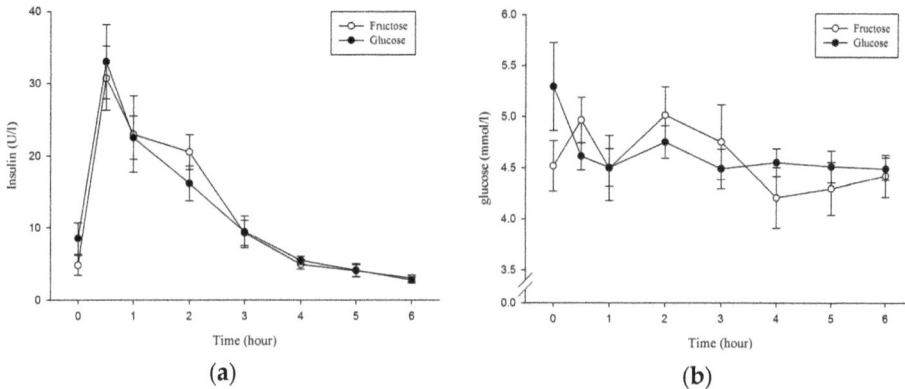

Figure 2. Serum insulin concentrations (**a**) and plasma glucose concentrations (**b**) during the 6 h postprandial period, $p < 0.05$.

Table 1. The plasma and serum sample concentrations area under the curve.

	Fructose	Glucose	p	ES
Insulin (µU/mL × 6 h)	74.06 ± 20.95	71.73 ± 17.88	0.717	0.12
TG (mmol/L × 6 h)	9.97 ± 3.64	10.91 ± 3.56	0.033 *	0.26
TG IAUC (mmol/L × 6 h)	6.57 ± 2.46	7.14 ± 2.64	0.004 *	0.22
Glucose (mmol/L × 6 h)	27.46 ± 3.30	27.56 ± 1.59	0.951	0.04
NEFA (mmol/L × 6 h)	3.05 ± 0.45	3.06 ± 0.54	0.962	0.02
Glycerol (µmol/L × 6 h)	395.84 ± 69.55	363.19 ± 64.67	0.192	0.49
HDL-C (mmol/L × 6 h)	8.02 ± 1.68	7.49 ± 1.52	0.003 *	0.33

* Significant difference between F and G ($p < 0.05$). Values are mean $\pm$ SD. TG: triacylglycerol; IAUC: incremental area under the curve; NEFA: non-esterified fatty acids; HDL-C: high density lipoprotein cholesterol.

Table 2. The plasma and serum sample concentrations before OFTT.

	Fructose	Glucose	p	ES
Insulin (µU/mL)	4.78 ± 3.86	8.50 ± 6.31	0.177	0.71
TG (mmol/L)	0.57 ± 0.21	0.63 ± 0.31	0.580	0.23
Glucose (mmol/L)	4.52 ± 0.69	5.29 ± 1.22	0.101	0.78
NEFA (mmol/L)	0.39 ± 0.10	0.24 ± 0.06	0.011 *	1.82
Glycerol (µmol/L)	168.8 ± 38.86	131.7 ± 45.30	0.015 *	0.88
HDL-C (mmol/L)	1.47 ± 0.31	1.33 ± 0.27	0.184	0.48

* Significant difference between F and G ($p < 0.05$). Values are mean $\pm$ SD. TG: triacylglycerol; NEFA: non-esterified fatty acids; HDL-C: high density lipoprotein cholesterol.

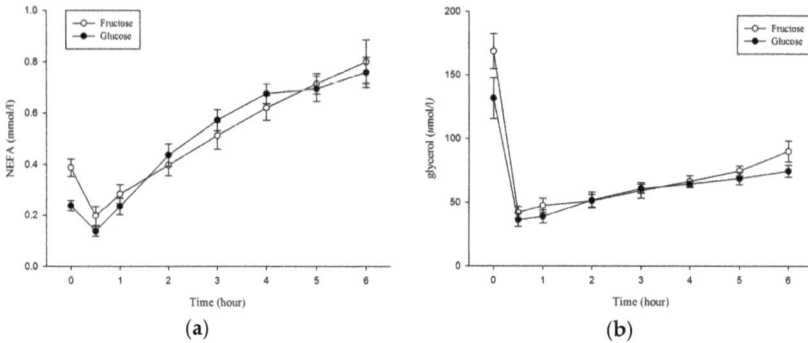

Figure 3. Plasma NEFA concentrations (**a**); and glycerol concentrations (**b**) during the 6 h postprandial period. NEFA: non-esterfied fatty acid.

3.4. Plasma High-Density Lipoprotein Cholesterol

Plasma HDL-C concentrations (Figure 4) showed no significant difference between trials by time interaction (condition × time, $p = 0.336$; condition, $p = 0.118$; time, $p = 0.021$). Plasma HDL-C AUC was significantly higher in the F trial compared to the G trial (8.02 ± 1.68 vs. 7.49 ± 1.52 mmol/L × 6 h, $p = 0.003$; Table 1).

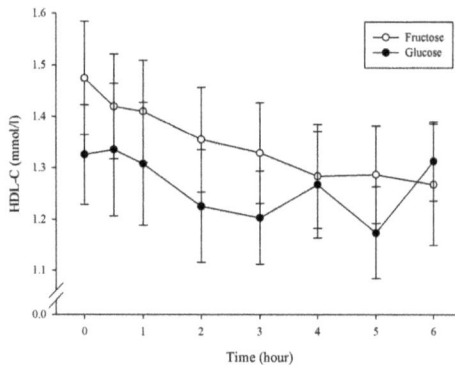

Figure 4. Plasma HDL-C concentrations during the 6 h postprandial period.

4. Discussion

The major finding of this study is that ingestion of fructose before endurance exercise lowered subsequent postprandial plasma TG concentrations compared to that of the glucose drink. Several studies demonstrated that endurance exercise effectively reduced postprandial lipemia [7,8,17,25,27]. To our knowledge, no studies have been conducted on how pre-exercise CHO with different GI and endurance exercises influence the subsequent lipid profile after oral ingestion of a high-fat meal.

A previous study reported that while a lower insulin level was induced by ingesting an LGI meal before exercise, a lower CHO oxidation rate was observed compared with when an HGI meal was ingested during exercise [20,22]. After ingestion of the HGI CHO meal, the rise in insulin level, in turn, decreases the fat oxidation rate during exercise, thereby inhibiting exercise-induced lipid metabolism [22,28]. Although we did not measure the RER to examine the rate of substrate utilization during exercise, the plasma NEFA and glycerol concentrations of the F trial was significantly higher than that of the G trial following 60 min of exercise (Figure 3). The current result is similar to previous studies, which might indicate a higher fat oxidation occurred in the F trial during exercise in the current study [20,22]. Therefore, the current study supported those of previous studies that, after ingesting a CHO diet with different GI and engaging in exercise, the LGI trial showed a significantly higher fat oxidation rate during the exercise than did the HGI trial [20–22]. This result verifies that ingesting an LGI CHO drink before exercise could depress lipid metabolism less during exercise compared with the HGI CHO drink. The higher fat oxidation occurring in the F trial during exercise might result in a higher plasma TG clearance rate during subsequent postprandial period.

Previous study demonstrated that the additional insulin administration after ingestion of high fat meal showed an improvement in postprandial TG disposal in type I diabetes [29]. The study indicated that the insulin concentration played an important role on postprandial TG removal. However, the current study did not find differences in insulin concentration between trials during OFTT. This might be due to the exercise before OFTT diminishing the difference in postprandial insulin response even though we fed different GI CHO before exercise.

In the present study, the postprandial TG AUC and IAUC in the F trial demonstrated significantly lower values than the G trial (Figure 1). An increase in the plasma TG removal rate is possibly the factor causing a reduction in the plasma TG level, including decreasing the release of liver TG and increasing the transport of TG into muscle cells, or storage and utilization [24,30,31]. Ingestion of fructose was thought to increase postprandial lipemia in a sedentary population. Chong and colleagues [32] concluded that ingestion of fructose induced lower insulin secretion and might result in less activation of lipoprotein lipase (LPL), which consequently leads to impairing TG clearance. However, the negative effect may be offset by increasing physical activities [33]. Interestingly, Egli and colleagues [34] reported exercise prevented short-term high-fructose diet induced hypertriglyceridemia and increased lipid oxidation. After exercise, muscle LPL activity is increased, stimulating the transport of TG into the muscle cell for storage and utilization, which may reduce plasma TG concentration [24,35]. A previous study determined that a single session of exercise significantly enhanced the muscle LPL activity [36,37]. Seip and colleagues [38] also found that the expression of LPL genes in fat tissues did not differ significantly after exercise, which further reflects the importance of muscle to blood lipid metabolism after exercise. However, the different insulin responses caused by ingestion of CHO solution with different GI were likely to influence the muscle LPL activity [36,39]. In addition, the levels of glycogen and insulin increased considerably following ingestion of CHO; however, LPL activity was significantly decreased [40,41]. Seip and colleagues [42] demonstrated that the LPL mRNA level was significantly increased after 4 h of exercise, while lower insulin level was observed. This result caused a rise in VLDL concentration and a reduction in HDL-C release [39,43]. Another study reported that glucose ingestion elicits an insulin response that is evidently more apparent than that of fructose ingestion [44,45]. Moreover, the fat oxidation rate during exercise is relatively higher after fructose ingestion [46], which also influences LPL activity, eliciting changes in the postprandial TG level. This probably partially explains why the F trial postprandial TG level was significantly lower than that of the G trial in the current study.

The postprandial HDL-C level is related to the metabolic rate of TG-rich lipoprotein [24], and exercise may promote a rise in HDL-C concentration [47,48]. However, HDL-C concentration could be influenced by the insulin level [49]. The result of the present study indicated that when the participants ingested the F drink before exercise, the significantly higher postprandial HDL-C level was observed compared with when ingesting the G solution (Figure 4). A previous study reported when CHO ingestion was controlled for four weeks, the fasting insulin level increased significantly, and the HDL-C level was significantly lower than the pretest value [50]. Another study compared ingestion of different concentrations of CHO beverages and found that low CHO intervention resulted in higher HDL-C concentration, which was effective for triggering a decrease in the postprandial TG level [51]. Compared with ingesting the F drink, ingesting the G drink induced higher insulin response which, in turn, caused a reduction in the postprandial HDL-C level, thereby weakening the TG removal capability.

Previous studies have largely explored the relationship between exercise intervention and postprandial lipid metabolism. No studies have investigated ingestion of different GI CHO before exercise generating an influence on the postprandial lipemia immediately after exercise. This is the first study to elucidate the interactive effects of different GI drinks and exercise intervention on lipid metabolism following OFTT. We found that intervention might exert a retention effect, suggesting that ingesting a CHO drink with different GI before exercise influences the rate of substrate utilization during exercise, as well as the postprandial lipid level when OFTT is ingested after exercise.

Limitations

One of the major limitation of the present study was that the participants were given 75 g glucose or fructose drinks which were not adjusted by their body weight, although the different body size may have resulted in different magnitudes of glycemia when ingesting of the same amount of CHO. However, the design of the present study was mainly to induce different glycemic and insulinemic responses via different GI carbohydrate prior to exercise in order to influence the subsequent substrate utilization. In addition, the present study design was a with-in subject design. Therefore, we speculate that even if we adjust the amount of pre-exercise CHO by the subject's body weight, the outcome will possibly be similar to the present results.

5. Conclusions

This study found that when fructose was ingested before endurance exercise, the fructose trial significantly lowered TG AUC and IAUC compared with the glucose trial after OFTT. This result is possibly related to the pre-exercise low insulin level that induced higher fat oxidation during exercise. However, the mechanism involved remains elusive and warrants further investigation in the future.

Acknowledgments: The study was funded by National Science Council in Taiwan (NSC 96-2413-H-028-003-MY2). We deeply appreciated Sport Science Research Center of National Taiwan University of Sport provided the technical supported for this study.

Author Contributions: Ching-Lin Wu and Cheng-Kang Chang conceived and designed the experiments; Mei-Hui Tseng and Tsung-Jen Yang performed the experiments; Chih-Hui Chiu and Tsung-Jen Yang analyzed the data; Ching-Lin Wu, Cheng-Kang Chang, and Chih-Hui Chiu contributed reagents/materials/analysis tools; and Ching-Lin Wu and Tsung-Jen Yang wrote the paper.

Conflicts of Interest: The authors declare no conflict of interest. The funding sponsors had no role in the design of the study; in the collection, analyses, or interpretation of data; in the writing of the manuscript; or in the decision to publish the results.

References

1. Lefebvre, P.J.; Scheen, A.J. The postprandial state and risk of cardiovascular disease. *Diabet. Med.* **1998**, *15*, S63–S68. [CrossRef]
2. Chan, D.C.; Pang, J.; Romic, G.; Watts, G.F. Postprandial hypertriglyceridemia and cardiovascular disease: Current and future therapies. *Curr. Atheroscler. Rep.* **2013**, *15*, 309. [CrossRef] [PubMed]
3. Nordestgaard, B.G.; Benn, M.; Schnohr, P.; Tybjaerg-Hansen, A. Nonfasting triglycerides and risk of myocardial infarction, ischemic heart disease, and death in men and women. *JAMA* **2007**, *298*, 299–308. [CrossRef] [PubMed]
4. Patsch, J.R.; Miesenbock, G.; Hopferwieser, T.; Muhlberger, V.; Knapp, E.; Dunn, J.K.; Gotto, A.M., Jr.; Patsch, W. Relation of triglyceride metabolism and coronary artery disease. Studies in the postprandial state. *Arterioscler. Thromb.* **1992**, *12*, 1336–1345. [CrossRef] [PubMed]
5. Sharrett, A.R.; Chambless, L.E.; Heiss, G.; Paton, C.C.; Patsch, W. Association of postprandial triglyceride and retinyl palmitate responses with asymptomatic carotid artery atherosclerosis in middle-aged men and women. The Atherosclerosis Risk in Communities (ARIC) Study. *Arterioscler. Thromb. Vasc. Biol.* **1995**, *15*, 2122–2129. [CrossRef] [PubMed]

6. Aldred, H.E.; Perry, I.C.; Hardman, A.E. The effect of a single bout of brisk walking on postprandial lipemia in normolipidemic young adults. *Metabolism* **1994**, *43*, 836–841. [CrossRef]

7. Gill, J.M.; Al-Mamari, A.; Ferrell, W.R.; Cleland, S.J.; Sattar, N.; Packard, C.J.; Petrie, J.R.; Caslake, M.J. Effects of a moderate exercise session on postprandial lipoproteins, apolipoproteins and lipoprotein remnants in middle-aged men. *Atherosclerosis* **2006**, *185*, 87–96. [CrossRef] [PubMed]

8. Tsetsonis, N.V.; Hardman, A.E.; Mastana, S.S. Acute effects of exercise on postprandial lipemia: A comparative study in trained and untrained middle-aged women. *Am. J. Clin. Nutr.* **1997**, *65*, 525–533. [CrossRef] [PubMed]

9. Hargreaves, M.; Hawley, J.A.; Jeukendrup, A. Pre-exercise carbohydrate and fat ingestion: Effects on metabolism and performance. *J. Sports Sci.* **2004**, *22*, 31–38. [CrossRef] [PubMed]

10. Wright, D.A.; Sherman, W.M.; Dernbach, A.R. Carbohydrate feedings before, during, or in combination improve cycling endurance performance. *J. Appl. Phys.* **1991**, *71*, 1082–1088. [CrossRef] [PubMed]

11. Parks, E.J. Effect of dietary carbohydrate on triglyceride metabolism in humans. *J. Nutr.* **2001**, *131*, 2772S–2774S. [CrossRef] [PubMed]

12. Shin, Y.; Park, S.; Choue, R. Comparison of time course changes in blood glucose, insulin and lipids between high carbohydrate and high fat meals in healthy young women. *Nutr. Res. Pract.* **2009**, *3*, 128–133. [CrossRef] [PubMed]

13. Siri-Tarino, P.W.; Sun, Q.; Hu, F.B.; Krauss, R.M. Saturated fat, carbohydrate, and cardiovascular disease. *Am. J. Clin. Nutr.* **2010**, *91*, 502–509. [CrossRef] [PubMed]

14. Abbasi, F.; McLaughlin, T.; Lamendola, C.; Kim, H.S.; Tanaka, A.; Wang, T.; Nakajima, K.; Reaven, G.M. High carbohydrate diets, triglyceride-rich lipoproteins, and coronary heart disease risk. *Am. J. Cardiol.* **2000**, *85*, 45–48. [CrossRef]

15. Hamilton, M.T.; Etienne, J.; McClure, W.C.; Pavey, B.S.; Holloway, A.K. Role of local contractile activity and muscle fiber type on LPL regulation during exercise. *Am. J. Physiol.* **1998**, *275*, E1016–E1022. [CrossRef] [PubMed]

16. Zhang, J.Q.; Thomas, T.R.; Ball, S.D. Effect of exercise timing on postprandial lipemia and HDL cholesterol subfractions. *J. Appl. Phys.* **1998**, *85*, 1516–1522. [CrossRef] [PubMed]

17. Katsanos, C.S.; Grandjean, P.W.; Moffatt, R.J. Effects of low and moderate exercise intensity on postprandial lipemia and postheparin plasma lipoprotein lipase activity in physically active men. *J. Appl. Physiol.* **2004**, *96*, 181–188. [CrossRef] [PubMed]

18. McLaughlin, T.; Abbasi, F.; Lamendola, C.; Yeni-Komshian, H.; Reaven, G. Carbohydrate-induced hypertriglyceridemia: An insight into the link between plasma insulin and triglyceride concentrations. *J. Clin. Endocrinol. Metab.* **2000**, *85*, 3085–3088. [CrossRef] [PubMed]

19. Olefsky, J.M.; Farquhar, J.W.; Reaven, G.M. Reappraisal of the role of insulin in hypertriglyceridemia. *Am. J. Med.* **1974**, *57*, 551–560. [CrossRef]

20. Stevenson, E.J.; Williams, C.; Mash, L.E.; Phillips, B.; Nute, M.L. Influence of high-carbohydrate mixed meals with different glycemic indexes on substrate utilization during subsequent exercise in women. *Am. J. Clin. Nutr.* **2006**, *84*, 354–360. [PubMed]

21. Wee, S.L.; Williams, C.; Tsintzas, K.; Boobis, L. Ingestion of a high-glycemic index meal increases muscle glycogen storage at rest but augments its utilization during subsequent exercise. *J. Appl. Physiol.* **2005**, *99*, 707–714. [CrossRef] [PubMed]

22. Wu, C.L.; Nicholas, C.; Williams, C.; Took, A.; Hardy, L. The influence of high-carbohydrate meals with different glycaemic indices on substrate utilisation during subsequent exercise. *Br. J. Nutr.* **2003**, *90*, 1049–1056. [CrossRef] [PubMed]

23. Kaviani, M.; Chilibeck, P.D.; Yee, P.; Zello, G.A. The effect of consuming low-versus high-glycemic index meals after exercise on postprandial blood lipid response following a next-day high-fat meal. *Nutr. Diabetes* **2016**, *6*, e216. [CrossRef] [PubMed]

24. Gill, J.M.; Herd, S.L.; Vora, V.; Hardman, A.E. Effects of a brisk walk on lipoprotein lipase activity and plasma triglyceride concentrations in the fasted and postprandial states. *Eur. J. Appl. Physiol.* **2003**, *89*, 184–190. [CrossRef] [PubMed]

25. Koutsari, C.; Hardman, A.E. Exercise prevents the augmentation of postprandial lipaemia attributable to a low-fat high-carbohydrate diet. *Br. J. Nutr.* **2001**, *86*, 197–205. [PubMed]

26. Chiu, C.H.; Burns, S.F.; Yang, T.J.; Chang, Y.H.; Chen, Y.L.; Chang, C.K.; Wu, C.L. Energy replacement using glucose does not increase postprandial lipemia after moderate intensity exercise. *Lipids Health Dis.* **2014**, *13*, 177. [CrossRef] [PubMed]
27. Wu, C.L.; Williams, C. A low glycemic index meal before exercise improves endurance running capacity in men. *Int. J. Sport Nutr. Exerc. Metab.* **2006**, *16*, 510–527. [CrossRef] [PubMed]
28. Gregory, S.; Wood, R.; Matthews, T.; Vanlangen, D.; Sawyer, J.; Headley, S. Substrate Utilization is Influenced by Acute Dietary Carbohydrate Intake in Active, Healthy Females. *J. Sports Sci. Med.* **2011**, *10*, 59–65. [PubMed]
29. Campbell, M.D.; Walker, M.; Ajjan, R.A.; Birch, K.M.; Gonzalez, J.T.; West, D.J. An additional bolus of rapid-acting insulin to normalise postprandial cardiovascular risk factors following a high-carbohydrate high-fat meal in patients with type 1 diabetes: A randomised controlled trial. *Diabetes Vasc. Dis. Res.* **2017**, *14*, 336–344. [CrossRef] [PubMed]
30. Magkos, F.; Wright, D.C.; Patterson, B.W.; Mohammed, B.S.; Mittendorfer, B. Lipid metabolism response to a single, prolonged bout of endurance exercise in healthy young men. *Am. J. Physiol. Endocrinol. Metab.* **2006**, *290*, E355–E362. [CrossRef] [PubMed]
31. Tsekouras, Y.E.; Yanni, A.E.; Bougatsas, D.; Kavouras, S.A.; Sidossis, L.S. A single bout of brisk walking increases basal very low-density lipoprotein triacylglycerol clearance in young men. *Metabolism* **2007**, *56*, 1037–1043. [PubMed]
32. Chong, M.F.; Fielding, B.A.; Frayn, K.N. Mechanisms for the acute effect of fructose on postprandial lipemia. *Am. J. Clin. Nutr.* **2007**, *85*, 1511–1520. [PubMed]
33. Bidwell, A.J.; Fairchild, T.J.; Redmond, J.; Wang, L.; Keslacy, S.; Kanaley, J.A. Physical activity offsets the negative effects of a high-fructose diet. *Med. Sci. Sports Exerc.* **2014**, *46*, 2091–2098. [CrossRef] [PubMed]
34. Egli, L.; Lecoultre, V.; Theytaz, F.; Campos, V.; Hodson, L.; Schneiter, P.; Mittendorfer, B.; Patterson, B.W.; Fielding, B.A.; Gerber, P.A.; et al. Exercise prevents fructose-induced hypertriglyceridemia in healthy young subjects. *Diabetes* **2013**, *62*, 2259–2265. [CrossRef] [PubMed]
35. Malkova, D.; Evans, R.D.; Frayn, K.N.; Humphreys, S.M.; Jones, P.R.; Hardman, A.E. Prior exercise and postprandial substrate extraction across the human leg. *Am. J. Physiol. Endocrinol. Metab.* **2000**, *279*, E1020–E1028. [CrossRef] [PubMed]
36. Kiens, B.; Lithell, H. Lipoprotein metabolism influenced by training-induced changes in human skeletal muscle. *J. Clin. Investig.* **1989**, *83*, 558–564. [CrossRef] [PubMed]
37. Kiens, B.; Richter, E.A. Utilization of skeletal muscle triacylglycerol during postexercise recovery in humans. *Am. J. Physiol.* **1998**, *275*, E332–E337. [CrossRef] [PubMed]
38. Seip, R.L.; Angelopoulos, T.J.; Semenkovich, C.F. Exercise induces human lipoprotein lipase gene expression in skeletal muscle but not adipose tissue. *Am. J. Physiol.* **1995**, *268*, E229–E236. [CrossRef] [PubMed]
39. Pollare, T.; Vessby, B.; Lithell, H. Lipoprotein lipase activity in skeletal muscle is related to insulin sensitivity. *Arterioscler. Thromb. Vasc. Biol.* **1991**, *11*, 1192–1203. [CrossRef]
40. Jacobs, I.; Lithell, H.; Karlsson, J. Dietary effects on glycogen and lipoprotein lipase activity in skeletal muscle in man. *Acta Physiol. Scand.* **1982**, *115*, 85–90. [CrossRef] [PubMed]
41. Lithell, H.; Jacobs, I.; Vessby, B.; Hellsing, K.; Karlsson, J. Decrease of lipoprotein lipase activity in skeletal muscle in man during a short-term carbohydrate-rich dietary regime. With special reference to HDL-cholesterol, apolipoprotein and insulin concentrations. *Metabolism* **1982**, *31*, 994–998. [CrossRef]
42. Seip, R.L.; Mair, K.; Cole, T.G.; Semenkovich, C.F. Induction of human skeletal muscle lipoprotein lipase gene expression by short-term exercise is transient. *Am. J. Physiol.* **1997**, *272*, E255–E261. [CrossRef] [PubMed]
43. Lithell, H.; Karlstrom, B.; Selinus, I.; Vessby, B.; Fellstrom, B. Is muscle lipoprotein lipase inactivated by ordinary amounts of dietary carbohydrates? *Hum. Nutr. Clin. Nutr.* **1985**, *39*, 289–295. [PubMed]
44. Decombaz, J.; Sartori, D.; Arnaud, M.J.; Thelin, A.L.; Schurch, P.; Howald, H. Oxidation and metabolic effects of fructose or glucose ingested before exercise. *Int. J. Sports Med.* **1985**, *6*, 282–286. [CrossRef] [PubMed]
45. Koivisto, V.A.; Karonen, S.L.; Nikkila, E.A. Carbohydrate ingestion before exercise: Comparison of glucose, fructose, and sweet placebo. *J. Appl. Physiol.* **1981**, *51*, 783–787. [CrossRef] [PubMed]
46. Massicotte, D.; Peronnet, F.; Allah, C.; Hillaire-Marcel, C.; Ledoux, M.; Brisson, G. Metabolic response to [13C]glucose and [13C]fructose ingestion during exercise. *J. Appl. Physiol.* **1986**, *61*, 1180–1184. [CrossRef] [PubMed]

47. Blazek, A.; Rutsky, J.; Osei, K.; Maiseyeu, A.; Rajagopalan, S. Exercise-mediated changes in high-density lipoprotein: Impact on form and function. *Am. Heart J.* **2013**, *166*, 392–400. [CrossRef] [PubMed]

48. Grandjean, P.W.; Crouse, S.F.; Rohack, J.J. Influence of cholesterol status on blood lipid and lipoprotein enzyme responses to aerobic exercise. *J. Appl. Physiol.* **2000**, *89*, 472–480. [CrossRef] [PubMed]

49. Laws, A.; Reaven, G.M. Evidence for an independent relationship between insulin resistance and fasting plasma HDL-cholesterol, triglyceride and insulin concentrations. *J. Intern. Med.* **1992**, *231*, 25–30. [CrossRef] [PubMed]

50. Volek, J.S.; Sharman, M.J.; Gomez, A.L.; DiPasquale, C.; Roti, M.; Pumerantz, A.; Kraemer, W.J. Comparison of a very low-carbohydrate and low-fat diet on fasting lipids, LDL subclasses, insulin resistance, and postprandial lipemic responses in overweight women. *J. Am. Coll. Nutr.* **2004**, *23*, 177–184. [CrossRef] [PubMed]

51. Volek, J.S.; Sharman, M.J.; Gomez, A.L.; Scheett, T.P.; Kraemer, W.J. An isoenergetic very low carbohydrate diet improves serum HDL cholesterol and triacylglycerol concentrations, the total cholesterol to HDL cholesterol ratio and postprandial lipemic responses compared with a low fat diet in normal weight, normolipidemic women. *J. Nutr.* **2003**, *133*, 2756–2761. [PubMed]

nutrients

MDPI

Article

ChREBP-Knockout Mice Show Sucrose Intolerance and Fructose Malabsorption

Takehiro Kato [1], Katsumi Iizuka [1,2,*], Ken Takao [1], Yukio Horikawa [1], Tadahiro Kitamura [3] and Jun Takeda [1]

[1]   Department of Diabetes and Endocrinology, Graduate School of Medicine, Gifu University,
      Gifu 501-1194, Japan; bado_aberu@yahoo.co.jp (T.K.); lamgerrpard@yahoo.co.jp (K.T.);
      yhorikaw@gifu-u.ac.jp (Y.H.); jtakeda@gifu-u.ac.jp (J.T.)
[2]   Gifu University Hospital Center for Nutritional Support and Infection Control, Gifu 501-1194, Japan
[3]   Metabolic Signal Research Center, Institute for Molecular and Cellular Regulation, Gunma University,
      Gunma 371-8512, Japan; kitamura@gunma-u.ac.jp
*   Correspondence: kiizuka@gifu-u.ac.jp; Tel.: +81-58-230-6564; Fax: +81-58-230-6376

Received: 31 January 2018; Accepted: 9 March 2018; Published: 10 March 2018

Abstract: We have previously reported that 60% sucrose diet-fed *ChREBP* knockout mice (KO) showed body weight loss resulting in lethality. We aimed to elucidate whether sucrose and fructose metabolism are impaired in KO. Wild-type mice (WT) and KO were fed a diet containing 30% sucrose with/without 0.08% miglitol, an α-glucosidase inhibitor, and these effects on phenotypes were tested. Furthermore, we compared metabolic changes of oral and peritoneal fructose injection. A thirty percent sucrose diet feeding did not affect phenotypes in KO. However, miglitol induced lethality in 30% sucrose-fed KO. Thirty percent sucrose plus miglitol diet-fed KO showed increased cecal contents, increased fecal lactate contents, increased growth of lactobacillales and *Bifidobacterium* and decreased growth of clostridium cluster XIVa. *ChREBP* gene deletion suppressed the mRNA levels of sucrose and fructose related genes. Next, oral fructose injection did not affect plasma glucose levels and liver fructose contents; however, intestinal sucrose and fructose related mRNA levels were increased only in WT. In contrast, peritoneal fructose injection increased plasma glucose levels in both mice; however, the hepatic fructose content in KO was much higher owing to decreased hepatic *Khk* mRNA expression. Taken together, KO showed sucrose intolerance and fructose malabsorption owing to decreased gene expression.

Keywords: carbohydrate-responsive element-binding protein; ketohexokinase; fructose; glucose transporter 5; glucose transporter 2

1. Introduction

Excess intake of high sucrose and fructose diet were thought to be associated with the development of obesity, metabolic syndrome, and diabetes [1,2]. Many experimental animal studies, for example, experiments feeding 70% fructose-containing water, supported this hypothesis [2]. However, recent human epidemic data suggest that there is little association between metabolic syndrome and consumption of sucrose and fructose [3,4].

Moreover, the mechanism of sucrose and fructose metabolism remains unclear. Sucrose is a disaccharide composed of glucose and fructose, and is digested by intestinal sucrase-isomaltase (SI), which is inhibited by miglitol, an α-glucosidase inhibitor [5]. Fructose is more potent and has higher capacity of protein glycation than glucose and, thus, is more harmful than glucose [6]. Fructose is metabolized in the intestine and liver. Previously, it has been considered that large amounts of fructose are metabolized mainly in the liver [7]. However, portal fructose levels are five times lower and plasma fructose levels are 100 times lower than plasma glucose levels [8,9].

Moreover, excess intake of fructose can cause dietary fructose malabsorption and thereby irritable bowel syndrome [10]. Taken together, we hypothesized that intestinal fructose absorption, but not hepatic fructose metabolism, regulates portal and plasma fructose levels [11].

To clarify the intestinal sucrose and fructose metabolism, we focused on the phenotypes of high-sucrose diet-fed carbohydrate-responsive element-binding protein (ChREBP)-knockout (KO) mice [12]. ChREBP is a glucose-activated transcription factor that regulates glucose and lipid metabolism. We have formerly reported that high-sucrose diet-fed KO mice showed body weight loss and eventual lethality, although high-glucose diet-fed and high-starch diet-fed KO mice did not [12]. As SI is induced by sucrose, we wondered whether SI expression is decreased in KO mice [13]. Moreover, high-fructose diet-fed KO mice showed similar phenotypes (body weight loss and appetite loss) [14–16]. ChREBP regulates the gene expression of glucose transporter 5 (*Glut5*) and ketohexokinase (*Khk*), which regulate fructolysis [12,17,18]. Taken together, we speculated that altered sucrose and fructose metabolism may contribute to the pathology of sucrose intolerance and fructose malabsorption seen in KO mice.

In this study, we focused on the effect of ChREBP on sucrose and fructose metabolism in the liver and intestine. We tested whether 30% sucrose plus miglitol (S + M) diet-fed KO mice show phenotypes similar to sucrose intolerance. Furthermore, by comparing the results of oral and peritoneal fructose injection, we tried to clarify the role of hepatic and intestinal ChREBP in fructose metabolism. This study will be beneficial for understanding the mechanism of sucrose and fructose metabolism.

2. Materials and Methods

2.1. Materials

Sucrose, fructose, and glucose measurement kits were purchased from Wako Pure Chemical Industries (Osaka, Japan). Lactate measurement kits were purchased from Kyowa Medex Co. (Tokyo, Japan). Triglyceride and cholesterol measurement kits were purchased from Wako Pure Chemical Industries. Glucose-6-phosphate dehydrogenase (G6PDH), phosphoglucose isomerase, hexokinase, and NADP were purchased from Roche Custom Biotech Inc. (Mannheim, Germany).

2.2. Animals, and Sucrose and Sucrose + Miglitol Diets

Animal experiments were carried out in accordance with the National Institutes of Health guide for the care and use of Laboratory animals (NIH Publications No. 8023, revised 1978). All animal care was approved by the Animal Care Committee of the University of Gifu (Approval number 27–31, Approval date 4 June 2015). Mice were housed at 23 °C on a 12-h light/dark cycle. KO mice were backcrossed for at least 10 generations onto the C57BL/6J background [19].

Mice had free access to water and were fed an autoclaved CE-2 diet (CLEA Japan, Tokyo, Japan). Wild-type (WT) and KO mice were housed separately with a total of three mice per cage. To examine mortality and body weight changes, 12 week old male WT and KO mice were fed a 30% sucrose diet (S; protein 17% kcal, carbohydrate 73% kcal, fat 10% kcal) or a 30% sucrose + 0.08% miglitol diet (S + M; protein 17% kcal, carbohydrate 73% kcal, fat 10% kcal, miglitol 0.08%) for eight weeks [20]. To examine phenotypes (tissue weight, tissue metabolites, plasma profile, mRNA levels), 18 week old male WT and KO mice were fed S or S + M diets for seven days. The diets were purchased from Research Diets Inc. (New Brunswick, NJ, USA). Miglitol was gifted by Sanwa Kagaku Kenkyuusho Co. (Nagoya, Japan).

2.3. Liver Glycogen, Triglyceride, Cholesterol and Fructose Contents, and Plasma Profile Measurements

The liver glycogen content was measured as previously reported [12,19]. Liver lipids were extracted using the Bligh and Dyer method [21], and measured using triglyceride (Wako Pure Chemical Industries) and cholesterol E-tests (Wako Pure Chemical Industries). Liver fructose contents were measured by enzymatic methods [22]. Briefly, freeze-clamped tissues (100 mg) were homogenized in 2 mL of cold 6% perchloric acid, neutralized, and centrifuged. The assay is based on the oxidation of glucose as glucose-6-phosphate (G6P) using G6PDH. Fructose-6-phosphate is converted to G6P by the phosphoglucose isomerase enzyme, and subsequently oxidized by the G6PDH in the assay mixture. The fructose concentration is determined as the difference in G6P concentration before and after phosphoglucose isomerase treatment. All enzymes were purchased from Roche Custom Biotech Inc. Blood plasma was collected from the retro-orbital venous plexus following ad libitum feeding or after a 6-h fast. Blood glucose levels were measured using a FreeStyle Freedom monitoring system (Nipro, Osaka, Japan). Plasma triglycerides and total cholesterol levels were determined using the commercial kits, triglyceride E-test (Wako Pure Chemical Industries), and cholesterol E-test (Wako Pure Chemical Industries), respectively.

2.4. Cecal Contests Weight, Cecal Lactate Contents, and Intestinal Bacterial Flora

Mice fed with S or S + M were sacrificed at 19 weeks of age by cervical dislocation. After tissue weight, length of intestine and cecal contents were measured, the intestine and liver were immediately snap-frozen in liquid nitrogen and stored at -80 °C until further analysis of hepatic triacylglycerol and cholesterol contents, and quantitative PCR. For measurement of cecal lactate contents, frozen cecal content (20 mg) was homogenized in 80 µL of cold 6% perchloric acid, neutralized and centrifuged. Supernatants were collected and measured by a lactate measurement kit (Kyowa Medex). Terminal restriction fragment length polymorphism (T-RLFP) flora analysis of cecal contents was performed by Techno Suruga Labo Inc. (Shizuoka, Japan) [23].

2.5. Oral and Intraperitoneal Fructose-Loading Test

Fructose (3 g/kg BW) was orally or intraperitoneally injected into 14 weeks old male WT and KO mice. Plasma glucose was measured at the indicated times. For liver fructose contents and mRNA expression analyses, mice were sacrificed at 0, 1, 2 or 4 h, and the liver and intestine were removed and stored at -80 °C until further analysis.

2.6. RNA Isolation and Quantitative Real-Time PCR

Total RNA isolation, cDNA synthesis and real-time PCR analysis were performed as previously described [12,19]. Real-time PCR primers for mouse/rat *ChREBP*, liver type pyruvate kinase (*Pklr*), glucose transporter 2 (*Glut2*), fibroblast growth factor-21 (*Fgf-21*) and RNA polymerase II (*Pol2*) have been previously reported [19]. Primers used for *Glut5*, *Khk*, and *Si* were as follows: *Glut5* forward, 5′-CGGCTTCTCCACCTGCCTC-3′, *Glut5* reverse, 5′-CGTGTCCTATGACGTAGACAATGA-3′; *Khk-C* forward, 5′-GCTGACTTCAGGCAGAGG-3′, *Khk-C* reverse, 5′-CCTTCTCAAAGTCCTTAGCAG-3′; *Si* forward, 5′-TTGATATCCGGTCCACGGTTCT-3′, *Si* reverse, 5′-CAGGTGACATCCAGGTTGCATT-3′. All amplifications were performed in triplicate. The relative amounts of mRNA were calculated using the comparative CT method. *Pol2* expression was used as an internal control.

2.7. Statistical Analysis

All values are presented as means $\pm$ SD. Data were analyzed using Tukey's test. A value of $p < 0.05$ was considered statistically significant.

3. Results

3.1. ChREBP Knockout Mice Show Intolerance to Modest Amounts of Sucrose and Miglitol Diet

We have reported that a high-sucrose diet (60% sucrose) caused decreased appetite and eventual lethality in KO mice [12]. First, we investigated whether KO mice have any problems with sucrose digestion. We tested whether a medium amount of sucrose (30%) feeding caused body weight loss.

A 30% sucrose diet was not lethal, although the body weight gain of 30% sucrose-diet-fed KO (KO S) mice was much lower than that of 30% sucrose-fed WT (WT S) mice (Figure 1A,B).

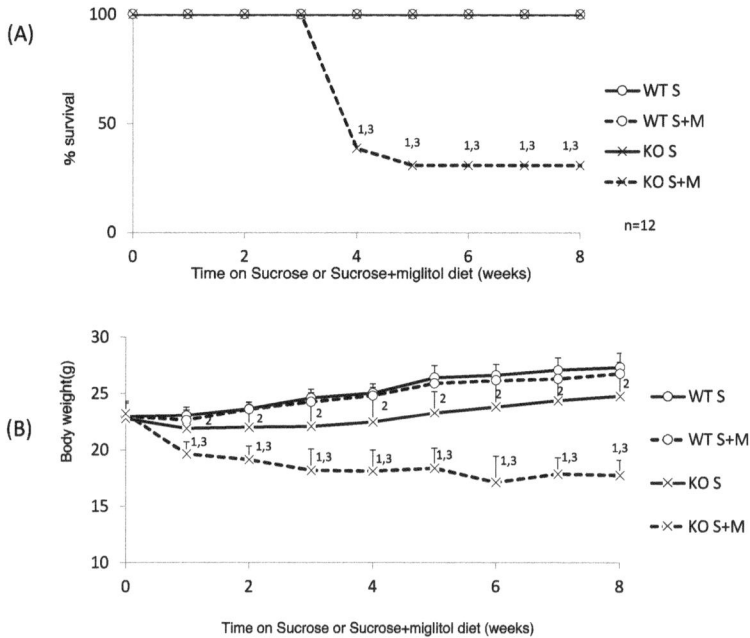

Figure 1. Thirty percent sucrose + 0.08% miglitol diet causes body weight loss and high lethality. Twelve week old male wild-type (WT) mice and *ChREBP* knockout (KO) mice were fed a 30% sucrose (S) or 30% sucrose plus 0.08% miglitol (S + M)-containing diet for eight weeks. (**A**) Survival rate. WT S, WT S + M, and K S, except KO S + M, survived. Data represented as % survival; (**B**) Body weight change. Data represented as mean ± SD (n = 12 per group). [1] KO S vs. KO S + M, $p < 0.05$, [2] WT S vs. KO S, $p < 0.05$, and [3] WT S + M vs. KO S + M, $p < 0.05$.

Interestingly, the addition of miglitol, which inhibits sucrose digestion in the upper intestine, caused decreased body weight and increased mortality (75 and 75%, six and eight weeks after feeding the specific diet, respectively; Figure 1A,B). Next, we examined the following parameters one week after feeding the specific diet. The body weight changes and food intake of KO S mice were similar to those of WT S mice (Table 1). However, the body weight and food intake of sucrose plus miglitol (S + M) diet-fed KO (KO S + M) mice were significantly decreased compared with WT S + M mice. Consistently, the liver, epidydimal fat tissue and brown adipose tissue weight was decreased in KO S + M mice compared with WT S + M mice (Table 1). In contrast, the locomotor activity was similar among the groups (Table 1).

Regarding the plasma profile, the plasma glucose levels were lowest in KO S + M mice. Plasma triglyceride and total cholesterol levels in KO S and KO S + M mice were lower than those in WT S and WT S + M mice (Table 1). The liver triglyceride and cholesterol contents in KO S and KO S + M mice were also lower than those in WT S and WT S + M mice (Table 1). The liver glycogen content in KO S mice was increased; however, in KO S + M mice it was decreased owing to appetite loss (Table 1). Thus, KO S + M mice showed sucrose intolerance similar to high-sucrose diet-fed KO mice.

Table 1. The effect of 30% sucrose and 0.08% miglitol diet on wild-type mice and ChREBP knockout mice.

	WT S	WT S + M	KO S	KO S + M
BW (g) before	31.0 ± 1.77	29.8 ± 1.83	27.7 ± 1.62 [3]	26.6 ± 1.21 [4]
BW (g) after	29.3 ± 1.22	27.6 ± 0.91 [1]	25.2 ± 1.07 [3]	20.9 ± 1.00 [2][4]
BW (%) Difference	−5.38 ± 2.47	−7.5 ± 3.88	−8.97 ± 3.21	−21.5 ± 2.14 [2][4]
Liver (%BW)	5.33 ± 0.30	5.23 ± 0.23	7.12 ± 1.79 [3]	4.97 ± 0.57 [2]
Epidydimal Fat Weight (%BW)	1.78 ± 0.55	1.69 ± 0.32	1.35 ± 0.30 [3]	0.47 ± 0.16 [2][4]
Brown Adipose Tissue (%BW)	0.40 ± 0.09	0.38 ± 0.05	0.30 ± 0.07 [3]	0.26 ± 0.06 [4]
Locomotor activity (counts/day)	14550 ± 3788	12778 ± 2984	12875 ± 2303	10800 ± 2066
Food Intake (g/day)	2.51 ± 0.63	2.33 ± 0.26	2.53 ± 0.17	1.77 ± 0.30 [2][4]
Plasma Glucose (mg/dL)	100.6 ± 9.6	96.3 ± 8.3	80.3 ± 10.8 [3]	57.6 ± 6.8 [2][4]
Plasma Triglyceride (mg/dL)	137.2 ± 49.4	181.7 ± 54.2	72.7 ± 17.5	70.2 ± 14.2 [4]
Plasma T-Chol (mg/dL)	127.5 ± 15.3	130.6 ± 4.4	60.3 ± 7.8	65.4 ± 6.46 [4]
Liver Glycogen (mg/g liver)	38.6 ± 14.3	50.4 ± 17.4	83.5 ± 36.2 [3]	56.9 ± 27.4
Liver Triglyceride (mg/g liver)	6.60 ± 1.97	5.54 ± 1.50	2.72 ± 0.84 [3]	1.35 ± 0.45 [4]
Liver Cholesterol (mg/g liver)	0.99 ± 0.32	1.54 ± 0.79	0.44 ± 0.14	0.56 ± 0.33 [4]

Thirty percent sucrose fed wild-type mice (WT S), 30% sucrose plus 0.08% miglitol fed wild-type mice (WT S + M), 30% sucrose fed ChREBP knockout mice (KO S), and 30% sucrose plus 0.08% miglitol fed ChREBP knockout mice (KO S + M); BW: body weight; T-chol: total cholesterol. [1] WT S vs. WT S + M, $p < 0.05$, [2] KO S vs. KO S + M, $p < 0.05$, [3] WT S vs. KO S, $p < 0.05$, and [4] WT S + M vs. KO S + M, $p < 0.05$.

3.2. Sucrose Plus Miglitol Diet-Fed KO Mice Show Cecum Enlargement

Next, we checked the intestinal changes in WT and KO mice. The length of the small intestine was comparable in WT S, WT S + M, KO S and KO S + M mice (Figure 2A). The cecal enlargement and cecal contents in KO S mice were higher than those in WT S mice (Figure 2B,C). Although the food-loading test was performed only for one week, the cecal content in KO S + M mice was about 3.5 times higher than that in WT S and WT S + M mice (Figure 2B,C). Moreover, analysis of the intestinal flora and cecal contents showed that the ratios of *Bifidobacterium* and lactobacillales, and the cecal lactate contents were the highest in KO S + M mice (Figure 2D,E). In contrast, the abundance of clostridium cluster XIVa was dramatically diminished in KO S + M mice (Figure 2D).

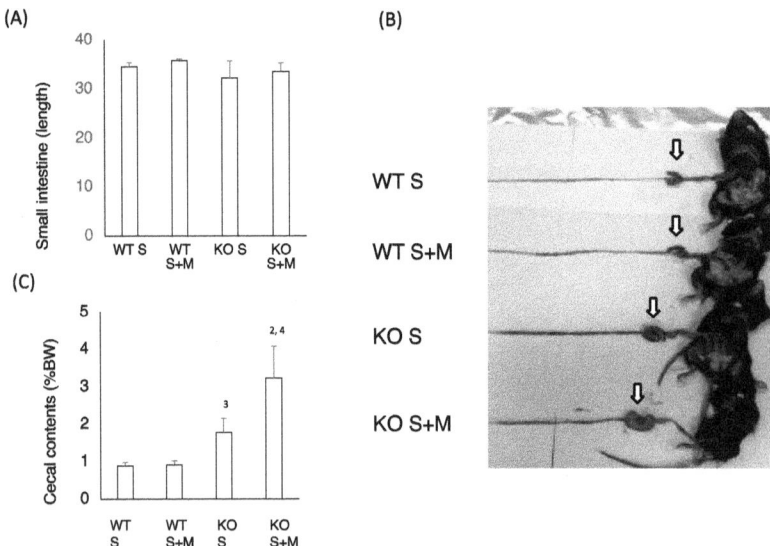

Figure 2. *Cont.*

(D) (E)

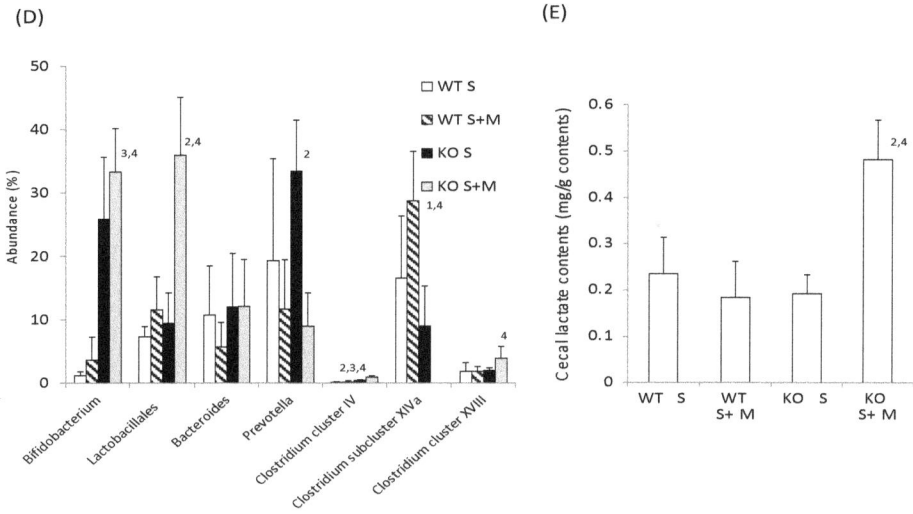

Figure 2. Sucrose plus miglitol diet-fed KO mice show cecal enlargement, higher lactate contents and altered intestinal flora. Eighteen week old male wild-type (WT) mice and *ChREBP*-knockout (KO) mice were fed a 30% sucrose (S) or 30% sucrose plus 0.08% miglitol (S + M)-containing diet for seven days. (**A**) Lengths (cm) of small intestine; (**B**) Representative image of intestinal enlargement; (**C**) Weight of cecal contents (% BW). Open arrows indcate cecum. Cecum in KO S and KO S + M were enlarged; (**D**) Gut microbes in cecum contents of WT and KO mice are expressed as a percentage of total DNA sequences; (**E**) Cecal lactate contents (mg/g). Data represented as mean ± SD (n = 6 per group). [1] WT S vs. WT S + M, $p < 0.05$, [2] KO S vs. KO S + M, $p < 0.05$, [3] WT S vs. KO S, $p < 0.05$, and [4] WT S + M vs. KO S + M, $p < 0.05$. BW: body weight.

3.3. Miglitol Affects the Expression of ChREBP Target Genes in the Intestine

Next, we tested the sucrose and fructose metabolism in relation to gene expression. In WT S mice, the expression of sucrose metabolism (*Si*), fructose metabolism (*Glut2*, *Glut5* and *Khk*), and *ChREBP* and its target genes in the upper intestine were higher as compared with those in the lower intestine (Figure 3). Upon addition of miglitol, the mRNA expression of these genes was highest in the middle and lower intestine. In the liver, the mRNA expression of these genes was not affected by the addition of miglitol. Interestingly, the expression of *Glut5* mRNA in the liver was much lower than in the intestine (Figure 3E). By contrast, the mRNA levels of the abovementioned genes were lower in the KO mice than in the WT and the effect of miglitol on these mRNA levels was suppressed in KO S mice (Figure 3A–F). As compared with *Glut5* expression, *SGLT1* mRNA levels were not affected by *ChREBP* gene deletion (data not shown). Thus, we concluded that *ChREBP* regulates sucrose and fructose metabolism through gene expression.

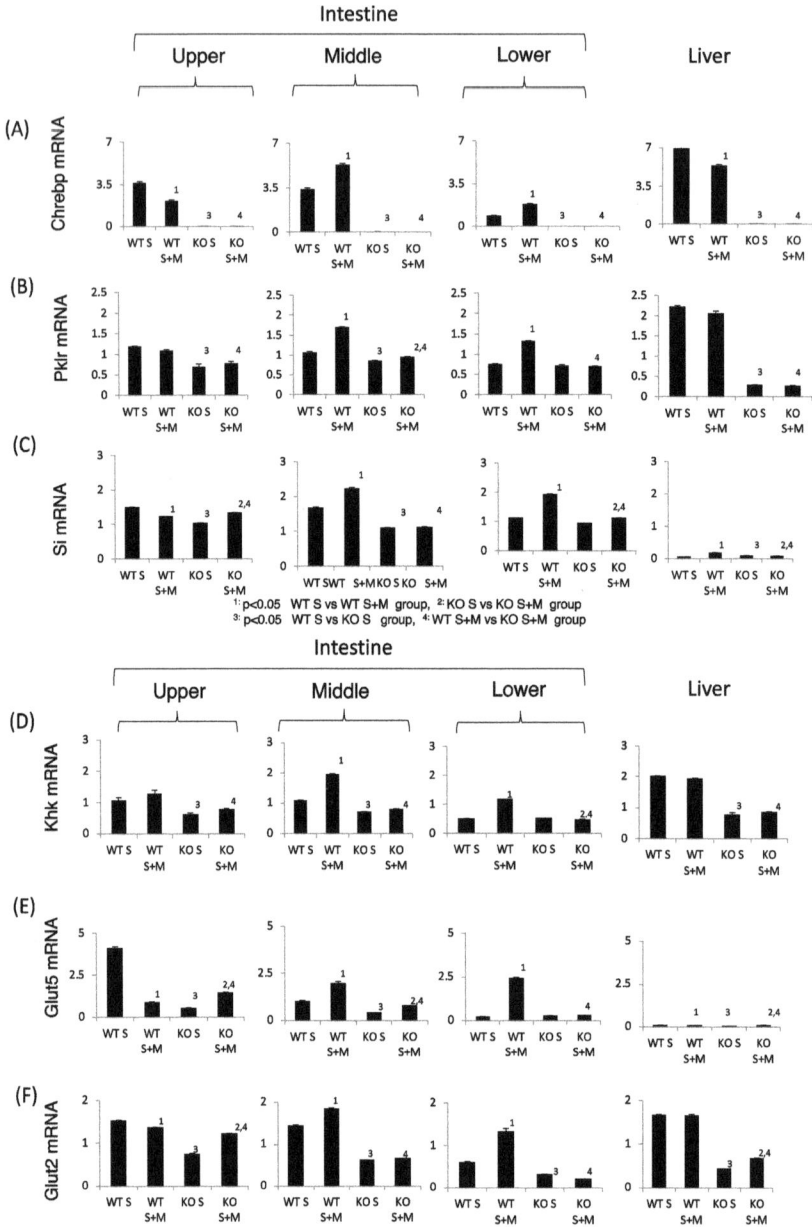

Figure 3. The effect of miglitol and the *ChREBP* gene deletion on genes related to ChREBP, fructose and sucrose metabolism. Eighteen week old male wild-type (WT) mice and *ChREBP*-knockout (KO) mice were fed a 30% sucrose (S) or 30% sucrose plus 0.08% miglitol (S + M)-containing diet for seven days. The intestine was divided into three parts (upper, middle and lower) and the mRNA levels were measured by real-time PCR. (**A**) *ChREBP*; (**B**) liver pyruvate kinase (*Pklr*); (**C**) sucrase isomerase (*Si*); (**D**) ketohexokinase (*Khk*); (**E**) glucose toransporter 5 (*Glut5*); and (**F**) glucose transporter 2 (*Glut2*). Data represented as mean ± SD (n = 3 per group). [1] WT S vs. WT S + M, $p < 0.05$, [2] KO S vs. KO S + M, $p < 0.05$, [3] WT S vs. KO S, $p < 0.05$, and [4] WT S + M vs. KO S + M, $p < 0.05$.

3.4. Fructose Is Difficult to Metabolize in the Intestine, but Not in the Liver

As KO mice showed disturbance not only in sucrose metabolism but also in fructose metabolism, we next tested the role of intestinal and hepatic ChREBP in fructose metabolism. After oral fructose injection, fructose is absorbed in the intestine (Figure 4A). After peritoneal injection, fructose is absorbed in the portal vein (Figure 4B) [24]. In the oral fructose-loading test (3 g/kg BW), the plasma glucose levels in WT mice only modestly increased to 120 mg/dL at 30 min (Figure 4A). In KO mice, the plasma glucose levels at 30 min were slightly lower than those in WT mice (Figure 4A). By contrast, in peritoneal fructose loading, the plasma glucose levels in WT mice increased to 200 mg/dL at 30 min (Figure 4B). In KO mice, the plasma glucose levels were lower than those in WT mice, and the peak time shifted right (Figure 4B). Consistent with these results, the hepatic fructose content in the oral fructose-loading test (at 0 and 1 h) was undetectable (Figure 4C). Therefore, we concluded that fructose is difficult to metabolize and absorb in the intestine. In contrast, the fructose content after the peritoneal fructose-loading test at 1 h was measurable. Moreover, in KO mice, the hepatic fructose content at 1 h was about three times higher than that in WT mice (Figure 4D). These results suggest that hepatic fructose metabolism was inhibited at the level of KHK in the liver of KO mice.

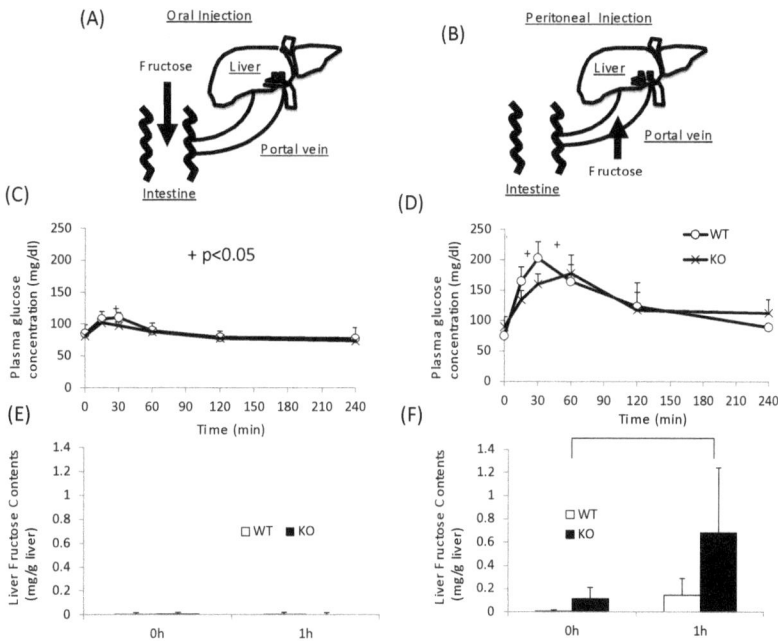

Figure 4. Oral and peritoneal fructose injection test. Oral (**A**) and perinoteal (**B**) injected fructose is absorbed in intestine and portal vein, respectively. Time course of glucose concentration after oral (**C**) or peritoneal (**D**) fructose injection. Liver fructose content at 0 and 1 h after oral (**E**) or peritoneal (**F**) fructose injection. Data are presented as means ± SD ($n = 6$ per group). $^{+}$ WT vs. KO, $p < 0.05$.

3.5. ChREBP Regulates the Expression of Genes Related to Fructose Metabolism in the Intestine

Finally, we examined whether fructose induces the expression of intestinal and hepatic ChREBP target genes. After oral fructose injection, the expression of intestinal ChREBP target genes (*ChREBP*, *Pklr*) and fructose metabolism genes (*Glut2*, *Glut5*, and *Khk*) in WT mice increased in a time-dependent manner, while the mRNA expression of these genes was much lower in KO mice (Figure 5A–F). Consistent with the plasma glucose levels, the mRNA expression of the hepatic ChREBP target genes (*ChREBP*, *Pklr*, and *Fgf-21*) and fructose metabolism genes (*Glut2*, *Glut5*, and *Khk*) was not affected by fructose (Figure 5A–F). After peritoneal fructose injection, the hepatic mRNA expression of *ChREBP*, *Pklr*, *Glut2*, *Glut5*, and *Khk* in WT mice increased in a time-dependent manner; however, this induction was diminished in KO mice. By contrast, the intestinal mRNA levels of these genes were not affected by fructose injection (Figure 5A–F). In the liver, *Fgf-21* mRNA levels in KO mice were lower than those in WT mice. However, the hepatic *Fgf-21* mRNA levels in WT mice were not induced by oral or peritoneal fructose injection (Figure 5C). Thus, we concluded that oral and peritoneal fructose injection mainly induced intestinal and hepatic fructose metabolism genes regulated by ChREBP, respectively.

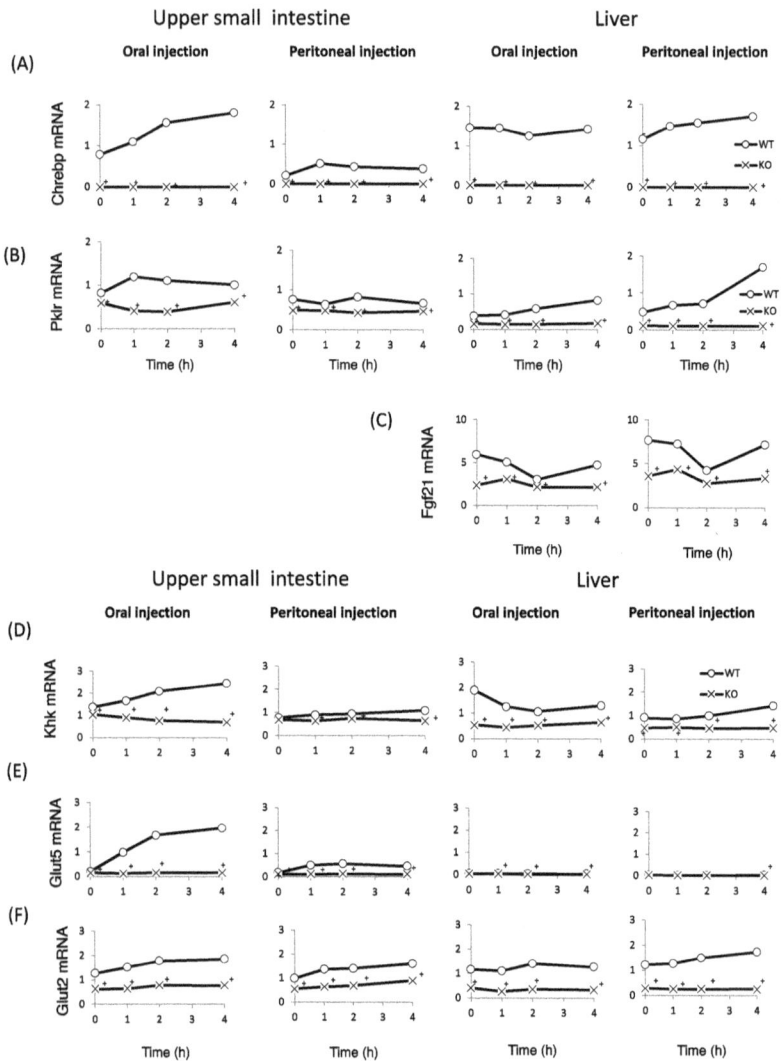

Figure 5. The effect of oral and peritoneal fructose injection on genes related to ChREBP and fructose metabolism. After oral or peritoneal fructose injection (3 kg/kg BW), the mRNA expression of *ChREBP* (**A**); liver type pyruvate kinase (*Pklr*) (**B**); fibroblast growth factor-21 (*Fgf21*) (**C**); ketohexokinase (*Khk*) (**D**); glucose transporter 5 (*Glut5*) (**E**); and glucose transporte 2 (*Glut2*) (**F**) in the intestine and liver was measured by real-time PCR analysis. $n = 3$ per group. [+]WT vs. KO, $p < 0.05$.

4. Discussion

In this study, we tried to identify the mechanism by which *ChREBP*-KO mice show sucrose intolerance. Thirty percent sucrose (30%) diet-fed KO mice did not present the body weight loss and lethality seen in 60% sucrose diet-fed KO mice; however, Si inhibition by miglitol successfully exhibited sucrose intolerance. Increased fecal lactate contents, and increased growth of lactobacillales and *Bifidobacterium*, consistent with increased lactate contents, was seen only in S + M fed KO mice. These findings were consistent with decreased expression of sucrose and fructose metabolism-related genes, which are regulated by ChREBP. Moreover, oral and peritoneal fructose injection mainly induced ChREBP-regulated intestinal and hepatic fructose metabolism genes, respectively. These results suggest that alternations in the expression of both sucrose and fructose-related genes contribute to sucrose intolerance and fructose malabsorption in KO mice (Figure 6).

(A) In 30% sucrose plus 0.08% miglitol diet fed wild-type mice (WT), sucrose was digested into glucose and fructose in upper intestine. Glucose was almost absorbed in upper intestine. In contrast, fructose was partly absorbed and unabsorbed fructose was used for intestinal bacterial growth.

(B) In 30% sucrose plus 0.08% miglitol diet fed ChREBP knockout mice (KO), owing to decreased sucrase-isomaltase (SI) expression or SI inhibition by miglitol, undigested sucrose was moving into the lower intestine. Moreover, fructose absorption in KO was also decreased due to decreased intestinal glucose transporter 5 (*Glut5*), glucose transporter 2 (*Glut2*), and ketohexokianse (*Khk*) expression. Undigested sucrose and fructose in lower intestine and cecum affected intestinal bacterial flora (increased growth of lactobacillales and *Bifidobacterium* and decreased growth of clostridium cluster XIVa).

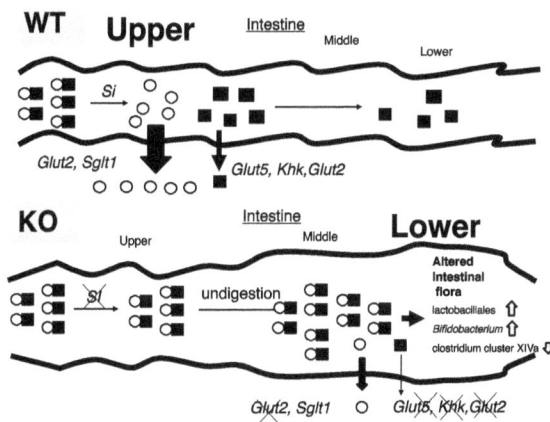

Figure 6. Schematic presentation of intestinal carbohydrate metabolism in wild-type and ChREBP knockout mice.

We have formerly reported that 60% sucrose diet-fed KO mice showed body weight loss and decreased food intake [12]. Despite the appetite loss, the cecum of dead 60% sucrose diet-fed KO mice was enlarged (unpublished data), hence, we wondered whether sucrose metabolism was disrupted in KO mice. As miglitol is a well-known Si inhibitor, the addition of miglitol caused an increased flux of undigested sucrose into the lower intestine. Consistent with these results, the addition of miglitol caused sucrose intolerance in KO mice fed a 30% sucrose diet, which, by itself, did not induce sucrose intolerance. Consistent with our hypothesis, KO S + M mice showed malabsorption (body weight, food intake, and diarrhea), similarly to the 60% sucrose diet-fed KO mice. Therefore, the increased flux of undigested sucrose into the lower intestine was partly due to the pathology of sucrose intolerance in KO mice.

S + M fed KO mice showed cecal enlargement in addition to body weight and appetite loss. Moreover, the ratios of lactobacillales and *Bifidobacterium* increased and the ratio of clostridium cluster XIVa reciprocally diminished in these mice. As the growth of these bacteria favors sucrose and fructose, these results suggest that undigested sucrose was moving into the lower intestine and cecum, and promoted the growth of lactobacillales and *Bifidobacterium* [25,26]. In contrast, the abundance of clostridium cluster XIVa increased in mice fed with high-fat diets [27]. Our data showed that Si inhibition did not change the gut microbiota in WT, which is consistent with the finding that Si inhibition by inulin-type fructans did not change the total number of bacteria in the cecal content and did not induce a bifidogenic effect [28]. However, the abundance of clostridium cluster XIVa was diminished in KO S + M mice. As these changes in KO mice were caused by a 60% sucrose diet and by an S + M diet, we concluded that sucrose intolerance was partly due to both Si suppression and a large amount of sucrose intake, resulting in an increased flux of undigested sucrose into the lower intestine.

These phenotypes were similar to those of human SI deficiency patients [29]. After weaning from breast-feeding, human congenital SI deficiency patients experienced stomach cramps, bloating, excess gas production, and diarrhea, resulting in failure to gain weight and malnutrition. Most affected children have improved tolerance to sucrose and maltose as they get older. Moreover, α-glucosidase inhibitors (miglitol, voglibose, and acarbose) have gastrointestinal side effects such as flatulence, diarrhea, soft stool, and abdominal discomfort [30]. As S + M KO mice were sucrose-intolerant, KO mice may have another important metabolic defect, such as fructose malabsorption.

Indeed, high-fructose diet-fed intestine-specific *ChREBP*-KO mice showed cecal enlargement and body weight loss similar to high-fructose diet-fed $GLUT5^{-/-}$ mice, a model of fructose malabsorption [18,31]. These phenotypes appear similar to those of S + M KO mice. GLUT5 is mainly expressed in the intestine and kidneys, and much less in the liver [32]. Fructose absorption in mice and humans appears to be limited at high fructose concentrations, which is consistent with the limited absorption capacity of a facilitated transport system [33,34]. Moreover, in these *GLUT5*-KO mice, fructose absorption was decreased by 75% in the jejunum and the concentration of serum fructose was decreased by 90%, compared with WT mice [31]. Therefore, decreased "intestinal" *Glut5* mRNA may contribute to the lower intestinal fructose absorption in KO mice, suggesting that S + M-fed KO mice have not only sucrose intolerance, but also fructose malabsorption. From a clinical viewpoint, metformin sometimes causes abdominal discomfort (diarrhea and vomiting) [35]. Considering metformin can inhibit ChREBP activity [36], abdominal side effects may be due to suppression of ChREBP, and thereby decreased *Glut5* mRNA expression. If excess amounts of carbohydrates are consumed by patients with diabetes mellitus, the combination therapy of metformin and α-glucosidase inhibitor may increase abdominal side effects.

SI has important roles in the regulation of intestinal sucrose absorption [37]. SI is an enzyme that digests sucrose into glucose and fructose. *Si* mRNA is induced by sucrose and fructose [13,38]. Moreover, it has been reported that glucose "negatively" regulates human *Si* gene expression through two HNF binding sites in Caco-2 cells [39,40]. Therefore, it is reasonable that ChREBP does not directly regulate SI. However, we found that *Si* mRNA levels in the intestine of KO mice were lower than those in WT. We considered some potential pathways through which ChREBP indirectly regulates *Si* mRNA expression. First, the amount of sucrose intake by KO mice may be lower than the intake by WT because of appetite loss in KO mice. Second, intracellular metabolites derived from sucrose may be a signal for induction of SI genes. As ChREBP regulates glucose and fructose metabolism, intracellular metabolites may be decreased in KO mice. Interestingly, it has been reported that, independently of ChREBP, fructose uniquely induces *SREBP1c* and fatty acid synthesis genes, resulting in impaired insulin signaling [41]. Although further investigation is still needed, decreased *Si* mRNA levels in KO mice also partly contribute to the pathogenesis of sucrose intolerance.

In addition to decreased sucrose metabolism, decreased fructose metabolism has a more important role in the pathogenesis of sucrose intolerance in KO mice. We and other groups have reported that ChREBP has an important role in regulating fructose metabolism [11,12,14–17]. Many of the fructose metabolism genes (*Glut2*, *Glut5*, *Khk*, and *Aldob*) are ChREBP-target genes [12,17,18]. The mRNA levels of *Khk*, *Glut2* and *Glut5* in intestine-specific *ChREBP*-KO mice were much lower than in WT mice after oral fructose injection [18]. Consistently, our data showed that the mRNA levels of *Khk*, *Glut2*, and *Glut5* in KO mice were much lower than in WT mice. Moreover, oral fructose injection induced *Khk*, *Glut2*, and *Glut5* mRNA levels in a time-dependent manner only in WT mice. Moreover, intestinal KHK has important roles in intestinal fructose metabolism [7,42]. Low doses of fructose are ~90% cleared by the intestine and high doses of fructose ($\geq$1 g/kg) overwhelm intestinal fructose absorption and clearance, resulting in fructose reaching both the liver and colonic microbiota [42]. Interestingly, Intestinal fructose clearance is augmented both by prior exposure to fructose and by feeding. These were compatible with our data. Intestinal *Khk* mRNA was induced by fructose and ChREBP gene deletion diminished *Khk* induction by fructose. Accordingly, these results reconfirmed that ChREBP coordinately regulates intestinal fructose metabolism by modulating *Khk*, *Glut2*, and *Glut5* gene expression.

Hepatic KHK has important roles in liver fructose metabolism [43,44]. It has been reported that the plasma fructose levels in $Khk^{-/-}$ mice were 10 times higher than those in WT and $Glut5^{-/-}$ mice [43]. Consistently, the hepatic fructose content in KO mice was much higher after peritoneal fructose injection, which is consistent with decreased *Khk* mRNA levels in the liver of KO mice. As with hepatic fructose transport, hepatic *Glut5* mRNA levels were much lower than in the intestine, which is consistent with a previous study [32]. Considering that the plasma fructose levels in $Glut5^{-/-}$ mice were much lower than in $Khk^{-/-}$ mice, other fructose transporters may regulate hepatic fructose uptake. Our data suggest that hepatic *Khk*, rather than *Glut5*, regulates hepatic fructose metabolism.

Fgf-21 is induced by starvation through PPAR alpha activation [45]. Dietary protein restriction causes Fgf-21 induction through the amino acid sensor GCN2 activation [46]. Moreover, we formerly reported that Fgf-21 is regulated by ChREBP [47]. Fructose feeding increase plasma fructose levels [48]. In this study, hepatic *Fgf-21* mRNA levels in KO mice were much lower than those in WT mice, however, fructose induction of *Fgf-21* mRNA were not seen in both mice, which were not consistent with other reports. PPARα is also required for the ChREBP-induced glucose response of Fgf-21 regulation [49]. Moreover, glucagon and insulin cooperatively stimulate fibroblast growth factor 21 gene transcription by increasing the expression of activating transcription factor 4 [50]. Therefore, in vivo regulation of Fgf-21 expression is complicated.

In this study, undigested excess fructose entered into the lower intestine, resulting in bacterial overgrowth. Fructose malabsorption causes irritable bowel syndrome. Moreover, excess fructose intake might increase colorectal cancer risk [51]. Interestingly, Aldolase B overexpression is associated with poor prognosis and promotes tumor progression by epithelial-mesenchymal transition in colorectal adenocarcinoma [52]. As Aldolase B is a ChREBP-target gene [53], colorectal ChREBP activation by undigested excess fructose might cause colorectal tumor progression. These suggested that intestinal fructose metabolism by ChREBP might be associated with irritable bowel syndrome and colorectal cancer.

5. Conclusions

In conclusion, both sucrose feeding and Si inhibitor caused sucrose intolerance and fructose malabsorption in *ChREBP*-KO mice. ChREBP coordinately regulates sucrose and fructose metabolism by modulating the mRNA expression of intestinal *Si* and *Glut5*, and hepatic *Khk*. Considering intestinal absorption of fructose is more difficult than that of glucose, intestinal ChREBP rather than hepatic ChREBP has an important role in the pathology of sucrose intolerance and fructose malabsorption.

Acknowledgments: We thank Hiromi Tsuchida (Gifu University) and Wudelehu Wu (Gifu University) for technical assistance. We thank Michal Bell, from Edanz Group (www.edanzediting.com/ac) for editing a draft of this manuscript. This work was supported in part by a Grant-in-Aid for Scientific Research from the Japan Society for the Promotion of Science (Iizuka K.: Nos. 17K00850, 26500005, Takeda J.: No. 17K19902), research grants from MSD (Tokyo, Japan), Novartis Pharma (Tokyo, Japan) and Sanwa Kagaku Kenkyusyo Inc. (Nagoya, Japan). (Iizuka K. and Takeda. J.).

Author Contributions: Katsumi Iizuka conceived and designed the experiments; Takehiro Kato and Ken Takao performed the experiments; Takehiro Kato and Katsumi Iizuka analyzed the data; Tadahiro Kitamura and Yukio Horikawa gave support in the literature review; and Katsumi Iizuka and Jun Takeda wrote and revised the paper. All the authors approved the final version of the manuscript.

Conflicts of Interest: The authors declare no conflicts of interest.

References

1. Elliott, S.S.; Keim, N.L.; Stern, J.S.; Teff, K.; Havel, P.J. Fructose, weight gain, and the insulin resistance syndrome. *Am. J. Clin. Nutr.* **2002**, *76*, 911–922. [CrossRef] [PubMed]
2. Samuel, V.T. Fructose induced lipogenesis: from sugar to fat to insulin resistance. *Trends Endocrinol. Metab.* **2011**, *22*, 60–65. [CrossRef] [PubMed]
3. Macdonald, I.A. A review of recent evidence relating to sugars, insulin resistance and diabetes. *Eur. J. Nutr.* **2016**, *55*, 17–23. [CrossRef] [PubMed]
4. Khan, T.A.; Sievenpiper, J.L. Controversies about sugars: Results from systematic reviews and meta-analyses on obesity, cardiometabolic disease and diabetes. *Eur. J. Nutr.* **2016**, *55*, 25–43. [CrossRef] [PubMed]
5. Mochizuki, K.; Hanai, E.; Suruga, K.; Kuranuki, S.; Goda, T. Changes in α-glucosidase activities along the jejunal-ileal axis of normal rats by the α-glucosidase inhibitor miglitol. *Metabolism* **2010**, *59*, 1442–1447. [CrossRef] [PubMed]
6. Delbridge, L.M.; Benson, V.L.; Ritchie, R.H.; Mellor, K.M. Diabetic Cardiomyopathy: The Case for a Role of Fructose in Disease Etiology. *Diabetes* **2016**, *65*, 3521–3528. [CrossRef] [PubMed]
7. Douard, V.; Ferraris, R.P. The role of fructose transporters in diseases linked to excessive fructose intake. *J. Physiol.* **2013**, *591*, 401–414. [CrossRef] [PubMed]
8. Sugimoto, K.; Hosotani, T.; Kawasaki, T.; Nakagawa, K.; Hayashi, S.; Nakano, Y.; Inui, H.; Yamanouchi, T. Eucalyptus leaf extract suppresses the postprandial elevation of portal, cardiac and peripheral fructose concentrations after sucrose ingestion in rats. *J. Clin. Biochem. Nutr.* **2010**, *46*, 205–211. [CrossRef] [PubMed]
9. Kawasaki, T.; Akanuma, H.; Yamanouchi, T. Increased fructose concentrations in blood and urine in patients with diabetes. *Diabetes Care* **2002**, *25*, 353–357. [CrossRef] [PubMed]
10. DiNicolantonio, J.J.; Lucan, S.C. Is fructose malabsorption a cause of irritable bowel syndrome? *Med. Hypotheses* **2015**, *85*, 295–297. [CrossRef] [PubMed]
11. Iizuka, K. The Role of Carbohydrate Response Element Binding Protein in Intestinal and Hepatic Fructose Metabolism. *Nutrients* **2017**, *9*, 181. [CrossRef] [PubMed]
12. Iizuka, K.; Bruick, R.K.; Liang, G.; Horton, J.D.; Uyeda, K. Deficiency of carbohydrate response element-binding protein (ChREBP) reduces lipogenesis as well as glycolysis. *Proc. Natl. Acad. Sci. USA* **2004**, *101*, 7281–7286. [CrossRef] [PubMed]
13. Broyart, J.P.; Hugot, J.P.; Perret, C.; Porteu, A. Molecular cloning and characterization of a rat intestinal sucrase-isomaltase cDNA. Regulation of sucrase-isomaltase gene expression by sucrose feeding. *Biochim. Biophys. Acta* **1990**, *1087*, 61–67. [CrossRef]
14. Zhang, D.; Tong, X.; VanDommelen, K.; Gupta, N.; Stamper, K.; Brady, G.F.; Meng, Z.; Lin, J.; Rui, L.; Omary, M.B.; et al. Lipogenic transcription factor ChREBP mediates fructose-induced metabolic adaptations to prevent hepatotoxicity. *J. Clin. Investig.* **2017**, *127*, 2855–2867. [CrossRef] [PubMed]
15. Fisher, F.M.; Kim, M.; Doridot, L.; Cunniff, J.C.; Parker, T.S.; Levine, D.M.; Hellerstein, M.K.; Hudgins, L.C.; Maratos-Flier, E.; Herman, M.A. A critical role for ChREBP-mediated FGF21secretion in hepatic fructose metabolism. *Mol. Metab.* **2016**, *6*, 14–21. [CrossRef] [PubMed]
16. Kim, M.S.; Krawczyk, S.A.; Doridot, L.; Fowler, A.J.; Wang, J.X.; Trauger, S.A. ChREBP regulates fructose-induced glucose production independently of insulin signaling. *J. Clin. Investig.* **2016**, *126*, 4372–4386. [CrossRef] [PubMed]

17. Ma, L.; Robinson, L.N.; Towle, H.C. ChREBP*Mlx is the principal mediator of glucose-induced gene expression in the liver. *J. Biol. Chem.* **2006**, *281*, 28721–28730. [CrossRef] [PubMed]

18. Kim, M.; Astapova, I.I.; Flier, S.N.; Hannou, S.A.; Doridot, L.; Sargsyan, A.; Kou, H.H.; Fowler, A.J.; Liang, G.; Herman, M.A. Intestinal, but not hepatic, ChREBP is required for fructose tolerance. *JCI Insight* **2017**, *2*. [CrossRef] [PubMed]

19. Wu, W.; Tsuchida, H.; Kato, T.; Niwa, H.; Horikawa, Y.; Takeda, J.; Iizuka, K. Fat and carbohydrate in western diet contribute differently to hepatic lipid accumulation. *Biochem. Biophys. Res. Commun.* **2015**, *461*, 681–686. [CrossRef] [PubMed]

20. Sasaki, T.; Shimpuku, M.; Kitazumi, T.; Hiraga, H.; Nakagawa, Y.; Shibata, H.; Okamatsu-Ogura, Y.; Kikuchi, O.; Kim, H.J.; Fujita, Y.; et al. Miglitol prevents diet-induced obesity by stimulating brown adipose tissue and energy expenditure independent of preventing the digestion of carbohydrates. *Endocr. J.* **2013**, *60*, 1117–1129. [CrossRef] [PubMed]

21. Bligh, E.G.; Dyer, W.J. A rapid method of total lipid extraction and purification. *Can. J. Biochem. Physiol.* **1959**, *37*, 911–917. [CrossRef] [PubMed]

22. Kunst, A.; Drager, B.; Ziegenhorn, J. UV methods with hexokinase and glucose-6-phosphate dehy-dro-genase. In *Methods of Enzymatic Analysis*; Bergemeyer, H.Y., Ed.; Verlag Chemie: Deerfield, IL, USA, 1983; Volume VI, pp. 163–172.

23. Nagashima, K.; Hisada, T.; Sato, M.; Mochizuki, J. Application of new primer-enzyme combinations to terminal restriction fragment length polymorphism profiling of bacterial populations in human feces. *Appl. Environ. Microbiol.* **2003**, *69*, 1251–1262. [CrossRef] [PubMed]

24. Lukas, G.; Brindle, S.D.; Greengard, P. The route of absorption of intraperitoneally administered compounds. *J. Pharmacol. Exp. Ther.* **1971**, *178*, 562–564. [PubMed]

25. Gänzle, M.G.; Follador, R. Metabolism of Oligosaccharides and Starch in Lactobacilli: A Review. *Front. Microbiol.* **2012**, *3*, 340. [CrossRef] [PubMed]

26. Pokusaeva, K.; Fitzgerald, G.F.; van Sinderen, D. Carbohydrate metabolism in Bifidobacteria. *Genes Nutr.* **2011**, *6*, 285–306. [CrossRef] [PubMed]

27. Yamada, S.; Kamada, N.; Amiya, T.; Nakamoto, N.; Nakaoka, T.; Kimura, M.; Saito, Y.; Ejima, C.; Kanai, T.; Saito, H. Gut microbiota-mediated generation of saturated fatty acids elicits inflammation in the liver in murine high-fat diet-induced steatohepatitis. *BMC Gastroenterol.* **2017**, *17*, 136. [CrossRef] [PubMed]

28. Neyrinck, A.M.; Pachikian, B.; Taminiau, B.; Daube, G.; Frédérick, R.; Cani, P.D.; Bindels, L.B.; Delzenne, N.M. Intestinal Sucrase as a Novel Target Contributing to the Regulation of Glycemia by Prebiotics. *PLoS ONE* **2016**, *11*, e0160488. [CrossRef] [PubMed]

29. Treem, W.R. Clinical aspects and treatment of congenital sucrase-isomaltase deficiency. *J. Pediatr. Gastroenterol. Nutr.* **2012**, *55* (Suppl. 2), S7–S13. [CrossRef] [PubMed]

30. Johnston, P.S.; Coniff, R.F.; Hoogwerf, B.J.; Santiago, J.V.; Pi-Sunyer, F.X.; Krol, A. Effects of the carbohydrase inhibitor miglitol in sulfonylurea-treated NIDDM patients. *Diabetes Care* **1994**, *17*, 20–29. [CrossRef] [PubMed]

31. Barone, S.; Fussell, S.L.; Singh, A.K.; Lucas, F.; Xu, J.; Kim, C.; Wu, X.; Yu, Y.; Amlal, H.; Seidler, U.; et al. Slc2a5 (Glut5) is essential for the absorption of fructose in the intestine and generation of fructose-induced hypertension. *J. Biol. Chem.* **2009**, *284*, 5056–5066. [CrossRef] [PubMed]

32. Rand, E.B.; Depaoli, A.M.; Davidson, N.O.; Bell, G.I.; Burant, C.F. Sequence, tissue distribution, and functional characterization of the rat fructose transporter GLUT5. *Am. J. Physiol.* **1993**, *264*, G1169–G1176. [CrossRef] [PubMed]

33. Jones, H.F.; Butler, R.N.; Brooks, D.A. Intestinal fructose transport and malabsorption in humans. *Am. J. Physiol. Gastrointest. Liver Physiol.* **2011**, *300*, G202–G206. [CrossRef] [PubMed]

34. Douard, V.; Ferraris, R.P. Regulation of the fructose transporter GLUT5 in health and disease. *Am. J. Physiol. Endocrinol. Metab.* **2008**, *295*, E227–E237. [CrossRef] [PubMed]

35. Sanchez-Rangel, E.; Inzucchi, S.E. Metformin: clinical use in type 2 diabetes. *Diabetologia* **2017**, *60*, 1586–1593. [CrossRef] [PubMed]

36. Li, X.; Kover, K.L.; Heruth, D.P.; Watkins, D.J.; Moore, W.V.; Jackson, K.; Zang, M.; Clements, M.A.; Yan, Y. New Insight into Metformin Action: Regulation of ChREBP and FOXO1 Activities in Endothelial Cells. *Mol. Endocrinol.* **2015**, *29*, 1184–1194. [CrossRef] [PubMed]

37. Gericke, B.; Amiri, M.; Naim, H.Y. The multiple roles of sucrase-isomaltase in the intestinal physiology. *Mol. Cell. Pediatr.* **2016**, *3*, 2. [CrossRef] [PubMed]

38. Kishi, K.; Tanaka, T.; Igawa, M.; Takase, S.; Goda, T. Sucrase-isomaltase and hexose transporter gene expressions are coordinately enhanced by dietary fructose in rat jejunum. *J. Nutr.* **1999**, *129*, 953–956. [CrossRef] [PubMed]

39. Boudreau, F.; Zhu, Y.; Traber, P.G. Sucrase-isomaltase gene transcription requires the hepatocyte nuclear factor-1 (HNF-1) regulatory element and is regulated by the ratio of HNF-1 alpha to HNF-1 beta. *J. Biol. Chem.* **2001**, *276*, 32122–32128. [CrossRef] [PubMed]

40. Gu, N.; Adachi, T.; Matsunaga, T.; Tsujimoto, G.; Ishihara, A.; Yasuda, K.; Tsuda, K. HNF-1alpha participates in glucose regulation of sucrase-isomaltase gene expression in epithelial intestinal cells. *Biochem. Biophys. Res. Commun.* **2007**, *353*, 617–622. [CrossRef] [PubMed]

41. Softic, S.; Gupta, M.K.; Wang, G.X.; Fujisaka, S.; O'Neill, B.T.; Rao, T.N.; Willoughby, J.; Harbison, C.; Fitzgerald, K.; Ilkayeva, O.; et al. Divergent effects of glucose and fructose on hepatic lipogenesis and insulin signaling. *J. Clin. Investig.* **2017**, *127*, 4059–4074. [CrossRef] [PubMed]

42. Jang, C.; Hui, S.; Lu, W.; Cowan, A.J.; Morscher, R.J.; Lee, G.; Liu, W.; Tesz, G.J.; Birnbaum, M.J.; Rabinowitz, J.D. The Small Intestine Converts Dietary Fructose into Glucose and Organic Acids. *Cell Metab.* **2018**, *27*, 351–361. [CrossRef] [PubMed]

43. Patel, C.; Sugimoto, K.; Douard, V.; Shah, A.; Inui, H.; Yamanouchi, T.; Ferraris, R.P. Effect of dietary fructose on portal and systemic serum fructose levels in rats and in KHK−/− and GLUT5−/− mice. *Am. J. Physiol. Gastrointest. Liver Physiol.* **2015**, *309*, G779–G790. [CrossRef] [PubMed]

44. Ishimoto, T.; Lanaspa, M.A.; Le, M.T.; Garcia, G.E.; Diggle, C.P.; Maclean, P.S.; Jackman, M.R.; Asipu, A.; Roncal-Jimenez, C.A.; Kosugi, T.; et al. Opposing effects of fructokinase C and A isoforms on fructose-induced metabolic syndrome in mice. *Proc. Natl. Acad. Sci. USA* **2012**, *109*, 4320–4325. [CrossRef] [PubMed]

45. Inagaki, T.; Dutchak, P.; Zhao, G.; Ding, X.; Gautron, L.; Parameswara, V.; Li, Y.; Goetz, R.; Mohammadi, M.; Esser, V.; et al. Endocrine regulation of the fasting response by PPARalpha-mediated induction of fibroblast growth factor 21. *Cell Metab.* **2007**, *5*, 415–425. [CrossRef] [PubMed]

46. Laeger, T.; Albarado, D.C.; Burke, S.J.; Trosclair, L.; Hedgepeth, J.W.; Berthoud, H.R.; Gettys, T.W.; Collier, J.J.; Münzberg, H.; Morrison, C.D. Metabolic responses to dietary protein restriction require an increase in FGF21 that is delayed by the absence of GCN2. *Cell Rep.* **2016**, *16*, 707–716. [CrossRef] [PubMed]

47. Iizuka, K.; Takeda, J.; Horikawa, Y. Glucose induces FGF21 mRNA expression through ChREBP activation in rat hepatocytes. *FEBS Lett.* **2009**, *583*, 2882–2886. [CrossRef] [PubMed]

48. Dushay, J.R.; Toschi, E.; Mitten, E.K.; Fisher, F.M.; Herman, M.A.; Maratos-Flier, E. Fructose ingestion acutely stimulates circulating FGF21 levels in humans. *Mol. Metab.* **2014**, *4*, 51–57. [CrossRef] [PubMed]

49. Iroz, A.; Montagner, A.; Benhamed, F.; Levavasseur, F.; Polizzi, A.; Anthony, E.; Régnier, M.; Fouché, E.; Lukowicz, C.; Cauzac, M.; et al. A Specific ChREBP and PPARα Cross-Talk Is Required for the Glucose-Mediated FGF21 Response. *Cell Rep.* **2017**, *21*, 403–416. [CrossRef] [PubMed]

50. Alonge, K.M.; Meares, G.P.; Hillgartner, F.B. Glucagon and Insulin Cooperatively Stimulate Fibroblast Growth Factor 21 Gene Transcription by Increasing the Expression of Activating Transcription Factor 4. *J. Biol. Chem.* **2017**, *292*, 5239–5252. [CrossRef] [PubMed]

51. Higginbotham, S.; Zhang, Z.-F.; Lee, I.-M.; Cook, N.R.; Giovannucci, E.; Buring, J.E.; Liu, S. Dietary Glycemic Load and Risk of Colorectal Cancer in the Women's Health Study. *J. Natl. Cancer Inst.* **2004**, *96*, 229–233. [CrossRef] [PubMed]

52. Cao, W.; Chang, T.; Li, X.Q.; Wang, R.; Wu, L. Dual effects of fructose on ChREBP and FoxO1/3α are responsible for AldoB up-regulation and vascular remodelling. *Clin. Sci. (Lond.)* **2017**, *131*, 309–325. [CrossRef] [PubMed]

53. Li, Q.; Li, Y.; Xu, J.; Wang, S.; Xu, Y.; Li, X.; Cai, S. Aldolase B Overexpression is Associated with Poor Prognosis and Promotes Tumor Progression by Epithelial-Mesenchymal Transition in Colorectal Adenocarcinoma. *Cell. Physiol. Biochem.* **2017**, *42*, 397–406. [CrossRef] [PubMed]

MDPI

St. Alban-Anlage 66

4052 Basel

Switzerland

Tel. +41 61 683 77 34

Fax +41 61 302 89 18

www.mdpi.com

Actuators Editorial Office

E-mail: actuators@mdpi.com

www.mdpi.com/journal/actuators

www.ingramcontent.com/pod-product-compliance
Lightning Source LLC
Chambersburg PA
CBHW051724210326
41597CB00032B/5590